Knowledge, Policy, and Expertise

Knowledge, Policy, and Expertise

The UK Royal Commission on Environmental Pollution 1970–2011

Susan Owens

OXFORD
UNIVERSITY PRESS

Great Clarendon Street, Oxford, OX2 6DP,
United Kingdom

Oxford University Press is a department of the University of Oxford.
It furthers the University's objective of excellence in research, scholarship,
and education by publishing worldwide. Oxford is a registered trade mark of
Oxford University Press in the UK and in certain other countries

© Susan Owens 2015
© Chapter 4 with Tim Rayner 2015

The moral rights of the authors have been asserted

First Edition published in 2015
Impression: 1

All rights reserved. No part of this publication may be reproduced, stored in
a retrieval system, or transmitted, in any form or by any means, without the
prior permission in writing of Oxford University Press, or as expressly permitted
by law, by licence or under terms agreed with the appropriate reprographics
rights organization. Enquiries concerning reproduction outside the scope of the
above should be sent to the Rights Department, Oxford University Press, at the
address above

You must not circulate this work in any other form
and you must impose this same condition on any acquirer

Published in the United States of America by Oxford University Press
198 Madison Avenue, New York, NY 10016, United States of America

British Library Cataloguing in Publication Data
Data available

Library of Congress Control Number: 2014960021

ISBN 978–0–19–829465–8

Printed and bound by
CPI Group (UK) Ltd, Croydon, CR0 4YY

Links to third party websites are provided by Oxford in good faith and
for information only. Oxford disclaims any responsibility for the materials
contained in any third party website referenced in this work.

Preface

Academic careers sometimes take unexpected turns. My interests in environmental governance meant that I was aware, from an early stage, of the work of the Royal Commission on Environmental Pollution, but I first attended closely to its reports while writing about integrated pollution control in the late 1980s. When I visited the Commission's offices at that time, it seemed to me a rather grand and august body. A few years later, my work on planning and environmental change led to my appointment as Special Advisor for the Commission's study of transport and the environment, and I attended many of its meetings in 1992–4. As the report took shape, I became more familiar with the Commission's working practices, and found myself asking questions about this intriguing advisory body. What was its role in the policy process? How did it choose topics, frame problems, arrive at conclusions, and promulgate its ideas? Why did it sometimes have substantial impact and sometimes (apparently) very little? I went away and drafted a research proposal. Happily, the Leverhulme Trust and the Commission itself were willing to fund a twenty-eight month project, with a research assistant (Tim Rayner). This enabled archives to be explored, many individuals to be interviewed, and initial findings about the Commission's role and influence to be presented (Owens and Rayner 1998, 1999).

In 1998, shortly after completion of the Leverhulme project, I was appointed as a member of the Commission—a turn of events that I could certainly not have predicted ten years earlier—and stayed until 2008, working on nine reports. This period, fascinating and enjoyable in equal measure, served also to intensify my interest in expertise, advisory practices, and processes of policy formation. My insights into the questions behind the earlier research were enriched, and new puzzles presented themselves regarding the ways in which the authority and autonomy of a body like the Commission could effectively be constructed and sustained. I was aware, too, of the rapidly increasing salience of science–policy relations, and of the controversies to which they could sometimes give rise. My enquiries into these issues deepened, and the plan to write a book about the Commission, hitherto eclipsed by other projects and commitments, was kept alive. On leaving the Commission, I was presented with a full set of minutes going back to its very first meeting

Preface

in 1970. I realized that I had become, quite unofficially, the Commission's historian.

Immediately afterwards, I was privileged to hold the King Carl XVI Gustaf Professorship of Environmental Science in Sweden. During a very fruitful year (2008–9) spent at the Stockholm Resilience Centre (Stockholm University) and the Royal Institute of Technology (KTH), I had time to think further about the book and to delve extensively into the diverse literatures which had some bearing on the interplay of knowledge, policy, and expertise. Political science, policy analysis, science and technology studies (STS), and the environmental social sciences all offered valuable insights into these issues, though they had tended, with honourable exceptions, to move forward along parallel tracks. There seemed to me to be enormous potential to draw these insights together in the context of an in-depth study of a particular advisory body; there had, in fact, been rather few investigations of this kind. I returned to Cambridge with renewed enthusiasm for my project but also—before too long—acquired a major administrative responsibility. This helps explain why the book itself took a further five years to complete.

In the Spring of 2011, the Royal Commission was abolished. With more than forty years of service, it had been one of Britain's longest-standing environmental advisory bodies. Ironically, perhaps—given the many years taken to complete this project—it seems particularly timely, now that a line has been drawn under the Commission, to reflect on its role, achievements, and ultimate demise. But my aim in this book has been to offer much more than the biography of a particular institution, however absorbing this might be in its own right. In interpreting the Commission's practices, roles, and influence within a wider frame, I have sought, too, to illuminate the complex and contingent relations among knowledge, expertise, politics, and policy formation. I am left in no doubt at all about the intricacies of these interactions, or their deep significance in modern democracies, or the need for further, 'forensic' research. I hope, nevertheless, that the insights offered by this study, grounded in the specific case of the Commission and the environmental policy domain, might help to further our understanding of knowledge–policy relations, and perhaps make a modest contribution to the project of 'good advice'.

Susan Owens
Newnham College, Cambridge
August 2014

Acknowledgements

Many people and organizations—too numerous to be named individually—have helped to bring this project to fruition. I most gratefully acknowledge the financial support from the Leverhulme Trust (grant no. F/753/A) and the Royal Commission on Environmental Pollution, as well as the excellent research assistance and contribution of Tim Rayner, during the initial phase of the study (1995–7). Tim (currently Senior Research Associate, School of Environmental Sciences, University of East Anglia) is the co-author of Chapter 4 of this book.

Archival material has been essential throughout, and I am grateful to the Royal Commission and to the Department of the Environment (and its successors) for granting access to their files and providing the necessary assistance. Thanks are due also to Imperial College London for facilitating the use of relevant material. The Commission's Secretariat was always extremely helpful, and a special word of thanks must go to David Lewis (Secretary 1992–2002) for his enthusiasm for the Commission to be scrutinized. And of course, I owe a substantial debt of gratitude to the many interviewees for this study, and to all the other individuals who provided insights on the Commission and on the policy processes to which it contributed. Interviews were transcribed, with patience and good humour, by Virginia Mullins at the University of Cambridge.

A year spent in Stockholm in 2008–9 enabled me substantially to develop this project, and I am deeply grateful to His Majesty King Carl XVI Gustaf for the opportunity to hold the King Carl XVI Gustaf Professorship of Environmental Science. My colleagues at the Stockholm Resilience Centre (SRC), Stockholm University, and the Environmental Strategies Group (fms) at the Royal Institute of Technology (KTH) were warmly welcoming and always willing to hear about, and offer valuable feedback on, my research. Special thanks go my hosts in each institution—Katarina Eckerberg of the Stockholm Environment Institute (then part of SRC) and Göran Finnveden at fms—for their intellectual and practical support.

A number of academic colleagues, as well as those with connections to the Royal Commission, read and commented on drafts of this book at various points. Invariably, they provided new insights and offered sound advice, and

Acknowledgements

I owe many improvements to Harriet Bulkeley, Rob Doubleday, Katarina Eckerberg, John Lawton, David Lewis, Miles Parker, and John Roberts. In addition, Amy Donovan, Emily Noah, James Palmer, and Elizabeth Rough, as well as reading and commenting on drafts, provided efficient research assistance at various stages of the project. Many others have contributed to the book's development in a variety of different ways; I am indebted, in particular, to Nigel Haigh, for introducing me to the Commission and furnishing a wealth of information about environmental policy; and to Sheila Jasanoff and Albert Weale, both of whom have been sources of inspiration, advice, and support.

My sincere thanks are extended to Dominic Byatt, my ever-patient Commissioning Editor; to Olivia Wells, Assistant Commissioning Editor, who worked so hard to get the manuscript into production; and to all other members of the team involved with this book at Oxford University Press. I am also very grateful to the production team at SPi Global and to Helen Cooper for her careful attention to copy-editing.

Last but definitely not least, I thank my academic colleagues in Cambridge and elsewhere, my fellow members of the Royal Commission, and the two Chairs under whom I served (Sir Tom Blundell and Sir John Lawton), for never giving up hope that this book would eventually appear.

Permission to reproduce the extract from *The Rubáiyát of a Royal Commissioner* by Gordon Fogg (which appears in Chapter 5 of this book) has kindly been granted by Timothy Fogg. I am also grateful for permissions to reproduce published material as chapter epigraphs, as follows. In Chapter 1, material from *Ozone Discourses*, by Karen Litfin, copyright © 1994 Columbia University Press, reprinted with permission of the publisher. In Chapter 2, material from: *The New Politics of Pollution*, by Albert Weale, Manchester University Press, 1992, reprinted with permission of the author; *The Politics of Environmental Discourse*, by Maarten Hajer, 1995, reprinted with permission of Clarendon Press; 'In from the cold', by Nick Rowcliffe, *ENDS at 30* (Anniversary Supplement), May 2008, reprinted with permission of the author. In Chapter 3, material from: 'Axing this Commission is a right royal shame', by Geoffrey Lean, *Telegraph Blogs*, 10 March 2011, © Telegraph Media Group Ltd 2011, reprinted with permission of Telegraph Media Group Ltd; *In the Public Interest*, by Stephen Wilks, Manchester University Press, 1999, reprinted with permission of Manchester University Press and the author. In Chapter 4, material from: John Ashworth, introduction in Royal Society, *Science, Policy and Risk*, 1997, reprinted with permission of The Royal Society. In Chapter 5, material from *The New Politics of Pollution*, by Albert Weale, Manchester University Press, 1992, reprinted with permission of the author. In Chapter 6, material from Kingdon, John W., *Agendas, Alternatives, and Public Policies* (Longman Classics Edition), 2nd, © 2003, printed and electronically reproduced by permission of Pearson Education, Inc., Upper Saddle River, New Jersey. In Chapter 7, material from *Government by Committee*, by Kenneth Wheare, 1955, reprinted with permission of Clarendon Press. This book contains public sector information (including, for example, extracts from Royal Commission reports and government and Parliamentary publications) licensed under the Open Government Licence v3.0. All sources are acknowledged.

Contents

1. Knowledge, Advice, and Policy 1
   Introduction 1
   Knowledge, Policy, and Expertise 5
     Technical Rationality 6
     Political Rationality 8
     Cognitive Perspectives 9
     Co-production and Boundary Work 13
   Positioning Advisory Bodies 16
   Investigating the Royal Commission: A Note on Methods
     and Sources 18
   Outline of the Book 21

2. The Environment in Politics 23
   Introduction 23
   Revolution and Beyond: The Late 1960s to the Late 1970s 25
   Changing the Frame: The 'Long 1980s' 29
   From Rio to Recession—The Era of Climate Change: 1992–2007 35
   Recession and Environmental Policy: Regression
     or Resilience? 2007–11 40
   Forty Years of Policy Evolution: A Brief Reflection 42

3. A *Standing* Royal Commission 45
   Introduction 45
   Origins 46
   A Delicate Balance? 49
   Appointments 52
   A 'Collegiate Body' 55
   Working Practices 57
     Choosing Topics 59
     Taking Evidence 61
     Deliberation and Drafting 63
   The Reports in Brief 65
   Survival and Demise 66
   Concluding Comment 70

Contents

4. Risk, Precaution, and Governance: An Evolution of Ideas   71
with Tim Rayner

    Introduction   71
    Risk Paradigms   73
    Risk and the Royal Commission: Early Perspectives   76
        Nuclear Power: Precaution, Ethics, and 'Deliberation in Public'   79
    Risk, Uncertainty, and Precaution: Towards an Ecologically
       Modern Perspective?   81
        Lead in Petrol: Framing the Problem   83
    Sliding Between Paradigms   85
        Public 'Perceptions'   87
        Precaution Without Deliberation: The Case of Genetically
          Modified Organisms   88
        Regression?   91
    Beyond 'Enlightened Technocracy'?   92
        An Uncharacteristic 'Spat'   95
        Nanomaterials: Precaution and Adaptive Governance   97
    From Enlightened Technocracy to Cautious Constructivism   99

5. The Quest for Integration   103

    Introduction   103
    New Institutions for Pollution Control   104
        An 'Elephantine Gestation Period'   106
        Changing the Frame   109
        Practising Integration   112
        Learning to Integrate?   113
    Environmental Planning   115
        Integrated Spatial Strategies   117
        A Diffuse Effect?   120
    Conclusions: The Contingency of Influence   121

6. The Circumstances of Influence   125

    Introduction   125
    Visible, Short-term Responses   127
    Dormant Seeds   132
    Slowly Changing the Frame   136
    'Doing Good by Stealth'   140
    The 'Dogs That Didn't Bark'   141
    Assessing Influence: Conclusions and Reflections   143

Contents

7. Giving Good Advice?	146
Introduction	146
Constructing Authority	147
Interdisciplinary Deliberation	148
The 'Independence Thing'	150
An Intersection of Networks	154
Continuity	158
Explaining the Commission's Demise	159
Learning from the Commission	164
Lessons for the Future?	169
Appendices	
1. Reports of the Royal Commission and Government Responses	173
2. Location of Environmental Functions in UK Government, 1970–2011	177
3. Interviewees (and Other Respondents) by Category	179
4. United Kingdom and European Legislation Mentioned in the Text	181
5. Chairs of the Royal Commission and Membership, 1971, 1991, and 2011	189
Endnotes	197
References	263
Index	301

1

Knowledge, Advice, and Policy

> We have not overthrown the divine right of Kings to fall down before the divine right of experts.
>
> Harold Macmillan, at Council of Europe Consultative Assembly, Strasbourg, France (1950)[1]

> The relationships between science and policy, technical experts and policymakers, are multidimensional and not reducible to an *a priori* understanding; indeed, the possibilities of interaction are innumerable.
>
> Karen Litfin, *Ozone Discourses* (1994: 51)

> Advisers advise, and Ministers decide.
>
> Margaret Thatcher, responding to question in the House of Commons (1989)[2]

Introduction

All those who govern seek advice. In modern democracies, governments do so extensively, calling regularly upon committees, commissions, think tanks, and individual advisors in the formation and implementation of public policy.[3] Expertise, it seems, is indispensable, though it can also be controversial, and its role in the policy process raises a number of critical questions. How much impact do advisors really have in important political domains? What is the nature of their influence and in what ways is it manifested? In complex democracies, how might the role of advisory bodies be conceptualized? And is it possible, in widely differing contexts, to characterize and delineate 'good advice'? Given the contemporary significance of advisory functions, answers to such questions remain surprisingly elusive. In her groundbreaking study of scientific advisory committees in the United States, Sheila Jasanoff (1990: 1) observed that, in spite of the emergence of such committees as a 'fifth branch' of government, their activities were poorly documented and their impacts on policy decisions difficult to understand and evaluate. These observations, which could be extended beyond purely 'scientific' committees, remain substantially true, at least in the sense that we lack a well-developed theory of policy advice. Yet they are also paradoxical because a great deal has in fact been written about advisory bodies, which have existed in considerable variety in terms of their nature, constitution, degree of permanence, and relation to governmental institutions. In a diverse literature, stretching back over many decades, we can find commentaries and case studies from a range of sectoral and national contexts,[4] as well as taxonomic, conceptual, and evaluative frameworks.[5] Government-appointed commissions and committees—of particular interest here—have been the subject of extensive commentary, with typological and often richly descriptive studies gradually giving way to more critical and analytical accounts.[6] Independent think tanks, which share some characteristics with government advisory bodies, have also received attention.[7]

The effectiveness of advice and expertise has exercised many observers, not only in a spirit of enquiry but also in response to the powerful urge in modern democracies to make better use of scientific knowledge in policy formation.[8] But those seeking straightforward answers to 'what works?' encounter some difficult issues. Ironically, the demand for 'evidence-based policy' seems to have intensified alongside declining public trust in expertise (Bijker, Bal, and Hendriks 2009), while those who study the connections among science, politics, and policy-making have repeatedly highlighted their complexity. Knowledge, it seems, doesn't simply flow from one sphere to another, and the very notion of rigidly separated spheres may be misleading, for reasons that will be considered below. Some observers have reached pessimistic conclusions about the capacity of scientific knowledge to make much difference, at least in complex and contested areas of public policy. According to Collingridge and Reeve (1986), we can expect science to have only the most marginal influence, while in 't Veld and de Wit (2009: 153) claim that '[l]arge quantities of knowledge produced for the benefit of policy are never used in that policy-making'. Others have found the mechanisms through which analysis and ideas exert influence to be subtle, indirect, and long term, as 'knowledge creeps into policy in diffuse ways' (Booth 1990: 85; Weiss and Bucuvalas 1980). Much, as Roger Pielke (2007) has argued in a thoughtful analysis of the role of science in policy and politics, is unavoidably context-dependent.

Effectiveness has most often been judged in terms of tangible impact, and some effort has been devoted, particularly in evaluative frameworks, to identifying the conditions under which this outcome might be achieved. But 'impact' as a criterion is insufficient, since even bad advice can (and does) have effect in these terms. We might argue that to be effective in a meaningful sense, advice should not only be heeded but should also be seen to influence public policy for the better, enabling governments to identify problems and pre-empt or address them through legislation, other interventions, and the creation of appropriate institutions. These effects are more complex to evaluate and require assessment over the longer term. Beyond impact, and the familiar appeal to 'sound science', relatively little has been said about what might constitute the intrinsic qualities of 'good advice', though independence, transparency, and defensibility in the public realm have been upheld as desirable characteristics, alongside fitness for purpose and the authority that derives from experience or expertise (see, for example, Clark and Majone 1985). Most often, in guidelines that have proliferated at all levels of governance, good advice has been associated with adherence on the part of advisors and those who engage them to certain principles and rules of conduct. The UK, for example, has a code of practice for scientific advisory committees (Government Office for Science 2011) and expects all members of advisory bodies to uphold the 'seven principles of public life', as first laid down by the Nolan Committee in the mid-1990s (Committee on Standards in Public Life 1995).[9]

There is, then, no shortage of commentary, analysis, and guidance when it comes to expertise and policy advice, and significant insights from this body of work will be outlined later in this chapter. But if we seek a deeper understanding of advisory practices and their interactions with the policy process, the existing material presents the analyst with a number of immediate problems. One is that it lacks coherence, in part because the literatures with most to offer on this subject have tended to run in parallel tracks, with surprisingly little cross-fertilization of ideas. Political scientists whose primary interest has been in policy evolution, for example, have often paid scant attention to the nature and practices of advice, other than, in some cases, to acknowledge its existence and significance. Conversely, those focusing on 'knowledge utilization', or approaching science and expertise from sociological and cultural perspectives, have not always engaged critically with theories of policy formation. There are missed opportunities, then, for a constructive synthesis of insights from different disciplinary traditions. A second limitation is that relatively few studies combine theoretical or conceptual frameworks with rich empirical evidence: those developing the former often tend towards abstraction, while many of the fascinating stories about advice and advisory bodies lack any coherent framing. Finally, although there are compelling arguments for adopting a long-term perspective in studies of policy evolution, few projects have considered the role and influence of expert advice over extended periods of time. In particular, in-depth, longitudinal studies of individual advisory bodies are rare.[10]

This book is the outcome of one such investigation. It focuses on a British advisory body of long standing—the Royal Commission on Environmental

Pollution—and explores its influence on environmental policy over its lifetime of more than forty years. Royal commissions have been a feature of the British political landscape for a long time, and are sometimes thought of as a rather quaint, out-dated style of institution. In fact, such bodies continued to be appointed throughout the twentieth century, though they fell out of favour under the Thatcher government (1979–90) and only three came into existence after that. The Royal Commission on Environmental Pollution was established as a standing body under Harold Wilson's Labour government in 1970, part of the wider institutional response to mounting public and political concerns about the environment at that time. Standing royal commissions have been rarer than their ad hoc counterparts (the latter being set up for a limited period to report on a specific issue) and, for the last few years of its existence, this particular Royal Commission had the distinction of being the only advisory body of its kind in the UK.[11] More will be said about its origins, constitution, and working practices in Chapter 3. For now, it is worth noting that the Commission was established as an interdisciplinary body of experts, distinguishable in important respects from a more specialized scientific expert committee.[12] Its remit, set out in its Royal Warrant in 1970, was 'to advise on matters, both national and international, concerning the pollution of the environment; on the adequacy of research in this field; and the future possibilities of danger to the environment', and the Commissioners were granted powers to enquire into any such matters referred to them by ministers, or any others on which they should 'deem it expedient to advise'.[13] In the period from 1970 to 2011 (when it was abolished) the Commission produced thirty-three reports, dealing with an extensive range of subjects, which were mostly of its own choosing (Appendix 1). It is widely perceived to have influenced successive governments and, in some cases, European and other international institutions, in the rapidly evolving policy domains concerned with pollution and the environment.

The Royal Commission is mentioned in a number of works on commissions and committees of enquiry (Bulmer 1980c; Cartwright 1975; Everest 1990), though none discusses its activities in any depth. It features more prominently, usually as an active participant, in numerous accounts of environmental politics and policy-making (for example, Baker 1988; Ball and Bell 1994; Berkhout 1991; Hajer 1995; Lowe 1975a; Lowe et al. 1997; Weale 1992, 1997; Weale, O'Riordan, and Kramme 1991) and a few authors have attended specifically—though, again, not at length—to its role in the policy process (Lowe 1975b; Williams and Weale 1996). Further insights are offered by former chairs and members of the Commission, some reflecting on particular studies or themes (Flowers 1974; Silberston 1993; Southwood 1985, 1992), and others writing about their experience of the Commission in a wider, often autobiographical, context (Holdgate 2003; Houghton 2013; Nathan 2003;

Zuckerman 1988). No study to date, however, has considered the practices, role, and influence of this body in the round or over the full period of its existence.[14]

In fulfilling its remit over a period of more than four decades, the Commission became intricately involved with developments in British, European, and sometimes wider international environmental policy. This is promising terrain in which to explore questions about the nature, practices, and impacts of policy advice. Environmental regulation invariably demands expertise and, in dealing frequently with issues that are high profile as well as 'trans-scientific',[15] it has been one of the most visible arenas in which experts have had to (re)negotiate their relations with other social actors (Grove-White, Kapitza, and Shiva 1992). Furthermore, policy in the environmental field has acquired immense political and economic significance and has developed rapidly, offering unusual scope to investigate the dynamics of change over a period of half a century or so. An additional advantage is that the environment has been the subject of voluminous official discourse as well as a great deal of scholarly attention. There is an extensive literature on the emergence and politicization of environmental issues, the evolution of specific conflicts, the motivations and influence of interest groups, and the ways in which policies have been shaped. Such work has both drawn upon and enriched a variety of theoretical and methodological perspectives on policy and political processes.[16] Indeed, the emergence and institutionalization of environmental concerns might be said to have shifted the scenery on the political stage, as new issues, ideologies, and interests have destabilized established relations.

Knowledge, Policy, and Expertise

Expert advice is one of the established mechanisms through which knowledge comes into interplay with policy and politics. A study of the practices of an advisory body, and the reception and influence of its advice, can therefore be located within a broader field of enquiry concerned with the 'interface' (as it is often called) between knowledge and policy. This important boundary is itself the subject of a large and somewhat unruly literature, which will be considered in more detail below. First, however, it will help to clarify the terms 'knowledge' and 'policy' as they are employed in this book, since both defy precise definition and their meanings are often vague in general use. A word is also necessary about the relationship between 'policy' and 'politics'.

Knowledge is interpreted generously in this book to include formal (scientific and specialist) as well as tacit and experiential knowledge, together with belief systems, analysis, and ideas. Evidence, argument, and representation are also included, as vital components of the practices of giving advice. Although the Royal Commission has often been categorized—not quite accurately—as a 'scientific' body (in the more restrictive meaning of the word[17]), all of these cognitive factors have been important in its interactions with wider policy communities. 'Policy' is taken to mean a set of commitments to particular ends, usually combined with a specification of some means of achieving them, and governed by certain (explicit or tacit) principles. In the case of public policy, with which this book is primarily concerned, such commitments are ultimately made by governments, though many more actors are normally involved in their formulation. 'The policy process' entails the practices, procedures, and interactions involved in defining phenomena as problems and in formulating and implementing public policies; typically this includes setting agendas; generating alternatives; making decisions; and designing, implementing, and evaluating specific programmes and regulations—though not necessarily in that order, or with the stages so clearly defined. In the literature concerned with these issues, policy and politics are sometimes treated as separate analytical categories, with the former seen as emerging from the latter.[18] But many contemporary commentators accept that politics is an integral part of the policy *process* (see, for example, Keeley and Scoones 2003; Kingdon 2003; Pielke 2007), and some go further, arguing that policy-making in modern democracies may itself be constitutive of politics (Hajer 2003a; see also Owens and Cowell 2011; Radaelli 1995). Certainly, in this study of the Royal Commission on Environmental Pollution, a clear distinction between policy and politics would have been extremely difficult to sustain.

Turning now to the literature that explores connections between knowledge and policy, we can see that is unruly for good reason. Different theories of the policy process (with their many variants) combine with disparate perspectives on what constitutes knowledge to create a complex and confusing picture, with many possible points of intersection. The focus can change sharply when the picture is viewed through different disciplinary lenses, and even within disciplines we find a diversity of starting points and primary objects of enquiry.[19] It is possible, nevertheless, to distil from this complexity four broad conceptualizations, or models (in the informal sense), of the interplay between knowledge and policy, and to relate these to the different ways in which advisory bodies, particularly those set up to provide expert advice to governments, have been imagined or represented.

Technical Rationality

In what is probably the most enduring conceptualization, the connection between knowledge and policy is seen as a linear, uni-directional one in which science and other forms of enquiry inform an apparently rational and ordered process of public policy formation. This is a model in which facts exist independently of values and can be tracked down, distilled, and conveyed to decision-makers at appropriate stages of the policy process—most obviously when problems have to be characterized and different possible strategies evaluated.[20] Policy analysis (in the sense of analysis *for* policy[21]) emerged in the second half of the twentieth century precisely to provide such an input, developing a multiplicity of methods and practices for assessing alternative courses of action. Experts, in this linear–rational model, have an analytic role, furnishing information in a dispassionate way for the guidance of those in power. This is, of course, a familiar representation of royal commissions and scientific advisory committees: both have long been portrayed as uncovering 'the facts' relating to policy problems, 'filtering and making usable expert knowledge' (Bulmer 1993: 48), and thereby providing authoritative, objective advice.

Perhaps the most striking feature of this representation is its sheer tenacity in the face of a sustained and often powerful critique. The linear–rational model has been challenged by alternative conceptions of the policy process,[22] by post-positivist and post-structuralist perspectives on knowledge and expertise,[23] and by its own frequent failure, in real-life policy controversies, to convince and reassure. Yet it constantly rebounds, never far below the surface in calls for 'sound science' or 'the facts' to settle contested issues, and reinforced by political rhetoric when advisory bodies are established or their findings invoked. Some see the endless reiteration of this model, with its insistence that advisors advise and politicians decide, as a 'democratic façade' for technocracy, in which power shifts markedly towards the experts (Hisschemöller et al. 2001: 2; Fischer 1990, 2000, 2009). Others interpret it as a form of political technology, in which power is expressed more subtly, through the very delineation of what is rational, what is routine, and what remains open for political negotiation.[24] It is conceivable, however, that the linear–rational model persists because it retains a certain intuitive appeal and a modicum of explanatory power; it might reasonably represent the relationship between expertise and policy, for example, when the issue at hand is highly 'structured'—that is, not beset by intrinsic uncertainties, deep value conflicts, or high political stakes (see, for example, Funtowicz and Ravetz 1985, 1993; in 't Veld 2009; Owens, Rayner, and Bina 2004; Pielke 2007). And despite the manifest failings of the model, we should not dismiss out of hand the possibility of a well-meaning quest for guidance, or a willingness on the part of advisors to provide it as best they can. It remains the case that governments, on some occasions, are 'in genuine doubt as to what to do' (Baker 1988: 178) and may seek expert advice 'simply because they want to make the right decisions' (Peters and Barker 1993b: 2). The origins of the Royal Commission on Environmental Pollution, discussed in Chapter 3, would seem to be attributable, at least in part, to such a motive.

Political Rationality

One of the obvious shortcomings of the linear–rational model is its presumption of technical rationality—of policy-makers being guided by truths delivered from outside. In contrast, many political scientists have seen knowledge (when they have attended to its use at all) as an instrument of *political rationality*—not so much a neutral input as a resource to be deployed selectively and strategically in the interplay of interests, institutions, and power. In these strategic perspectives, knowledge features as an intervening variable, while the dynamics of policy processes are variously attributed to the jostling of pressure groups, to systemic structures in society, or to the workings of bureaucracies or institutional norms.[25] The point is not that knowledge is insignificant. It is treated, however, as an 'epiphenomenal expression of material interests' (Litfin 1994: 3), rather than as a variable whose role is worthy of analytical attention in its own right.

Like the linear–rational model, this 'strategic' account of the role of knowledge aligns with a familiar, even popular, representation of advisory bodies. But this time it is one in which they have an instrumental or symbolic function, and arguably it has a stronger grounding in real experience. Setting up a commission or committee of enquiry is, after all, a time-honoured device for removing difficult issues from the political arena: Gladstone was aware, in 1869, that '[a] committee keeps a cabinet quiet' (Morley 1903: 289),[26] while Macmillan, nearly a century later, saw deflecting the question of air pollution as an excellent way for government to 'seem to be very busy'.[27] Harold Wilson, in the 1960s and early 1970s, was particularly fond of establishing royal commissions, and not averse to rigging them to produce the desired results (Hennessy 1989). In the contemporary academic literature, too, it is widely recognized that governments in power use advisory bodies to placate and depoliticize, sometimes casting political issues into the realm of 'science', and always appealing to the ideal of dispassionate analysis. On several occasions, as we shall see, governments attempted to use the Royal Commission on Environmental Pollution in this way, though such attempts were never particularly successful.

Advice, in this strategic perspective, can be used in either manipulative or opportunistic ways. In the former sense, an advisory body might be constituted, through careful choice of members and terms of reference, with placatory or legitimizing intentions in mind;[28] or it might be captured by bureaucratic or vested interests. In extreme cases, the 'right' advice might even be bought— hence the emphasis on financial probity in modern codes of practice. But even if not controlled or captured in such crude ways, advisory bodies can be used or ignored when expedient. Thus an authoritative report can help legitimize a favoured policy, or enable ministers 'to change their minds without losing face or having to admit error' (Boehmer-Christiansen 1995: 197), while political actors both within and outside government deploy findings and recommendations as ammunition in policy controversies. Dorothy Nelkin (1975: 54) noticed such selectivity in planning conflicts, for example, when protagonists drew on expert advice in ways that 'embodie[d] their [own] subjective construction of reality'. Advisory bodies can also be used as conduits, particularly if (like royal commissions) they receive and interpret evidence from a wide variety of sources; indeed, Weale (1997: 100) specifically refers to the Royal Commission on Environmental Pollution in this way. Such a role might be seen as an analytic one if advisors are deemed capable of filtering out 'the facts', but as strategic when they enable interest groups (including, perhaps, less powerful ones) to gain leverage and promote their favoured objectives (Richardson and Jordan 1979; see also Bulmer 1993; Cartwright 1975).

The linear–rational and strategic models of relations between knowledge and policy are themselves traded in complex controversies, with the former persistently invoked while the latter is widely suspected. But neither of these conceptions, even in their more nuanced and subtle variations, can account in a satisfying way for the diversity of processes and outcomes actually observed. Over several decades, developments in two (largely separate) intellectual traditions have offered alternative approaches, each attending more explicitly (though in different ways) to the 'knowledge' component of the knowledge–policy interface. The first—mainly an extension of work on policy- and decision-making in political science and organizational theory—has produced a set of cognitive and discursive perspectives on the policy process in which knowledge (broadly defined) is afforded a more substantive, and more complex, role than traditional perspectives would allow. The second, building on a long tradition in science and technology studies (STS),[29] has focused on the production and validation of (primarily scientific) knowledge in the regulatory processes of modern states. Although these developments have, to a surprising extent, proceeded independently of one another, their convergence offers a potentially powerful framework for investigating expert advice and its influence in the environmental domain. This potential will become clearer if the two bodies of work—providing the third and fourth of our models of knowledge–policy relations—are considered in turn, and their points of intersection identified.

Cognitive Perspectives

One of the first political scientists to emphasize the cognitive and discursive dimensions of the policy process was Hugh Heclo (1974), in his seminal study of social policy in Britain and Sweden. Dissatisfied with overly rational or interest-based accounts of the policy process, Heclo argued that public policy had to be seen as a product of many different forces. Moreover, its development involved a process of 'learning', which was essentially social and interactive in nature. Governments, he insisted, 'not only "power"...; they also puzzle. Policy-making is a form of collective puzzlement on society's behalf; it entails both deciding and knowing' (ibid.: 306). Others have taken forward and developed these ideas, producing a body of work in which knowledge might best be regarded as 'an autonomous variable, characterized by a dialectic link with power' (Radaelli 1995: 164). In what Weale (1992: 57) describes as the 'idiom of policy discourse', cognitive and intellectual components of policy-making, including belief systems, are significant in themselves and not reducible to 'the push and pull of mechanical forces' (see also Haas 1992; Kingdon 2003; Litfin 1994; Majone 1989; Roqueplo 1995; Sabatier 1987; Weale 2010). Thus the critical question, from this perspective, is not whether knowledge or power predominates, but *'when and how* knowledge matters' (Radaelli 1995: 160) in the evolution of public policy. For many authors, the answer has something to do with the capacities of political actors, institutions, and even whole societies to learn.

Learning is therefore a central concept in cognitive models of the policy process, and it has been much developed and extended since Heclo (1974: 306) defined it as 'a relatively enduring alteration in behaviour that results from experience'.[30] But two distinctive processes stand out in otherwise diverse accounts. One is a form of learning that enables actors to become more effective within an existing policy paradigm: drawing on evidence or experience, for example, they might work out how to design better policy instruments in pursuit of a given objective. The other is a more profound and transformative process in which policy norms and objectives, and even problems themselves, come to be challenged and redefined. These two forms of learning have variously been described as 'simple' and 'complex' (Jachtenfuchs 1996, after Nye 1987), 'instrumental' and 'social' (May 1992), 'single-' and 'double-loop' (Argyris and Schön 1996), and 'technical' and 'conceptual' (Fiorino 2001, after Glasbergen 1996).[31] Both forms are of considerable interest if we seek to understand the evolution of public policy, because it seems that different forms of learning can be associated with different levels of policy change.

The nature of this association remains in dispute, however, especially when accounting for fundamental shifts in policy, which are relatively infrequent. In the 'advocacy coalition framework' developed by Paul Sabatier and his colleagues such shifts require 'perturbations in non-cognitive factors external to the [policy] subsystem' (Jenkins-Smith and Sabatier 1994: 183). 'Policy-oriented learning' is driven in this framework by the interactions between rival coalitions, but it is mostly of the simple, or technical, kind and its effects are largely confined to the peripheral aspects of belief systems and policies.[32]

Peter Hall (1993), in his study of British macro-economic regulation, similarly links 'normal policymaking' (ibid.: 279), involving what he calls first and second order change,[33] with the simpler form of learning, which (in his account) takes place within relatively closed policy communities. But Hall goes further, to implicate a more complex form of learning in radical, 'third order change' (ibid.), through which a new policy paradigm comes to replace the old. What shifts in this process is the whole interpretive framework of policy, 'embedded in the very terminology through which policy-makers communicate...about their work' (ibid.). Something similar is envisaged in Baumgartner and Jones' (1991, 1993, 2002) model of 'punctuated equilibrium', when changes to the 'policy image' are involved (see also Baumgartner 2006),[34] and in work that emphasizes the significance of frames and framing, through which particular problem definitions, knowledge claims, and policy options are brought to the fore while others are removed from consideration.[35] Baumgartner and Jones (1991) have shown, in addition, that the site or 'venue' within which policies are shaped can be of considerable significance in these processes (the shift of venue from the UK to European institutions in the 1980s, discussed in Chapter 2, provides a good example). In focusing on interpretive frameworks, such studies turn attention to the discursive dimensions of policy-making, and show that learning can be less of a rationalistic process than 'a struggle for cultural or discursive hegemony' (Hisschemöller et al. 2001: 6).

Cognitive perspectives take us beyond the rational, sequential models in which expert knowledge simply 'informs' the policy process, and they add a new dimension, at least, to theories that privilege 'the balance of organized forces' (Kingdon 2003: 163). They sit most comfortably with conceptions of policy-making as a messy, non-linear affair, which, as Kingdon (ibid.: 225) reflects in his analysis of agenda setting in the United States, 'doesn't seem incremental...or hierarchical, or tightly rational, or driven simply by power and pressure politics, or tracked into simple chronological stages'.[36] The point is that within this complex system, different forms of knowledge and processes of learning are of considerable, sometimes fundamental, significance, though always in combination with other factors. Kingdon himself envisages three 'process streams'—involving problems, policies, and politics—which develop, for the most part, independently; policy ideas can 'float around' (ibid.: 122), for example, without necessarily becoming attached to problems. But at 'critical junctures', which might be brought about by exogenous events or engineered by 'policy entrepreneurs', the streams merge, and such convergence can be a precursor to substantial policy change (ibid.: 19, 20). Cognitive factors, in Kingdon's model, are at work predominantly in the problem and policy streams. Ideas, and the framing of problems and policy options, are important, as is the 'gradual accumulation of knowledge and perspectives among the specialists in a given policy area and the generation of policy proposals by such specialists' (ibid.: 17).

Given the complex and interactive role of knowledge in cognitive perspectives on the policy process—the third of our models of knowledge–policy interactions—we might expect to find expert advice serving neither a purely analytic nor a purely symbolic function. Surprisingly, perhaps, experts and advisors are not always the subject of explicit analytical attention in these accounts, and sometimes they are present in a rather shadowy capacity. Still, their role is acknowledged, and generally deemed to be important. They are seen, for example, as occupying 'privileged positions at the interface between the bureaucracy and the intellectual enclaves of society' (Hall 1993: 277), from which they can act as brokers of knowledge and ideas,[37] as policy entrepreneurs, developing and promoting particular proposals (Kingdon 2003), or as advocates, attempting 'to redirect the...attitudes, preferences, or cognitive beliefs' of policy-makers (Majone 1989: 38; Pielke 2007). All of these roles were recognized by earlier observers of commissions and committees of enquiry, even if they expressed them in different terms. Chapman (1973a: 187), for example, noted how commissions could have indirect effects 'by breaking the ice so that changes, including those not previously mentioned at all, become discussable', while for Cartwright (1975: 217) such bodies could 'go where ministers and their officials might hesitate to tread'. And, as Wheare argued as long ago as the 1950s, these incursions could have lasting influence, even if a body had been established with rational or symbolic intent:

> [T]he value of the work of a committee to inquire goes beyond [fact finding]. It is intended to do something to educate opinion. Its report, and often the evidence submitted to it, are a contribution to the study of the subject and the basis of discussions about it both in the [government] department and outside. Even, therefore, when a department has no very strong intention of acting upon a committee's report, it by no means follows that the committee's work is wasted. It may perform an educative function, which in the end, perhaps, may lead to some action.
>
> (Wheare 1955: 89)

In effect, in cognitive models as well as in earlier, more descriptive accounts, advisory bodies are portrayed as agents of learning, in both its rationalistic and discursive forms. Of course, it might be argued that learning in the former sense is precisely what expert advisory bodies as 'rational analysts' are established to bring about, so that the linear–rational and cognitive models have something in common at this point. But it is the role of advisors in the more complex, discursive form of learning that is of particular interest in the context of substantial policy change. This too might be intentional: a body like a standing royal commission is set up at least in part with an 'educative function' in mind and, as Heclo (1974: 313) observed in his analysis of social policy, institutions like enquiries and investigatory committees can be 'consciously designed to aid in political learning'. Beyond such intent, however, advisory bodies can aid learning in less purposeful ways, when their work helps 'gradually to change the...vocabulary and interpretative frames of policy-makers' over time (Radaelli 1995: 164). In this sense, they might be said to exert 'atmospheric influence' (James 2000: 163), and to contribute to the slow and diffuse processes of 'enlightenment' (Weiss 1977: 531; see also Weiss 1975, 1991).[38] These effects are not easily planned or controlled, and nor should we expect them to be readily or directly observable.

Co-production and Boundary Work

Our fourth and final perspective on knowledge and policy has a number of features in common with the third: it challenges the familiar accounts in which either technical or political rationality is seen as dominant, and it treats knowledge itself as an object of critical attention. More specifically, this representation draws on a long tradition of fine-grained enquiry in historical and sociological studies of science, which has exposed the 'contingent and negotiated character' of scientific knowledge (Jasanoff 1990: 2; also Irwin 2008); and, importantly, it extends these insights into an analysis of contemporary science–policy relations. It might be seen, therefore, as broadening Radaelli's (1995) question of 'when and how knowledge matters' in the policy process to ask, in addition, how knowledge-making itself is influenced by social and political formations. Significantly, it problematizes the very concept of a 'knowledge–policy interface', with its implication of discrete and distinctive spheres on either side.

A central tenet of this approach is that the boundaries separating 'science' from other human endeavours—in this case politics and policy-making—are not readily defined by any essential characteristics of these categories; instead they must actively be constructed, negotiated, and defended by scientists and decision-makers alike. Such processes, which Thomas Gieryn (1983, 1995) first referred to as 'boundary work', have been especially apparent in the domain of 'post-normal' (Funtowicz and Ravetz 1985, 1993) or 'regulatory' (Jasanoff 1990) science, where controversies are characterized by high stakes, profound uncertainties, and deep value conflicts (see also Fischer 2009; Guston 2001; Hilgartner 2000; Hoppe 2005; Miller 2001). In such circumstances, it is not difficult to see how boundary work might help to sustain the linear–rational model in which science is separated from policy, removing important issues from the political agenda in the process. But, as Sheila Jasanoff (1990) has shown, boundary work can *re*-politicize as well as depoliticize, casting 'scientific' questions firmly back into the sphere of democratic deliberation. As we shall see, the Royal Commission performed this role on a number of significant occasions.

In this fourth perspective, some envisage science and politics influencing each other in a reciprocal relationship, or 'coupling' (Weingart 1999: 157), across a boundary that is never fixed. Others describe a more complex process in which facts (the realm of knowledge) and values (the realm of politics) 'interpenetrate' (Ezrahi 1980: 120), becoming almost impossible to tease apart: risk controversies (explored in Chapter 4) are among the best-documented examples of this phenomenon (Funtowicz and Ravetz 1985, 1993; Irwin 2001; Ravetz 1990; Wynne 2002). Jasanoff argues that science and politics in such domains are not usefully represented as separate spheres at all: the point is not that these worlds are indistinguishable, but that they are mutually constitutive—a clear case, in contemporary knowledge societies, of the natural and social orders being 'co-produced' (Jasanoff 1990, 2006a, b, c).[39]

Of all the perspectives on knowledge and policy considered here, it is this fourth, built around concepts of boundary work and co-production, that pays the most explicit attention to the role of expert advice; indeed, it is the only one that interrogates the very notion of expertise and its functioning in modern democracies (see, for example, Funtowicz and Ravetz 1990, 1993; Jasanoff 1990; Renn 1995; Weingart 1999). Jasanoff's (1990) seminal study of expert scientific committees advising US Federal agencies is of particular interest here, since it was the first to look in depth, from an STS perspective, at a scientific advisory system.[40] Its findings are important not only because they challenge conventional representations, but also because they are ultimately positive and optimistic about the role of expert advice. Jasanoff argues that neither a technical nor a political rationality dominates the interactions of advisory committees with policy-makers. The advisors do nothing so predictable as speaking truth to power, still less 'seizing the reins of decisionmaking from political institutions' (ibid.: 250). But nor can they simply be seen as pawns in the political system. Instead, and in a range of circumstances, they engage in negotiation and role playing, and perform 'skilled boundary work' (ibid.: 237) through which their expertise comes to be upheld as authoritative and 'scientific'. For Jasanoff, this is a flexible and often valuable process, helping to produce 'serviceable truth[s]' which meet hybrid criteria to satisfy the needs of both science and politics (ibid.).[41]

Bijker, Bal, and Hendriks (2009) reached similar conclusions in their analysis of the work of the *Gezondheidsraad* (the Health Council of the Netherlands)— one of a small number of detailed investigations of individual advisory bodies, and therefore also of special interest here.[42] The authors found that the *Gezondheidsraad* succeeded 'most of the time, in preserving the role of an independent and credible scientific institution' (ibid.: 163), and that it did so largely by 'draw[ing] boundaries and then relat[ing] the newly created domains in specific ways' (ibid.: 4). Like Jasanoff, they emphasize that 'science' and 'politics' are not simply collapsed into one another in this process, but rather that boundary work provides a means of both 'distinguishing and co-ordinating' these social worlds in ways that are beneficial to both (ibid.: 149). Seen in this light, scientific advisors perform a critical function in the governance of technological societies, becoming 'part of a necessary process of political accommodation among science, society, and the state' (Jasanoff 1990: 250).

A related, but broader concept is that of the 'boundary organization' (Guston 1999, 2001)—significant here because it embraces bodies that may not be purely scientific in purpose and composition. Indeed, it is characteristic of these organizations that they exist at the frontier of science and politics and typically involve actors from both sides; their boundary work consists in part in the creation of 'boundary objects' (such as reports, models, or ideas), which can be interpreted differently by individuals within each social world 'without losing their own identity' (Guston 2001: 400, after Star and Griesemer 1989).[43] International institutions within the global climate change regime have, for example, been interpreted as boundary organizations (Hoppe, Wesselink, and Cairns 2013; Miller 2001), as have the European Environment Agency (EEA) (Dammann and Gee 2011; Scott 2000), the Dutch Advisory Council for Research on Spatial Planning, Nature and the Environment (*Raad voor Ruimtelijk, Milieu- en Natuuronderzoek*, RMNO) (de Wit 2004, 2011), and the Office of Technology Transfer of the US National Institutes of Health (Guston 1999). In some cases (those of the EEA and the RMNO, for example), boundary work is treated not simply as a (perhaps unconscious) product of the structures and practices of such bodies, but as a proactive, intermediating activity, capable of being planned and managed.

There is an interesting affinity between these contemporary insights on boundary work, emphasizing its constructive dimensions, and the understanding of earlier commentators that the strategic and symbolic functions of policy advice could be part of the necessary and normal processes of governing (see, for example, Rhodes 1975; Wheare 1955).[44] These insights also chime with cognitive perspectives on the policy process, in which legitimizing functions (which have an important discursive component) are interpreted positively as a sign that political actors cannot get their way simply by 'powering', to use Heclo's term, but find it necessary to 'persuade and convince' (Radaelli 1995: 174). We should be wary, then, of an undue emphasis on manipulation, and recognize, with Litfin (1994: 15), that the power of scientific discourse—and by extension, advice and expertise—can have 'positive, capacity-giving dimensions' (see also Flyvbjerg 1998; Majone 1989; Weale 2010).[45]

Before turning to consider the role of advisory bodies in more detail, one more point should be made about the third and fourth models of knowledge–policy interactions. As noted earlier, the cognitive and co-productionist perspectives derive from different intellectual traditions, between which there has been a somewhat limited conversation. There are, however, some important exceptions. In particular, close encounters between these perspectives can be found within the post-positivist critique of classical policy analysis,[46] and in the interpretive studies of the policy process to which it has given rise. Many such studies emphasize both the discursive dimensions of policy-making and the mutually constitutive nature of knowledge and power, and empirically they have often focused on risk and environmental controversies. Prominent examples include Maarten Hajer's (1995) work on the politics of acid rain in Britain and the Netherlands, Karen Litfin's (1994) analysis of the international negotiations leading to the Montreal Protocol, and Bent Flyvbjerg's (1998) study of the interplay of rationality and power in transport planning in the city of Aalborg, Denmark. All attend to the production of knowledge, to the intricacies of policy and political processes at different scales, and to the complex and contingent intermingling of these spheres.

Positioning Advisory Bodies

All of the perspectives on knowledge and policy outlined above allow for expert enquiry and advice to be implicated in the genesis and evolution of public policy, though they tell different stories about the interplay of rationalities, interests, and power. As rational analysts or political symbols, advisory bodies serve the purposes of others, and their function might be characterized as *instrumental*. As cognitive and discursive agents, they can develop a degree of *autonomy*, in the sense that they may not only (or not even) serve the purposes for which they were established.[47] As participants (conscious or otherwise) in boundary work and co-production they have hybrid functions, in part technocratic but also constructive in stabilizing relations among science, society, and the state. It will be clear, of course, that these categories are neither tidy nor mutually exclusive: they merge and overlap, both conceptually and in practice, as do the broader models of knowledge–policy interactions to which they relate. An intended analytic role can become symbolic, for example, and advisors have discursive functions (including those of boundary work) when acting ostensibly in either of the former capacities. Moreover, as Chapters 3–7 will demonstrate, autonomy (or perceived autonomy) can serve an instrumental purpose, while hoped-for legitimizers can and do go 'off remit', acquiring autonomy despite any strategic intent.[48] We should expect the various roles to be performed in different measure according to the advisory body concerned and the nature of the issue in question, and it is both possible and productive for advisors to perform multiple roles, sometimes simultaneously. Indeed, Jasanoff (1990: 237) sees such versatility as an important advantage of scientific advisory committees, whose members, '[p]rotected by the umbrella of expertise... in fact are free to serve in widely divergent professional capacities: as technical consultants, as educators, as peer reviewers, as policy advocates, as mediators and even as judges.'

The most promising framework for the analysis of expert advice would seem to be one that combines the cognitive and co-productionist models of knowledge–policy relations. The former allows for knowledge to be used 'rationally' or strategically, and to have a greater or lesser effect depending on circumstance; in this sense, cognitive perspectives might be said to embrace both the linear–rational and strategic models as 'special cases'. But the co-productionist idiom brings much-needed critical attention to scientific knowledge and to the possibility (and the practices) of its separation from politics. Combining these perspectives enables us to see how an advisory body might perform different, and sometimes multiple, roles, such that careful examination of its work in specific contexts might help illuminate both the functioning of expert advice and relations between knowledge and policy more broadly defined.

There is, however, one final issue to be addressed in establishing the conceptual framework: if advisory bodies have agency, through what mechanisms (beyond the formal procedures of investigating, reporting, and making recommendations) might we expect their agency to be expressed? To understand such processes, especially in their more subtle forms, we need to look to the embedding of advisory bodies (and their members and secretariats) within personal, professional, epistemic, and policy networks. In the models outlined above, networks are attended to most explicitly in cognitive and discursive perspectives on knowledge–policy interactions, where they are treated both as structures within which transfers of ideas and information take place and as agents of learning in themselves. Kingdon's (2003) 'specialists', for example, would probably be part of an epistemic (or knowledge) community, circulating and promoting ideas in particular policy domains; Sabatier's (1988) advocacy coalitions are seen as central to the process of 'policy-oriented learning'; and Hajer's (1993, 1995) discourse coalitions, held together by 'storylines', as significant actors in the construction of policy frames.[49]

Network structures vary between policy domains, and different kinds of network can co-exist within a given domain at any one time. The structures are also dynamic. Since the mid–late twentieth century, for example—the period of interest in this book—tightly knit policy communities in many domains have been challenged, and sometimes replaced, by looser, more complex, and potentially less stable formations: such changes have certainly been characteristic in the field of environmental policy. The significance for our analysis is that emergent networks, in which power and interests may be 'redefined and relocated' (Hajer and Wagenaar 2003b: 5; see also Bulkeley 2005; Hajer and Versteeg 2005b), can provide important new settings for policy.[50] Hall (1993), for example (in the macro-economic domain), found that a loosely structured issue network was deeply implicated in the process of third order policy change (the shift from Keynesianism to monetarism), in part because it enabled 'outsiders' to influence a previously closed policy process (Hall 1993: 287, 289). Similarly, Saward (1992) associated the opening up of a tightly drawn nuclear policy community with substantial changes in British nuclear energy policy—of special note here because this was a process in which the Royal Commission on Environmental Pollution was closely involved.

Changes in network structures have been associated by some observers with a shift from government to governance, in which policy-making in modern democracies has become less state-centred and hierarchical, and more inclusive of a range of non-state actors. This theory presents something of a paradox for a study of the Royal Commission, which in some ways might be seen as a vestige of what Hajer and Wagenaar (2003b: 10) call 'classical–modernist government'. Yet the Commission was itself instrumental in the opening up of policy communities, it occupied a powerful position at the intersection of different networks, and it was sometimes an active agent within advocacy or discourse coalitions. These different roles, and the implications for the Commission's influence, will be considered in more detail in Chapters 3–7.

Investigating the Royal Commission: A Note on Methods and Sources

The previous sections give a sense of the extent and diversity of material that engages in some way with relations between knowledge and policy–political processes, and with the roles of advice and expertise. One notable point of agreement in this large body of work is that fine-grained analysis in a range of different contexts is vital if we are to understand how 'powering' and 'puzzling' interact (Hall 1993: 292; see also Bennett and Howlett 1992; Jasanoff 1996; Litfin 1994; Pielke 2007). Many analysts further emphasize the need to study these relationships over extended periods of time. The investigation of the Royal Commission on Environmental Pollution on which this book is based satisfies both of these criteria. It draws on the range of disciplinary literatures and traditions identified in this chapter, seeking both a synthesis of different insights and a constructive connection between theory building and 'forensic' empirical research.

The discussion so far points to a number of important questions about the Commission, its working practices, and its role and influence in the environmental policy process. We should ask how the functions of this advisory body were defined and perceived, which (if any) of the characterizations of advisors identified in this chapter it played out in practice, and whether and how these roles varied across policy contexts and over time. As with any advisory body, critical questions arise about the sources of legitimacy and trust, and about the maintenance of these attributes as circumstances change. Such questions are all the more interesting given the decreasing deference, over the Commission's lifetime, towards scientific and other forms of authority, including expert advice. Reporting on this issue at the turn of the century, the House of Lords Select Committee on Science and Technology (2000) found that public interest in science co-existed with deep unease about certain science-based technologies, as well as 'increasing scepticism about the pronouncements of scientists on science-related policy issues' (para. 2.2; see also Cabinet Office 2002). The Committee noted, in addition, that 'public confidence in scientific advice to Government ha[d] been rocked by a series of events, culminating in the BSE fiasco' (ibid.: para. 1.1).[51] In parallel with this 'crisis of trust' (ibid.: para. 2.2), the late twentieth century (in the UK and elsewhere) saw a growing demand for the democratization of expertise, and a shift, as some would see it, towards more dispersed and accountable processes of knowledge production (Nowotny, Scott, and Gibbons 2001, 2003; see also Gibbons et al. 1994). Such changes suggest that the Commission's working practices as an advisory body merit special scrutiny: it is through these practices that topics were selected, problems framed, written and oral evidence interpreted, and recommendations constructed. And it is necessary, of course, to scrutinize activities 'behind the scenes', as well as those that were formally and more visibly conducted.

A central question for this book is that of the nature and extent of the Commission's influence. But influence is a complex concept, and discerning it is unlikely to be straightforward. Because of the intricacies of the policy process (and of its relations with expertise), there is a danger of false positives—of crediting the Commission with changes that might have happened anyway. Conversely, when the line of influence has been indirect or obscure, significant longer-term impacts might be overlooked. There are two implications: first, that we have to look critically at the classic sequence of 'report, recommendation, and action'; and second, that we must look beyond this sequence to more subtle ways in which the Commission might have exerted influence over time. Did it, for example, operate quietly behind the scenes, or invisibly within wider networks, to promote particular ideas or frames? We must consider, too, why some of its reports were highly influential while others had little or no effect: in this respect, Chapter 6 will seek to identify 'the circumstances of influence', in the short to medium term as well as over extended periods of time.

In seeking to achieve these ends, the study of the Royal Commission drew its evidence from four main sources. First, there was a substantial amount of material in the public domain, including the Commission's own reports, government responses, parliamentary papers, policy documents, and analyses and commentaries of various kinds. These were vital not only in providing general background, but also in tracing the development and fate of the Commission's ideas and recommendations. Unpublished material was also available, particularly in the archives of the Commission and those of the government's environment department.[52] In addition, a number of individuals provided correspondence and personal papers. A third source, for the period up to the mid-1990s, consisted of around 100 in-depth interviews with people who had first-hand experience of the Commission (as chairs, members, or employees within the Secretariat), or had been otherwise engaged with its work: the latter group included civil servants, politicians, regulators, special advisors, academic observers, and members of wider environmental policy communities.[53] Interviews proved an extraordinarily rich source of information, revealing much about the routine practices of the Commission, the reception of its advice in policy communities, and the beliefs, values, and experiences of those involved. Finally, from 1998 to 2008 the author was herself a member of the Commission, providing direct experience and a perspective that is not available from any of the other sources. While several former members have written about the Commission, previous accounts have generally been brief, and none has combined insights from participant observation with an intensive study based on interviews and documentary sources. This unusual opportunity to look at the Commission both from the outside and from within offered the potential for a rich and reflexive analysis.

All of these sources, individually, have limitations. It is well known, for example, that documents can be selectively written and selectively archived, while oral accounts are coloured by personal experience and depend on recollections of variable reliability. As one senior civil servant admitted (and historians know very well), there is a tendency with hindsight to re-write the past: 'My basic conception is that anything good that happens, it turns out that practically everyone was playing a crucial part in formulating the ideas...and anything bad that happens they were somewhere else at the time, or advising against it.'[54] Insider accounts, though they offer privileged access to experiences, knowledge, and events, cannot, by definition, be detached or objective (Booth 1990). Nevertheless, a combination of these approaches can be powerful; in this study it provided invaluable insights into the internal practices of the Commission, the emergence and development of arguments and ideas, and the interplay of knowledge with other important variables in policy–political processes. Documentary evidence and interviews together, for example, could show that many Commission reports were cited as a basis for 'normal policymaking', indicative of instrumental, or 'single-loop' learning, while some could be implicated in the reframing of problems and policies over extended periods of time: these findings will be considered in Chapters 3–6. Accounts did indeed vary, but there was a reassuring degree of convergence. When the recollections of a wide range of individuals are triangulated with published reports, unpublished material, and personal experience, the picture that emerges, if not exactly the 'truth', is one in which we can have a reasonable degree of confidence.[55]

As well as exploring the practices and impacts of a unique institution, the study was concerned with wider questions about the role of expertise and advice in the policy processes of modern democracies. Given the complexity of the interactions involved, the variety of contextual factors, and the unpredictability of random events, it would be inappropriate to search for grand generalizations[56]—and even if this were the object, the scrutiny of a single advisory body could hardly aspire to fulfil it. There are good reasons, nevertheless, to feel that the study can contribute to theory-building through interpretive analysis, paying close attention to language, practices, and outcomes in particular contexts. One is that the Commission, although distinctive in many ways, exemplified a familiar format in which bodies set up by government 'consider specific policy problems... gather evidence about them, and... report and make recommendations for action' (Bulmer 1980a: 1). Another is that the study embraced multiple investigations conducted by the Commission over a period of more than forty years; in effect, therefore, it comprised not just one analysis in context, but a series of interconnected case studies in circumstances that varied widely over time. This long view adds depth to the analysis and enables reasonably robust conclusions to be drawn. And if, as some analysts maintain, even a messy and unpredictable policy process is likely to be characterized by at least some regularities when assessed over a sufficiently extended period (Baumgartner and Jones 1991; Kingdon 2003),[57] we can be optimistic that the findings of the study will be of interest and applicability beyond the specific subject of the Royal Commission and the environmental policy domain.

Outline of the Book

This chapter has set out the broad conceptual framework and background for the analysis in the remainder of the book. The next two chapters provide the historical and policy–political context. Chapter 2 looks at the evolution of environmental policy, identifying the norms, pre-occupations, trends, and external events that provided a shifting backdrop for (and were sometimes influenced by) the Commission's work. Chapter 3 then turns to the Commission itself, exploring its origins, constitution, working practices, and positioning within policy communities. It also considers the broad sweep of the Commission's enquiries over time.

To cover the whole of the Commission's work and its impacts on environmental policy at the same level of detail would require many volumes. Whilst retaining the 'big picture', therefore, certain key themes are developed in greater depth in Chapters 4 and 5. The former traces the Commission's changing perspectives on risk and precaution—key issues for environmental policy, which featured prominently in many of its reports. Chapter 5 then turns to its long-term advocacy of 'integration', considering first its proposals for integrated pollution control, which ultimately came to fruition, and, second, its more radical (but less successful) case for the better integration of land use and environmental planning. The focus expands again in Chapter 6, to identify the roles that the Commission played, its spheres of influence, and the circumstances that pertained when, at different times, its advice was accepted gratefully, used strategically, rejected (at least initially), or ignored.

Finally, Chapter 7 draws together the findings of the study and revisits the wider questions posed above about knowledge, policy, and the practices of expert advice. It seeks to identify those attributes of the Royal Commission that enabled it, on many occasions, to be effective, as well as the factors that ultimately led to its demise. This final chapter reflects, too, on the challenges of generating 'good advice' within changing structures of governance and increasingly complex policy domains.

2

The Environment in Politics

> When laws and regulations, or organisations and advisory bodies were established...in the 1960s and 1970s, what was established was not simply a configuration of policy institutions, but also a *process* of policy exploration and development. This process has itself contributed to the new politics of pollution.
>
> Albert Weale, *The New Politics of Pollution* (1992: 211)

> [P]eople understand the bigger problem of the ecological crisis through the example of certain emblems.
>
> Maarten Hajer, *The Politics of Environmental Discourse* (1995: 5)

> It is no longer weird to worry about the future of the planet.
>
> Nick Rowcliffe, in *ENDS at 30* (2008: 3)

Introduction

Since the 1960s, when 'the environment' first rose to prominence as a category for political action and concern, environmental politics and policies have changed almost everywhere beyond recognition. This transformation presents something of a dilemma. It must clearly provide the setting for any discussion of the Royal Commission on Environmental Pollution, whose activities were woven into the fabric of environmental policy at different scales. At the same time it is impossible to do justice to the vast array of political, institutional, and conceptual developments that shaped and were shaped by policy changes over a period of around half a century.[1] This chapter has more modest ambitions: it provides an overview of environmental policy from the late 1960s to 2011, focusing on key developments and trends affecting the UK and connecting in some way with the remit of the Royal Commission.[2] Many of the policy processes in which the Commission has been implicated will be explored in more depth in Chapters 3–6.

For most of its life, the Commission's advice was directed primarily at the UK government, and we might expect its effects on policy to be most apparent on a UK-wide, national level. Other influences, however, transcend national boundaries. In particular, Britain's membership of the European Union (formerly the European Economic Community) from 1973 resulted in 'a major shift in the locus of environmental policy making towards the European level' (Lowe and Ward 1998a: 25). By the mid-1990s, some 80 per cent of British environmental legislation was estimated to originate in Europe,[3] and this is of singular significance if 'hierarchically imposed reform' is among the exogenous factors that can destabilize a policy core (Jenkins-Smith and Sabatier 1994: 191) or influence the formation of networks (Marsh and Rhodes 1992a). Nor has the relationship been all one way: as Lowe and Ward (1998a: 3) point out, European politics and policy-making have themselves been shaped 'by the culture, agendas and actions of the individual member states'. We must look, therefore, to European institutions and policies when considering both the context for, and the impacts of, the Royal Commission's work. Developments on a wider international stage must also form part of the framework, since over the period concerned many environmental issues with transnational or even global ramifications became the subject of international negotiations, treaties, and conventions.

At times, then, the Royal Commission's work intersected in significant ways with European and wider international agendas, though its recommendations were not explicitly directed towards bodies beyond national level.[4] At the other end of the scale, significant aspects of environmental policy within the UK became the responsibility (after 1997) of the devolved administrations in Scotland, Wales, and Northern Ireland, to which the remit of the Royal Commission extended.[5] Policy developments within the constituent parts of the UK are not the main focus of analysis in this book, but it is important to acknowledge that devolution affected the Commission's working practices and that the devolved administrations sometimes responded separately (and with different perspectives) to its post-1997 reports. It is arguable, too (see Chapter 7), that the complexity of looking simultaneously to multiple levels of governance was one of a number of factors that contributed to the Commission's demise.

In tracing the evolution of ideas and policies, it seems appropriate to treat the material chronologically. To that end, four eras of environmental policy, of somewhat unequal length, are addressed in turn in this chapter: they cover the late 1960s to the end of the 1970s; the 'long 1980s' (roughly 1979–91); the period from 1992 (the year of the UN Conference on Environment and Development in Rio de Janeiro) to the end of 2007; and from then until 2011, the last year in which the Royal Commission was active. The rationale is that certain distinctive characteristics can be attributed to each period, though any such division is 'fraught with peril' (Haigh 1995: 7) and the device is used here as one of narrative convenience rather than rigid classification. Another point should be stressed at the outset: in a rapidly expanding field, it becomes tricky to tease out even the most significant and far-reaching developments in thinking and legislation. It has been impossible to do justice, for example, to the huge volume of activity at European level, the many initiatives spawned by the concept of sustainable development, or the multiple developments in the area of climate change, particularly in the second half of the period under consideration.[6]

In addition, we must take account of the longer-term trends that have often been gathered together under the heading of 'ecological modernization'. This influential concept, initially developed by West German social scientists in the 1980s, has both normative and descriptive–analytical dimensions, but in essence it has been concerned with a reconciliation of economy and environment through 'the institutional restructuring of late modernity' (Mol 1996: 303; see also Christoff 1996; Hajer 1996; Mol and Sonnenfield 2000).[7] Its significance in the present context is that the associated shifts in environmental policy—from reaction and fragmentation towards anticipation, precaution, and integration, and from closed to more open policy networks—have some of the characteristics of paradigmatic, or 'third order' policy change, as described by Hall (1993) and discussed in Chapter 1.[8] Changing conceptions of environmental risk, connecting with a move towards greater precaution and growing scepticism about science and expertise, have also been of great importance; they are touched upon here but addressed more specifically in Chapter 4. All of these developments constitute the backdrop against which we must assess the Commission's activities; but they could equally well be interpreted in terms of ideas and practices to which the Commission itself made persistent and significant contributions. We need to start, however, a few years before the Commission's appointment, with the stirrings of an environmental revolution.

Revolution and Beyond: The Late 1960s to the Late 1970s

The 1960s witnessed a sharp upturn in public and political concerns about the environment, focusing on pollution, population, and resources, and questioning the global capacity to sustain indefinite growth. By the end of that decade, Western governments were under pressure to respond to this new phenomenon, and to do something to tackle the many instances of environmental degradation that were now being framed as 'problems'. The pressure was intensified by seminal publications like *The Limits to Growth* (Meadows et al. 1972) and *Blueprint for Survival* (*Ecologist* 1972), and by the need to prepare for the first United Nations Conference on the Human Environment, initiated by Sweden and held in Stockholm in 1972. Most of all, governments were spurred into action by the emergence of a distinctive and politically orientated environmental movement, one of the clearest examples in the twentieth century of a challenge to established policy communities.[9]

In many countries, early institutional responses to these pressures included realignment of departments of state and establishment of quasi-independent bodies charged with environmental protection or the provision of scientific advice.[10] There was a powerful appeal to administrative and technical rationalities: new or reorganized ministries would be better positioned to coordinate environmental responsibilities and develop the requisite policies, while the formulation and implementation of the latter would be informed and guided by 'sound science'. For the most part, however, the *content* of environmental policy was not itself the subject of radical readjustment: changes were of the first or second order, to use Hall's (1993) terminology, rather than paradigmatic. Pollution control, for example, continued to focus on individual pollutants and was organized according to the medium (air, water, or land) into which they were discharged.[11] In many cases, a 'dilute and disperse' philosophy depended on the assumed assimilative capacities of the environment, and in those instances where emissions had to be reduced at source, an 'end-of-pipe' approach was favoured over changes to processes or products. In the bigger picture, environmental degradation had not yet been identified as a 'structural problem' (Hajer 1995: 25), nor was there much concern for the integration of environmental considerations into wider domains of public policy, though the US *National Environmental Policy Act* of 1969 was an early, and notable, exception (Weale 1992).

In the UK, amidst a generally heightened environmental awareness, the issue attracting sustained attention around 1970 was predominantly that of gross pollution, an emphasis clearly reflected in the Royal Commission's early reports. Problems were highlighted by campaigning journalists, who played an important and effective role in an emergent policy network: for example, Jeremy Bugler, environment correspondent for *The Observer*, graphically portrayed the 'grievous damage' inflicted by industrial discharges in a compact but influential book, *Polluting Britain* (Bugler 1972: 32; see also Tinker 1975).[12] The environmental agenda was also influenced by a series of particularizing, or focusing, events in what Kingdon (2003) would call the problem stream,[13] including pesticide-induced deaths of wildlife, the 1969 'seabird wreck' in the Irish sea (attributed to polychlorinated biphenyls [PCBs]),[14] and marine and coastal oil spills, most notably the massive pollution caused by the wreck of the supertanker, the *Torrey Canyon*, off the south-west coast of Britain in 1967. Responses included the merging of several ministries into a new Department of the Environment (the world's first cabinet-level environment ministry), the formation of a Central Scientific Unit on Pollution, and the appointment of the Royal Commission on Environmental Pollution, all taking effect in 1970. A White Paper on pollution (UK Government 1970) was also published in that year, and a flurry of legislation followed in the first half of the 1970s.

These changes were grafted onto an existing, fragmented system for environmental protection in the UK and at first barely challenged the 'traditional pragmatist' (Hajer 1995: 112) approach to setting and enforcing standards in pollution control.[15] Policy communities were tight and relatively closed, with industrial interests closely engaged in policy and regulatory processes, but environmental groups involved to a much more limited extent; this remained the case for some time, even as the movement gathered momentum (Jordan 1998; McCormick 1991; Weale 1997). Neither emission standards nor quality objectives for receiving environments were normally set down in national legislation. Rather, the regulatory agencies, operating on a decentralized basis, interpreted legislative intent in a discretionary and non-coercive manner, allowing for 'negotiated compliance' (Lowe and Ward 1998a: 8) in close consultation with affected parties.[16] Prosecutions for infringements were rare: the Alkali Inspectorate—the body then responsible for regulating major air polluters—brought only three prosecutions between 1920 and 1967.[17] In *Polluting Britain*, Bugler noted that the rate of prosecution had subsequently increased to about two per year, but that fines had remained small.

In 1972, a high-profile incident involving the illegal dumping of toxic waste led rapidly (after many years of procrastination) to the *Deposit of Poisonous Waste Act*:[18] the sequence of events is nicely described by Eric Ashby (1978), the first Chair of the Royal Commission, in his book *Reconciling Man with the Environment*. The more comprehensive *Control of Pollution Act* (CoPA) reached the statute book in 1974.[19] But the oil crisis of 1973—an archetypal 'external (system) event' (Sabatier 1987: 653)—intersected with the emergent environmental agenda in different and contradictory ways. While the shock of a fourfold increase in the price of oil focused attention on energy resources and brought issues such as nuclear power to the fore, the recession that followed, and the severe economic difficulties faced by the British government, tempered the nascent political enthusiasm to 'do something' about the environment. Implementation of key provisions of CoPA was much delayed (an issue that exercised the early Royal Commission) and, with the exception of an act in 1976 dealing with the import and export of endangered species, there was a long gap before further significant environmental legislation. In the view of one official from the environment department at that time, 'pollution as a whole was very much a backwater' for most of the 1970s.[20] But if the environment became less of an overt political priority, a number of developments in that decade nevertheless sowed seeds for more substantial and radical reform in later years.

There were, for example, emergent pressures relating not only to specific pollutants or control measures, but also to more fundamental aspects of the policy core—pressures, in effect, for policy to move in a more ecologically modern direction. One of the most important was concerned with the opening up of policy communities, which, as we saw in Chapter 1, can be a precursor to significant policy change. The cosy relationship between regulators and regulated, reflected in the low rate of prosecution, became a particular target: while bodies such as the Alkali Inspectorate presented the absence of legal proceedings as a indicator of success, their critics saw only a closed and unaccountable system, in which environmental abuses were tolerated.[21] A closely associated campaign pressed for public access to environmental information at a time when, as one senior official put it, 'virtually nothing was accessible'.[22] Other important ideas beginning to crystallize at this time included the notion that pollution prevention might make good economic sense; the concept of a less sectoral, more integrated regulatory approach; and the principle that the polluter should pay. The Royal Commission engaged with all of these issues at various times, absorbing, developing, promoting, and sometimes initiating the new ideas.

Britain's incorporation in 1973 into what was then the European Economic Community (EEC) was also of lasting significance for environmental policy, though its full implications were not immediately apparent; the environment had not been much of an issue in the vigorous debate about whether Britain should join the EEC, nor was it especially prominent within the Community itself.[23] Nevertheless, membership exposed British policy-makers to the more centralized and prescriptive continental approach to pollution control, shaped in part by the necessity of dealing routinely with transboundary problems and regulation (Lowe and Ward 1998a). They were not impressed, as Martin Holdgate recalls in his account of the early days of the Central Scientific Unit on Environmental Pollution, of which he was the first director (see Chapter 3):

> [I]t soon became clear that [civil servants] judged foreigners as the principal environmental sinners. There was a deep satisfaction with British achievements.... The Central Unit soon got the message that we were expected to proclaim our national way of doing things abroad. For there was a deep hostility to the foreign cult of emission standards—of fixing in law how much of a specified pollutant would be permitted in a particular kind of discharge.
>
> (Holdgate 2003: 179)

The 'deep satisfaction' was tested, soon after Britain joined the Community, in a series of conflicts over water directives (Ward 1998).[24] Initially, at least, British pragmatism hardly wavered, sometimes with strange effects such as the designation in 1979 of only twenty-seven British beaches as 'bathing beaches' (a minimalist interpretation of the Bathing Waters Directive, gently mocked by the Royal Commission in its tenth report[25]). Nevertheless, exposure to a different philosophy constituted a challenge to the British approach, demanding at the very least new powers of argument and persuasion. In fact—and even if the differences between the British and continental approaches were to become somewhat exaggerated (RCEP 1984)—the need to defend the status quo helped lay the foundations for substantial longer-term change, through policy refinement and expansion of the science base. Also foundational, though not immediately apparent, was the way in which membership of the European Community 'fundamentally altered the position and the powers of the environmental lobby' (McCormick 1991: 147; see also Lowe and Flynn 1989), enabling it to exploit a new venue in its promotion of particular policy images or frames, much as envisaged by Baumgartner and Jones (1991) in their analysis of policy change.

Another important venue, the UK's Parliamentary select committee system, was established at the end of the first era, in 1979. Committees of both houses were relatively accessible (through the evidence system) to a wide range of interests and perspectives, and would come to play a crucial role in environmental policy formation. At certain times, as we shall see, there was particular affinity (reinforced by shared membership) between the Royal Commission and select committees in the House of Lords. The significance of such connections within wider policy networks is discussed further in Chapters 3 and 7.

Changing the Frame: The 'Long 1980s'

A Conservative government, with Margaret Thatcher as prime minister, was elected in May 1979—a year in which oil prices once again rose dramatically, this time as a result of the Iranian revolution. Thatcher's emphasis was on economic growth, competitiveness, individualism, and 'rolling back the frontiers of the state'. There were to be tight controls on public expenditure. On coming into office, the new government apparently sought to 'reduce over-sensitivity to environmental considerations',[26] and a number of advisory bodies, including the Clean Air Council and the Commission on Energy and the Environment, were abolished in an enthusiastic cull of quangos. Significantly, the Royal Commission on Environmental Pollution survived, not least because of the tendency to locate it in a 'scientific' niche—a useful instance of 'boundary work' (Gieryn 1983, 1995), as far as the Commission was concerned.

In terms of Whitehall priorities, the environment continued to occupy a relatively lowly position: one senior official described his remit in the environment department at this time as 'yes, to do the work but to keep it all fairly quiet and not [to] raise the profile'.[27] But the profile of environmental issues was, inexorably, being raised. A succession of controversies over lead additives in petrol, acid rain, dangerous discharges to water, and stratospheric ozone depletion attracted considerable attention. The spotlight on nuclear power was intensified by a series of focusing events: leaks from the Sellafield reprocessing plant in 1983 (after which discharge consents were tightened); the complex public inquiry into Britain's first pressurized water reactor at Sizewell (which sat for 340 days in 1983–5); and the disastrous Chernobyl reactor accident in Ukraine in April 1986. What distinguishes the 1980s, however, is the emergence of a more systemic view of environmental problems. Whole policy domains, including agriculture, energy, and transport, now began to be scrutinized for their environmental implications, and the concept of 'environmental policy integration' was finding its way into official discourse—notably in the European Community's third and fourth Action Programmes on the Environment (Council of the European Communities 1983, 1987) and in the *Single European Act* of 1986, which made integration a requirement.[28]

Meanwhile, in the field of pollution, pressures for tighter controls on emissions increasingly ran into conflict with British faith in the assimilative capacities of the environment, and with reluctance on the part of both regulators and regulated to take costly action in the absence of proof of harm. A former director of the Central Unit on Environmental Pollution (CUEP, as the Central Scientific Unit came to be called) remembered the storyline of resistance: 'There was an absolutely consistent pattern of argument that came at us from industry..."it isn't necessary, you haven't proved that there is any problem, the science is not done or isn't good enough, it's too expensive and [it] would damage business".'[29]

But British pragmatism, and the tightly knit policy communities associated with it, came under siege from a number of different directions. Environmental campaigners were energetic and 'much more nimble' than the government.[30] Scandinavian countries were agitating about the effects of acid rain. The European Commission was not only initiating a growing volume of environmental legislation, but becoming more strategic and preventative in its overall approach. And the Royal Commission on Environmental Pollution, while defending some aspects of 'the British philosophy', argued that its success depended 'on the public being satisfied that environmental improvement [was] not being too readily delayed by commercial pressures' (RCEP 1984: para. 3.5). Within this diffuse issue network, there was a degree of mutual reinforcement: so, for example, green groups came to see the European Commission as 'an environmental beacon' (Lowe and Ward 1998a: 21), and actively sought European venues to pursue their differences with the UK government. In short, Britain began to find itself in a position of 'defensiveness and isolation', in which it 'seemed more and more at odds with other member states over the direction, pace and substance of environmental policy' (ibid.: 19, 23). A former Chief Pollution Inspector recalled a sense of dislocation: while the government was trying to hold its position, '[a]ll the time...the political environment outside [the] UK was changing. Public awareness was changing.'[31] His words call to mind Hugh Heclo's (1974: 304) observation that policy sometimes evolves because '[s]ocial and political events *move*'.

Policy was also shaped by factors unconnected with the environment, some of which constrained while others accelerated the pace of change. Among the former was the prevailing Euro-scepticism of the party in power, a 'deep core' belief of the kind that infuses all policy domains (Jenkins-Smith and Sabatier 1994: 180). One senior civil servant commented that in Britain during the 1980s, '[y]ou couldn't really separate attitudes...to the environment from attitudes to Europe, because if something was propounded by Europe...there was a slight—not always spoken, but some—presumption that we ought to be against it'.[32] Significant events in the political stream, however, added to the pressures for environmental policy reform. In a move that was to have lasting significance for British electoral politics, four senior Labour politicians broke away in 1981 to form a new Social Democratic Party (the SDP). Subsequently, the SDP formed an alliance with the Liberals—a potentially powerful third force in the political system—which survived until 1986.[33] In national and European election campaigns in 1983 and 1984, the SDP/Liberal Alliance moved rapidly to colonize what looked like fertile environmental ground, and the other major parties hastened to claim green (or greener) credentials (Owens 1986, 1987; Pepper 1987). In the event, the general election of 1983 returned the Conservative administration to power, but an interesting effect of party–political splintering had been to raise the profile of the environment still further, and propel it towards the political mainstream.

We have seen that European pressures in particular meant that there was a need for the British pragmatic approach to be rationalized and justified, sometimes even defined: political actors in this case certainly found themselves having to 'persuade and convince' (Radaelli 1995: 174). One consequence was 'to make explicit principles that were previously implicit or merely rhetorical' (Lowe and Flynn 1989: 266). Another was that environmental policy in the UK did in fact change to become more formal and more centralized (Haigh 1992).[34] But most interestingly, there was a change of mindset, or interpretive framework, in this particular policy domain. Gradually, despite powerful resistance, traditional pragmatism was giving way to a more precautionary approach, with a shift in the burden of proof about environmental harm. Reversal of the government's stance on lead in petrol, strongly influenced by the Royal Commission's ninth report (RCEP 1983), was among the first tangible manifestations of this change (discussed later, see Chapter 4). Another came in 1986 when the then Central Electricity Generating Board was given the go-ahead to retrofit 6,000 MW of generating plant with flue gas desulphurization technology (see Hajer 1995 for a detailed account). Two years later, the government signed up to the European Large Combustion Plant Directive, agreeing to a 60 per cent reduction in sulphur dioxide emissions by 2003. This undoubtedly represented a substantial movement in the British position, though in the end it was overtaken by electricity privatization and the subsequent shift towards cheaper, gas-fired generation—the so-called dash for gas.[35]

The announcement, at the second North Sea Conference held in London in 1987, that Britain would accept uniform emission standards based on 'best available technology' for certain dangerous discharges to water (the 'red list') was a further milestone. It followed a long and acrimonious dispute between the European Commission and the UK, in which the latter had fought hard to defend its preferred system based on quality objectives in the receiving environment (Haigh 1992, 1995).[36] The breakthrough on water facilitated another substantial change—the move towards an integrated approach to pollution control across different media, and the formation (in 1987) of Her Majesty's Inspectorate of Pollution (HMIP) to implement the new system. This important development, recommended more than ten years earlier by the Royal Commission (RCEP 1976a), will be analysed in Chapter 5. Interestingly, Nigel Haigh (1995: 9), a close observer of such matters, attributed the 'turning tide' in part to the appointment of William Waldegrave as Parliamentary Under-Secretary (1983–5), and subsequently Minister of State (1985–7), for the Environment. Not only was Waldegrave something of a policy entrepreneur—he initiated the moves on dangerous discharges to water and on HMIP—but it was he who attended meetings in Brussels throughout this period while successive Secretaries of State were occupied with non-environmental issues.

The privatization of key utilities in the late 1980s also impinged on environmental policy, albeit in unexpected ways. Preparations for electricity privatization exposed serious misgivings about the costs of nuclear power (notwithstanding the fact that the Sizewell B pressurized water reactor had been granted consent in 1987), and led to a moratorium on new nuclear construction (a full account can be found in Rough 2011). The proposed privatization of the water industry raised significant concerns about institutional arrangements for pollution control, resulting in the formation, in 1989, of the National Rivers Authority (NRA) as the main regulatory body for the water environment.[37] Thus, as John McCormick (1991: 13) notes, Margaret Thatcher's determination to privatize the utilities in line with core neo-liberal principles 'achieved the unintentional and unanticipated effect of making pollution control a major public issue'.

By the end of the decade, there could be little doubt that the environment occupied a prominent position on the political agenda once again. It had been helped by the Brundtland Report (WCED 1987), which popularized and politicized the concept of sustainable development. It had also received an unexpected boost from the prime minister herself. Addressing the Royal Society in September 1988, Thatcher declared herself convinced about the need for action in key areas such as ozone depletion and climate change.[38] Her speech 'reverberated through Whitehall'[39] and was credited by some observers with finally establishing the environment as a mainstream political issue (see McCormick 1991), though with hindsight we can see it more as a focusing event, contributing to the positive feedback that occurs at times of rapid policy change (Baumgartner 2006). Among the public, too, there was not only awareness but also an apparent readiness to act and to vote. The *Green Consumer Guide* (Elkington and Hailes 1988) sold the best part of two print runs within a fortnight, and in the European elections of 1989 the Green Party came from relative obscurity to take 15 per cent of the British vote. For a while it seemed that (green) politics might indeed change policy. Sensing that the power of the green lobby 'was starting to become something substantial',[40] the government introduced major new legislation in the form of an Environmental Protection Bill. This was to be taken forward by a new Secretary of State for the Environment, Chris Patten, who was generally regarded as more sympathetic to the environmental cause than his predecessor, Nicholas Ridley.

In short, we might say that the 1980s, in spite of unpromising beginnings, was a decade in which the environment acquired a new political significance. Long-established norms and practices were challenged, policy networks disrupted and reconstructed, and policies themselves adjusted (often substantially) as governments responded to pressures from above and below. While the causes of change were complex (as students of the policy process might expect), it is clear that those involved had a strong sense of the ground shifting beneath their feet as the 1980s progressed. One interviewee, who had held senior positions in both the UK environment department and the European Environment Directorate (in the 1970s and 1980s respectively), remembered the changes as follows:

> [W]hilst I was in CUEP [in the UK environment department] I had to go [to Brussels] and be difficult, because that was the policy line.... I ... didn't much like the 'dirty man of Europe' sobriquet and I tended to despair of the attitude that I encountered from London over some things... [But later] I felt, sitting in Brussels, [that the British position] was shifting, and lead [in petrol] was a sort of catalyst... things had begun to move... and it became a bit less disheartening as the atmosphere did seem to change... imagine the relief in Commission officials now that [the British had] a different tone, saying, 'yes, we're not a pushover, but let's cooperate and talk about it, instead of saying *"niet, niet!"'*[41]

A former Chief Inspector with the Industrial Air Pollution Inspectorate (IAPI) had similar recollections, and his account is strongly suggestive of a process of policy learning, at least of an instrumental variety and possibly of a more fundamental kind:

> [W]e went into Brussels to teach these foreigners something about it, and I think by and large we came back with our tails between our legs because when it comes to it there are a lot of instances where pollution control was not nearly as good as we thought it was and we had in fact something to learn from some of the continentals.[42]

Developments were particularly rapid in the second half of the decade—'a sort of sea change in the attitude of government to the environment', as described by one official who had been in the environment department at that time.[43] The new mindset was apparent in the wide-ranging *Environmental Protection Act* of 1990, which included provisions on integrated pollution control, public registers of information, the 'duty of care' in waste management,[44] and the control of genetically modified organisms (all of which had been influenced by the work of the Royal Commission—see Chapter 6). In a debate during the passage of this legislation Lord Clinton-Davis (a former European environment commissioner) commented on 'the utterly remarkable changes' affecting environmental policy, and its movement into the political mainstream: the environment was now 'at the very top of the political agenda', he said, no longer the preserve of 'the slightly eccentric, the sandalled and the hirsute'.[45] Still, by the early 1990s, the transformation was far from complete and, in spite of many gains, 'there was a widespread feeling...that the problems of pollution were becoming more serious and that existing policy strategies failed adequately to deal with them' (Weale 1992: 26). This sense of policy failure—by no means confined to the UK—was in turn to act as a stimulus for further, and more fundamental, reform.

For the immediate future, however, 'green thinking...went off the boil'[46] again as the electoral threat of 1989 receded, and the environment slid downwards once more in the hierarchy of political concerns. A White Paper on the subject (UK Government 1990) disappointed many in its failure to set out genuinely new policies (see, for example, Friends of the Earth 1990; Weale 1992, 1997) but was nevertheless significant in signalling broader changes in philosophy. Its attention to economic instruments,[47] for example, and its emphasis on the roles and responsibilities of non-governmental actors, hinted at the beginnings of not only a new style of environmental policy, but also a shift from government to governance in environmental affairs. Another signal of things to come was the publication of the first of the scientific assessments made by the Intergovernmental Panel on Climate Change (IPCC) (Houghton, Jenkins, and Ephraums 1990), a body established in 1988 by the World

Meteorological Organization (WMO) and the United Nations Environment Programme (UNEP). The report was presented as 'the most authoritative and strongly supported statement on climate change that has ever been made by the international scientific community', which would allow '[a]ppropriate strategies' to be firmly based on a scientific foundation (ibid.: Preface, iii).

From Rio to Recession—The Era of Climate Change: 1992–2007

Although the second era ended on a relatively quiet note in policy terms, this was soon to change. Attention to the environment was stimulated again by the landmark UN Conference on Environment and Development held in Rio de Janeiro in 1992,[48] and by the rapid diffusion and take up of a new discourse—that of sustainable development—which the conference itself did much to formalize and promote. Indeed, at around this time the credibility of political actors worldwide seemed to require them 'to draw on the ideas, concepts and categories' of this particular discourse, providing evidence for the condition that Hajer (1995: 60) has described as 'discourse structuration'. At Rio itself, *Agenda 21* set out a plan of action for sustainability at every level of governance, while in Europe the Maastricht Treaty of the same year moved tentatively towards embedding sustainable development within the objectives of the European Union.[49] The UK was one of the first countries to honour its Rio commitments by producing a national strategy for sustainable development (UK Government 1994c), together with a biodiversity action plan (UK Government 1994a) and a climate change programme, with the latter aiming to reduce greenhouse gas emissions to 1990 levels by 2000 (UK Government 1994b). All of these initiatives were to be significantly refined and developed during this third era.

Hajer has argued that in any given period certain issues ('emblems') dominate public and political attention, functioning as a metaphor for 'the environmental problematique at large' (Hajer 1995: 5). The third era, much more so than those preceding it, was dominated in this way by a single issue—that of global climate change. Concerns (and controversies) about climate steadily gathered momentum, driven in part by reports from the IPCC (Houghton, Jenkins, and Ephraums 1990; IPCC 1992, 1995, 1996, 2001, 2007). Major international policy developments (fully documented elsewhere[50]) included the Framework Convention signed at Rio, the Kyoto Protocol (under which many industrialized countries, and the European Union, adopted legally binding targets for the reduction of greenhouse gas emissions[51]), and the introduction, in 2005, of the European Emissions Trading Scheme for greenhouse gases, the largest of its kind in the world.

In the UK, the 'dash for gas' made it relatively easy to achieve the national targets set out in the first climate change programme (UK Government 1994b) and helped the UK to take a proactive position in international environmental diplomacy.[52] Kyoto required further reductions (to 12.5 per cent below 1990 levels by 2008–12), but Tony Blair's 'New Labour' Government, elected in 1997, adopted the more ambitious target of 20 per cent reduction in CO_2 emissions by 2010, and embarked on a series of initiatives that would mark a significant convergence of energy and climate policies (Lovell, Bulkeley, and Owens 2009; Owens 2010). A new climate change programme was set out in 2000 (DETR et al. 2000a),[53] and in 2003 an Energy White Paper was substantially framed around the climate issue (DTI 2003). Significantly for our discussion of policy advice, the White Paper took up a radical recommendation from the Royal Commission that 'the UK should put itself on a path' towards a reduction in CO_2 emissions of some 60 per cent by 2050 (ibid.: para. 1.10; RCEP 2000: para. 10.10)—a precursor to the even more demanding target (80 per cent for greenhouse gas emissions) eventually adopted in the *Climate Change Act* of 2008. This ratcheting up of policy objectives owed much to the Treasury-commissioned Stern Review (Stern 2006), which argued strongly that increasingly ambitious targets were affordable, and that *not* to act urgently on climate change would ultimately cost economies much more.[54] Adopting a distinctly ecomodernist perspective, Stern in effect reframed the need to act on climate as an economic as much as an environmental imperative, and in doing so ensured that the Review would have a high profile both within and beyond the UK. An updated climate change programme was published in the same year (UK Government 2006).[55]

Although at times it seemed almost synonymous with 'environment' in policy discourse, climate change was never in fact the sole issue commanding attention in the third era. Policies in 'traditional' areas continued to evolve, with particular controversies, and sometimes whole sectors, moving up or down on the policy agenda in the 'issue–attention cycle' that Anthony Downs (1972) had long observed. As always, a good deal of policy evolution involved what Hall (1993) would define as first or second order change—the tightening of regulations and the adoption of new instruments to achieve established ends. So, for example, in the UK the *Environment Act* of 1995 mandated the preparation of national strategies for air quality and waste, which appeared in due course and were subsequently revised and tightened. In both of these areas, policy and regulation came to reflect a new, multilayered system of environmental governance, in which Europe was still a key driver,[56] and the UK government remained an important actor, but many aspects of policy had become the responsibility of the devolved administrations. Interestingly, because of the transboundary nature of air pollutants, the air quality strategy continued to be presented in a single document, 'with common aims covering all parts of the UK' (Defra et al. 2007: para 9); for waste, however, separate publication became the norm.[57] There were significant developments, too, in the regulation of chemicals, with a new European system for the registration, evaluation, and authorization of chemical substances (REACH) coming into force towards the end of the period.[58] All of these issues were of interest to the Royal Commission, and were addressed in a number of its reports.

Specific policy domains attracted attention at different times. Road transport became a prominent issue in the 1990s, when the dominant frame ('predict and provide') was challenged by the so-called 'new realism', emphasizing environmental constraints, traffic management, and 'greener' modes of transport (Goodwin et al. 1991).[59] The challenge was such that something akin to third order policy change seemed possible after the 1997 election, when the new discourse was clearly reflected in a White Paper (DETR 1998), but the dominant frame proved resilient and the radicalism of the late 1990s gradually ebbed away (Owens and Cowell 2011; T. Rayner 2003). In the new century, the discursive struggle shifted towards aviation, a sector in which rapid growth and multiple environmental impacts provoked intense political conflict (see, for example, DfT 2003; EAC 2003, 2004a, 2004b, 2004c; House of Commons Transport Committee 2009; SDC 2002, 2004). In both cases, the Royal Commission was actively involved, producing two reports on ground transport in the 1990s (RCEP 1994, 1997c)—effectively becoming part of a 'new realist' discourse coalition—and later a special report on the climate change implications of aircraft in flight (RCEP 2002b). The energy sector became prominent, too, in part because of its links with climate change but also as a result of renewed concerns about energy security. Policy developed rapidly in the first decade of the new century (DTI 2003, 2006, 2007), and strategies for energy supply became as contentious as they had been in the 1970s and 1980s, with battle lines being drawn (or re-drawn) over the merits, costs, safety, and environmental impacts of different options for electricity generation, especially nuclear power and renewables.

There were many other active issues. They included protracted controversy about genetically modified crops;[60] tensions between infrastructure provision and environmental sustainability, bound up with reforms of the planning system;[61] and mounting concern about the impacts of fisheries and other human activities on marine ecosystems (Berry 2000; Defra 2007a; Prime Minister's Strategy Unit 2004). The Royal Commission anticipated or contributed to all of these debates (RCEP 1989, 2002a, 2004), with effects that will be discussed in more detail in Chapters 4–6. The point to be made here is that the environmental agenda in the third era was a dynamic one, quite apart from the many initiatives directly or indirectly related to climate change.

At least as important as the coming and going of specific issues, however, was a deeper process of change which might best be understood as one of 'discourse institutionalization' (Hajer 1995: 61), through which emergent discourses are translated into policies and institutional arrangements. Arguably, in the third era, this was happening to the key, interrelated concepts of sustainable development and ecological modernization, with their associated storylines of integration, precaution, and anticipation. Institutionalization does not, of course, begin at a particular moment, nor does it necessarily achieve its objectives; but many developments from the early 1990s onwards reflected a different, more structural, framing of environmental problems and a more systemic approach to the search for appropriate solutions.

These changes were manifest in a variety of ways. There were earnest attempts, for example, to put the principle of environmental policy integration into practice, both at the European level and within the UK. Integration was an explicit objective of the fifth and sixth European Action Programmes on the environment (Council of the European Communities 1993; European Commission 2002) and of the 'Cardiff process' launched in 1998, and was central to a strategy for sustainable development adopted in 2001 (CEC 2001).[62] In the UK, initial efforts (following the 1990 White Paper) included the nomination of 'green ministers' in all government departments and a requirement that new policies be scrutinized for their environmental implications, though these moves were widely seen as tokenistic.[63] More enduring were the production and publication of a wide-ranging set of indicators of sustainable development (Defra 2004a, 2010a; DETR 1999; DoE 1996); the creation of a Sustainable Development Unit, located within the environment department but with a remit to work across Whitehall; and the establishment of a new Parliamentary watchdog, the Environmental Audit Committee.[64] In a more specific sense, there was further integration of pollution control, with UK policy in this case preceding developments at the European level (see Chapter 5). So, for example, the Environment Agency for England and Wales, set up under the *Environment Act* of 1995, brought together the functions of Her Majesty's Inspectorate of Pollution and the National Rivers Authority with those of waste regulation (Carter and Lowe 1994), while the European Directive on Integrated Pollution Prevention and Control (IPPC) was adopted soon afterwards, in 1996.[65]

There was also a discernible shift at the European level from the older, prescriptive (often 'end-of-pipe') approach in environmental policy towards legislation and regulations that were more systemic and structural in character.[66] The Directive on Strategic Environmental Assessment (SEA), extending the scope of ex ante assessment from individual projects to public plans and programmes,[67] was a case in point, as was the Water Framework Directive, which adopted a 'river basin' approach, combined emission limits and quality standards (rather than seeing them as alternatives), and sought to bring coherence to a fragmented water policy system.[68] REACH and IPPC, already mentioned, would also fit into this more systemic category. Of course, none of these measures was achieved without contestation.

As theorists of ecological modernization might expect, the third era also saw greater use of market-based instruments in pursuit of environmental objectives. The carbon trading system mentioned above was one of the clearest examples at the European level, and the European Commission actively sought to extend such thinking to other areas (CEC 2007a); various measures were also adopted by individual member states and other industrialized countries.[69] In the UK, a landfill tax was introduced in 1997 and a 'climate change levy'—effectively a tax on industrial energy consumption—in 2001. Britain devised its own carbon emissions trading scheme, too, launched in 2002 (but superseded by the European scheme a few years later). Progress was by no means uniform, however: green taxes were always controversial, and sometimes proved too much of a political liability. The 'fuel duty escalator', introduced by the Conservatives in 1993 and initially increased under Labour, was hastily abandoned in the wake of vociferous protest in 2000, leaving a good deal of residual nervousness about this particular form of taxation.[70] Pesticides were also a candidate for green taxes, but after lobbying from the plant protection and agricultural industries a voluntary initiative aiming to reduce pesticide use was introduced instead in 2001.

Two further important trends need to be mentioned here. The first— ecomodernist in character and indicative of a move from government towards governance—involved a loosening of policy networks, the engagement and empowerment of new actors, and a partial shift from technocratic to more deliberative and inclusive styles of policy-making. Public rights of access to environmental information were strengthened by European legislation and by the Aarhus Convention of 1998, bringing to fruition campaigns that went back at least as far as the 1960s (and which had been reinforced, on numerous occasions, by interventions from the Royal Commission).[71] Environmental groups also became more routinely involved in the policy process, in stark contrast to their position in the 1970s, and indeed were included in the membership of new-style advisory bodies, such as the Round Table on Sustainable Development, created in 1994, and its successor, the Sustainable Development Commission (SDC), which became a significant player in sustainability politics in the period 2000–11.[72] The second trend, reinforcing (and reinforced by) these shifting structures of governance, was the diminishing deference—or at least, diminishing *automatic* deference—towards science and other forms of expertise. This change had particular significance for the framing of environmental risk, as we shall see in Chapter 4.

The above account of developments in the third era points to a high, if fluctuating, level of interest in and action on the environment, with governments paying close attention, even if not consistently affording the highest priority, to environmental goals. While climate dominated, and other issues often came to be viewed through a 'climate lens', the third era brought important, sometimes bold, developments across a range of different policy domains. It was also an era in which the emergent discourses of the 1980s became institutionalized to a considerable degree. Paradoxically, however, by the end of the period it was becoming increasingly apparent that neither ecological modernization nor sustainable development could readily deliver the promised reconciliation of social, economic, and environmental goals. In spite of multiple achievements, governments everywhere struggled with the conflicting demands of globalization, growth, and competitiveness on the one hand, and protection and enhancement of the environment on the other. By the end of 2007, such tensions were set to become more acute, as the harbingers of a deep financial crisis began to make themselves felt.

Recession and Environmental Policy: Regression or Resilience? 2007–11

In November 2007, long queues of anxious customers formed outside branches of a British bank, Northern Rock, which had been forced to seek emergency support from the Bank of England because of its exposure to the sub-prime mortgage problem in the United States.[73] The run on Northern Rock was one of the earliest signs in the UK of the global financial crisis and economic downturn that would come to dominate the fourth era. The crisis itself—much analysed and widely considered to have been the worst since the 1930s—is not the focus of attention here, but its interplay with environmental policy might be seen as the defining characteristic of this final period, covering the last few years of the Royal Commission's existence.

As in previous economic downturns, the environment lost traction as a political priority, though with environmental concerns now more deeply institutionalized than in earlier eras, sheer momentum could carry forward a range of important initiatives. On climate, for example, medium- and long-term targets were adopted both in the UK and internationally. The *Climate Change Act* of 2008 included the stringent target for 2050 discussed above and established an independent committee to set legally binding carbon budgets for five-year periods; it made the UK the first country in the world to adopt such a framework.[74] In the following year, an energy and climate change package looking forward to 2020 was adopted in the European Union,[75] while the G8 Summit in L'Aquila, Italy, agreed to targets of an 80 per cent reduction in greenhouse gas emissions in developed countries, and 50 per cent for the world as a whole, by 2050. The marine environment—rising rapidly on national and international policy agendas—was also the subject of new legislation;[76] in the UK, the *Marine and Coastal Access Act* of 2009 introduced a system of marine spatial planning, provided for the designation of Marine Conservation Zones, and recognized that the latter needed to be part of a wider network of Marine Protected Areas.[77] There was incremental progress, too, as seen in the further development and scrutiny of waste and air quality strategies, and the explicit linking of both to climate change (see, for example, Defra 2011; Defra et al. 2010; EAC 2010a, 2011). Promising (perhaps rashly) to be 'the greenest government ever',[78] in 2010 the incoming Conservative–Liberal Democrat coalition made a number of environmental commitments, including the cancellation or refusal of additional runways at major airports in the south east.

There were strong countervailing forces, however. Economic imperatives reinforced an already powerful backlash against climate change mitigation, including the affair that came to be known as 'Climategate'.[79] The high-profile Copenhagen conference of December 2008 (COP-15) failed to produce a successor to the Kyoto Protocol, as did subsequent meetings in Cancun (2009) and Durban (2011), where the parties could agree only that a legally binding treaty should be negotiated by 2015.[80] There were effects at national and European level too. In 2011, the UK's fourth carbon budget (2022–7) was agreed subject to the proviso that it might be reviewed if other European countries failed to follow a similar path,[81] while the Chancellor of the Exchequer seemed to revert to an ecologically pre-modern discourse, criticizing environmental policies for 'piling costs' onto companies and households, and declaring that the planet would not be saved 'by putting our country out of business'.[82] Concerns about competitiveness similarly drove renewed determination to facilitate the provision of 'nationally important' infrastructure; changes to this end were achieved in the *Planning Act* 2008, and featured prominently in a White Paper on nuclear power in the same year (BERR 2008).[83]

In a sense, then, during the fourth era a policy paradigm shaped by eco-modernist ideas was itself being challenged, as recession intensified the disciplining effects of the competitiveness and 'better regulation' agendas. If the pattern was familiar—an economic downturn dampening enthusiasm for environmental commitments—it was all the more disappointing when the 'green economy' had apparently held out such promise (Feindt and Cowell 2010). But there were additional reasons for the fragility of this discourse. Long-standing problems such as urban air pollution were becoming less tractable,[84] while the 'new' environmental issues, exemplified by climate change, demanded unprecedented levels of international negotiation and co-operation. Both, in their own ways, pointed to the limitations of technical fixes and the need for behavioural (and perhaps more fundamental) change.[85]

So the final period, paradoxically, saw both an intensified growth imperative and diminishing confidence that 'doing more with less' could sufficiently ameliorate environmental impacts. This paradox may have contributed to another interesting development towards the end of the Commission's life—the re-emergence of a 'limits' discourse, with renewed attention to relations among consumption, population, human well-being, and environmental degradation (Beddington undated, 2009; Rockström et al. 2009; Royal Society 2012; SRU 2012; Owens 2013, 2014). The Royal Commission entered the debate with its very last report, which explored the subject of demographic change and the environment (RCEP 2011).

Forty Years of Policy Evolution: A Brief Reflection

There is much in this chapter to support the view that policy evolution involves non-linear processes with important cognitive and discursive dimensions. Taking the long view in the environmental domain, we can see how different issues rose to prominence and faded, while problems that were alleviated (as well as many that were not) were periodically replaced by new concerns. No convincing account of specific developments or long-term trends in this area could omit the roles of scientific knowledge and research—both of which expanded enormously in the time period considered—or of ideas, persuasive argument, framing, and venue. Policy learning, of both the rationalistic and discursive kinds, does indeed seem to have occurred. But the somewhat erratic progress over the four eras considered here, with periods of inertia, spurts of legislative activity, shifting emblems, and sometimes dramatic change, supports theories in which knowledge (broadly defined) interacts in complex and contingent ways with interests, institutions, power, and events. As Kingdon (2003) and many others have argued, the seeds of change can take a long time to come to fruition, and knowledge can lie dormant until shifting circumstances instil it with meaning and authority.

There were many instances, throughout the period considered, of what Hall (1993) describes as first and second order policy change: in the UK and elsewhere, environmental controls were progressively tightened and policy instruments modified and refined. These changes were invariably contested and could often be slow to take effect, but over time they undoubtedly helped to mitigate, or even eliminate, the gross pollution of the 1960s and 1970s. 'Third order change', involving a fundamental reframing of problems and potential solutions, is also discernible, especially in the bigger picture and over the longer term. At the beginning of the period, for example, pollution was attributed (and was in many cases attributable) to the poorly regulated externalities of production, and the solution was seen to lie in 'end-of-pipe' adjustments, which, even if costly, could be effected without substantial disruption to processes or products. But by the beginning of the third era, environmental degradation was being reframed as a structural problem, pointing to the need for greater efficiencies, cleaner technologies, and 'environmental policy integration' (Nilsson and Eckerberg 2007). Though the different orders of policy change are not always easy to disentangle,[86] we can conclude that their combined effect in the second half of the twentieth century was to reduce the 'environmental impact co-efficient' of economic activity (Jacobs 1991: 55), much as envisaged by theorists of ecological modernization. Yet the limitations of this paradigm, too, began to be exposed during the third and fourth eras. Not only was progress becoming more difficult in traditional areas of concern, but 'new' environmental problems, materially and discursively different from the 'grievous damage' of the 1960s and 1970s, seemed to demand a wholly different level of response (Beck 1992; Hajer 1995; RCEP 1998b). By the end of our period it was increasingly apparent that policy-makers would have to confront deeply embedded practices if they wished to reduce absolute pressures on the planet, and fractures between growth and environment were beginning to be in evidence once again. It hardly needs to be said that this is politically challenging terrain, but it is worth noting that the changing nature of problems and potential solutions created difficulties for the Royal Commission too, as we shall see in Chapter 7.

Throughout the period considered, proponents of change must often have felt that the dominant force was inertia, that progress in environmental policy was maddeningly slow, and that their goals might even be unattainable. Yet in spite of obstacles and conflicts, and the constant reconstruction of problems, we can see that the transformation over forty or fifty years was profound. Elaborate institutional structures evolved, legislation expanded at every level of governance, policy networks grew in number and complexity, norms and practices were redefined—and these changes were apparent not only in the 'environmental' field, narrowly conceived, but in virtually all major domains of public policy that connected to the environment in some way. By the second decade of the twenty-first century (when the Royal Commission was abolished), the environment was conceptualized differently, and governed more systematically, than it had been half a century before. Latterly, and in part because of the changing nature of 'environmental problems', these changes were coming to impinge ever more closely on social and economic life.

The Royal Commission on Environmental Pollution—the main subject of investigation in this book—was itself involved in many of the key developments reviewed above, so that exploring its role, as argued in Chapter 1, should help to elucidate 'the interactive and multiplicative effects of the variables that contribute to policy change' (Baumgartner, 2006: 40). That will be an important task for Chapters 4–6, which examine certain developments, and the Commission's role and influence, in greater detail. First, however, it is necessary to look more closely at the constitution and practices of the Commission itself, and at its positioning within policy communities. It is to these issues that Chapter 3 will turn.

3

A *Standing* Royal Commission

[T]he existence of a standing Royal Commission with such broad terms of reference implies a judgment that pollution will continue to raise issues of such consequence to the nation as to justify an independent 'watch-dog' body.

RCEP *Pollution Control: Progress and Problems* (1974: 2)

As a standing Royal Commission, it was a rare beast and, as one that had lasted for four decades, it was even stranger.

Geoffrey Lean, *Telegraph Blogs* (10 March 2011)

Organisations mould the behaviour of those who work within them... They do so by regulating behaviour through rules; by imposing social obligations through normative pressures; and by providing automatic expectations through a cognitive framing of reality.

Stephen Wilks, *In the Public Interest* (1999: 116)

Introduction

The existence of the Royal Commission on Environmental Pollution for more than forty years offers rich empirical material for a study of relations among knowledge, advice, and policy. But exploring the role of any particular body in policy–political processes demands some understanding of its history, constitution, and working practices. This chapter provides the necessary background and is therefore essentially biographical. It offers an account of the Commission's origins, then examines its composition; its relations with others; and its methods for selecting topics, gathering evidence, and drafting reports. There is also a brief commentary on the thirty-three reports that the Commission delivered during its lifetime. This chapter serves, too, as an introduction to certain attributes of the Commission which helped to endow it with authority and engender trust in its advice. Finally, it considers the Commission's durability and its eventual abolition in 2011. We begin, however, in the late 1960s, when the idea of an independent environmental advisory body was gathering momentum within an embryonic policy community in the UK.

Origins

The end of the 1960s, as we have seen in Chapter 2, was a period of turmoil in environmental politics. In the UK, there was a widespread feeling that 'something had to be done', not least because the fragmentation of responsibilities for environmental matters made it difficult for governments to respond to emergent concerns in any co-ordinated way.[1] The mood of the time is encapsulated in a memorandum sent to the Secretary of State for Housing and Local Government by Lord Kennet, then a junior minister in that department, early in 1970. The note began: 'It may be useful to you to have some general observations about how we should handle environmental matters, and especially pollution, during this year.' Under the heading of 'General Policy', the memorandum is crisp: 'Do we really want to do anything, as opposed to looking busy? I think we do, because a) it needs doing, and b) everybody is now convinced it needs doing.'[2]

It was in this general climate that the idea of a committee or commission on environmental pollution was conceived. Both Lord Kennet and Sir Solly (later Lord) Zuckerman, the government's Chief Scientific Advisor, had been advocating something along these lines and encouraging the prime minister, Harold Wilson, to establish an appropriate body (Holdgate 2003).[3] Zuckerman recalls in his autobiography that after the disastrous oil spill from the *Torrey Canyon* he had enquired into the responsibilities of different government departments for environmental matters: 'In the summer of 1969, I brought the matter of environmental protection and conservation to the attention of the Central Council for Science and Technology[4]... After surveying the problem, the Council recommended that a standing commission on environmental pollution should be established' (Zuckerman 1988: 414).

In October 1969, Wilson appointed Anthony Crosland (then President of the Board of Trade) to the newly created post of Secretary of State for Local Government and Regional Planning. As well as directing the work of two ministries (Housing and Local Government, and Transport), Crosland's portfolio included coordination and promotion of government action in the field of environmental pollution.[5] His first and urgent task as 'environmental overlord', as he came to be called, would be to recommend improvements to the machinery of government in this field, taking advantage of the 'considerable amount of preparatory work' carried out by the Chief Scientific Advisor (Wilson 1971: 732). On 11 December, having received Crosland's report, Wilson told the House of Commons that the Queen had agreed to the appointment of a Royal Commission on Environmental Pollution, 'to act as a watchdog'.[6] A 'short but vigorous debate' on environmental pollution followed—probably the first of its kind in the UK (Johnson 1973: 99). According to Johnson, '[n]either Socialists nor Tories believed that the "environment issue" would be crucial to the forthcoming election. But there was an outside possibility that it might. Like the Rann of Kutch,[7] it seemed a flat and unpromising land but if one side was there, the other had to be there too' (ibid.: 101).

A second important initiative announced by the Prime Minister was the establishment of a new Central Scientific Unit on Pollution, a move that had also been under consideration for some time (Kennet 1970; Zuckerman 1988). The Unit was to be headed by a biologist, Dr (later Sir) Martin Holdgate, who would report to the head of Crosland's office, but have a 'strong dotted line' to the government's Chief Scientific Advisor.[8] In a Fabian pamphlet, based on a lecture delivered in November 1969, Lord Kennet stressed that the Royal Commission and the Central Scientific Unit were to have distinct roles:

> [The Royal Commission] will provide that outside focus of inquiry and information, and that outside stimulus to government, which are needed. The central unit will coordinate action among all the ten or twelve ministries concerned, and will also act as a secretariat to the royal commission. It will provide that internal oversight and consistency which is needed.
>
> (Kennet 1970: 12)

Kennet presented the tripartite solution—Crosland's new co-ordinating role, the Central Unit, and the Royal Commission—as preferable 'to the facile but disruptive alternative of tearing the statutory functions away from all the ministers who at present have them, and lumping them together in a capriciously conceived new ministry...you cannot have a ministry of everything' (Kennet 1970: 12). Nevertheless, planning for a new 'superministry' went ahead, though it was a Conservative government (returned in June 1970) that formally created the Department of the Environment, with Peter Walker as its first Secretary of State.[9] The Central Scientific Unit on Environmental Pollution (with a new name—the Central Unit on Environmental Pollution, or CUEP) was then transferred from the Cabinet Office to the environment department, where it continued to provide support for the Royal Commission. Thus the Commission's Secretariat became staff of the new department, and its Secretary now reported to the head of the Central Unit.

All of this constituted something of an environmental revolution in Whitehall. Setting up a royal commission, as we have seen in Chapter 1, is a time-honoured way of being 'seen to be busy', and it was a device towards which Harold Wilson was particularly inclined (Hennessy 1989). But we should not underestimate the genuine desire on the part of government to be informed and advised in this newly prominent and often perplexing area of public policy. Nor should we forget the important point made by Weale (1992: 211), that 'what was established was not simply a configuration of policy institutions, but also a *process* of policy exploration and development'. In this context, the constitution of the Commission as a *standing* body was highly significant. As the first director of the Central Unit explained: 'much was made of the word "standing" because of course these [bodies] were quite rare, and this implied government recognition that environmental pollution was not a topic that could be dealt with by a once-off three years royal commission: it required a continuing independent authoritative voice because it was going to be a very long term issue'.[10]

The Commission held its first meeting in London on 25 February 1970.[11] In the chair—on Zuckerman's recommendation (Holdgate 2003; Zuckerman 1988)—was Sir Eric (later Lord) Ashby, botanist, influential thinker on pollution, and at that time Master of Clare College, Cambridge. Other possible chairs had been considered, as Holdgate (2003: 177) records:

Anthony Crosland had originally favoured the Duke of Edinburgh and had been 'warned off' at Harold Wilson's express behest. His next choice, Dr Beeching, axeman of the railways, was also vetoed by the Prime Minister as 'too ICI, too industrialist'. We were fortunate in the final selection. Eric Ashby was not only a good scientist but a man of immense common sense, balance, sound judgement and consideration for others.

There were eight other members, from a range of backgrounds, including Sir Solly Zuckerman himself.[12] The Secretary of State was there in person to welcome them to their first meeting, observing that 'there was no simple answer to the question of what should be the purpose of the Commission: the Commission would themselves provide the answer as the work progressed'.[13] The membership grew to seventeen under the chairmanship of Sir Brian (later Lord) Flowers, who succeeded Ashby in February 1973, prompting one commentator to observe that: 'The inclusion of a physicist, a lawyer, professors of medicine and planning and a woman represents welcome broadening of the membership of the Commission: but its size raises doubts about its potential efficiency' (Lowe 1975b: 91).[14] In fact, the size had fallen to fourteen by the early 1980s, and remained at around that level for the next thirty years. But a diversity of membership, in terms of experience and expertise, became a lasting and important feature of the Commission, and if anything it increased over time. More will be said on this issue below.

A Delicate Balance?

In the early days of the Royal Commission, it was suggested that its potential effectiveness derived from 'two antithetical elements:... the degree of its independence from, and its co-operation with government' (Lowe 1975b: 92). Balancing these elements did indeed turn out to be crucial. On the one hand, the Commission's independence—a quality almost universally associated with 'good advice'—was always regarded as paramount. On the other, its links to Whitehall, varying in intensity and formality over the years, played a vital role in its function and operation (and, at the end, a change in the nature of this relationship was among the factors contributing to the Commission's demise—see Chapter 7). The delicacy of the balance will become apparent later in this book, but here it is worth outlining key developments in the Commission's interactions with Whitehall, and in particular with the department that housed the main environmental functions of government.[15]

Shortly after its establishment, the small Secretariat of the Royal Commission became closely integrated into the newly established Department of the Environment. Concern to safeguard the Commission's independence soon led to different arrangements, however, as the then Director of the Central Unit on Environmental Pollution recalled:

> Eric Ashby and I sat down and had a chat—I initiated it but Eric would have done if I hadn't—saying, 'look this is too cosy, too close, the independence of the Commission...could appear to be compromised...' And I went to the Permanent Secretary and got very ready agreement for this—and [from] the Minister—that the Commission Secretariat should be on secondment, because we wanted the long term career of the individuals concerned to be safeguarded but...when they were on secondment they would answer to the Chairman and nobody else. They were not subject, in other words, to my policy direction.[16]

Regular and close contact between the Commission and the Central Unit—'a very special relationship'[17]—was maintained for several years, with heads of the Unit attending many Commission meetings and being encouraged to participate in the discussion (Holdgate 2003: 178). The practice changed in 1974, when (at the September meeting) it was reported that: 'pressure of Commission business had been the reason for suggesting to Mr Fairclough [then Director of the Central Unit] that he should not as a matter of course attend Commission meetings. It was agreed that nevertheless he should be invited to attend periodically to report on important topical issues.'[18]

Afterwards, the relationship between the department and the Commission became one of 'progressive detachment...while maintaining good communication'.[19] The Chair (sometimes accompanied by other members) would meet senior civil servants and ministers from time to time, the Secretariat and departmental officials would liaise, and officials (occasionally ministers) would still attend certain meetings, though this had become unusual by the end of the Commission's life.[20] One former Secretary described the relationship from the 1980s onwards as an 'arm's length' one:

> Yes, there's co-operation with government on a sort of informal networking basis at various levels...but at the end of the day I think the Commission, when it comes to report, [has] to be seen to be independent and it...must avoid at all costs appearing to be in the pocket of the Department of the Environment, so it co-operates, but on the understanding that it makes its own mind up and it may in the end say things which are inconvenient to the government.[21]

Initially, in an environmental policy community that was still quite small and tightly defined, interpersonal networks involving the Chair, senior civil servants, and ministers were important in maintaining relations and also, as we shall see, in advancing the Commission's ideas. One member of the Secretariat recalled how Richard Southwood (Chair 1981–5) cultivated links with ministers 'so that they could trust him and the Commission not to... embarrass them.... [I]n building up confidence... [he] was... doing his best to keep both the Commission's independence and the Commission's life'.[22] The Secretariat exploited its own professional networks within Whitehall, often to test ideas and alert the Commission to opportunities and sensitivities. Coming predominantly from the environment department, they would 'know how the machinery works...they'd come along with a long background of seeing the world in a certain way'.[23]

When, on occasions, the 'delicate balance' tipped towards interference, the Commission's reaction was robust. There was polite resistance, for example, to too much direction on the few occasions when studies were requested by ministers. In 1974, at the outset of an enquiry into air pollution control, members expressed '[m]isgivings...about the degree to which Ministers were attempting to guide the Commission in the way it would conduct the study',[24] and, some thirty years later, the Commission modified the terms of reference proposed by the environment department for its study of bystander exposure to the spraying of agricultural pesticides (RCEP 2005). The Commission's 'sponsorship' within Whitehall, which had implications for the nature of its independence, was also a sensitive issue. In 1980, when it was proposed that financial responsibility for the Commission should pass from the Civil Service Department to the environment department, the Chair wrote to the prime minister to express concern that this 'would weaken the Commission's independence from any Departmental affiliations, which distinguishes it from the role of other advisory bodies and which is its greatest strength'. The Prime Minister replied that the change was one of Vote responsibility only (that is, accountability for expenditure) and that the status and powers of the Commission would remain unchanged.[25] The move went ahead, though the issue was not quite laid to rest, as we shall see. More anecdotally, on the subject of relations with the department, one long-serving member recalled that when a senior official was, in the Commission's view, becoming too closely involved, 'we had to tell him to bugger off'.[26]

As a body funded from the public purse, the Commission needed to be accountable, at least in the sense of using its resources responsibly, and was therefore subject to periodic scrutiny; after 1980, this was conducted by teams within the environment department, including its Internal Audit unit. Significantly, given the awkward relationship between accountability and independence, the scope and seriousness of these exercises increased during the second half of the Commission's lifetime. As the Secretary explained to members in 1991, there was by that time 'a trend in Government...for such audits to look beyond questions of financial regularity to an assessment of value for money'.[27] Members were characteristically resistant, expressing 'strong concern at the implication...that the Royal Commission needed to demonstrate to [the department] that it was providing value for money. This would compromise its independence.'[28] Feathers were smoothed at a meeting soon afterwards between the Chair and Secretary of the Commission and the Director General of Environmental Protection within the department; the latter subsequently wrote a reassuring letter, and not a great deal happened to change the Commission's practices at that time.[29] From the late 1990s, however, in line with the modernizing agenda of Tony Blair's 'New Labour' government, the Commission became subject to a different style of 'Financial Management and Policy Reviews' (FMPRs).

The first FMPR, conducted by the department's In House Policy Consultancy, was generally positive in its assessment, concluding that (in spite of divergent views on individual reports) the Commission's advice was scientifically authoritative and made a valuable long-term contribution to policy development (DETR 2000a). The review included a number of recommendations about working practices, noting that the Commission itself had recently introduced guidelines for the conduct of its studies (RCEP 1998a),[30] but its most significant proposal was that the working relationship between the Commission and the department should be set out in a framework document, describing the roles and responsibilities of the different parties. Produced soon afterwards, this 27-page document not only codified long-standing practices, but also introduced or consolidated a number of new procedures (DETR 2001). For example, it was agreed that the Commission would produce a report on activities as well as a 'corporate plan' each year, and would evaluate each of its main studies after three years—a surprisingly short period, given the timescales generally considered necessary by students of policy learning and change (see Chapter 1). At around this time the Commission also agreed to comply with the various guidelines and codes relating to scientific advisory bodies and public appointments that had appeared in the post-Nolan era (see Chapter 1).

The second FMPR, which was conducted in 2006–7 by external consultants, was different from the first in important respects; it will be discussed later, in the context of the Commission's survival and ultimate abolition. The important point to note here is that by the early twenty-first century, the requirements of accountability had already marked a subtle shift in the Commission's relations with government, serving to emphasize the extent to which its autonomy was relative, and of a 'scholarly' rather than a financial kind (Stone 2004: 3–4).

Appointments

One of the more immediate effects of modernization, as reflected in the framework document, occurred in the process of making appointments. Formally, this remained unchanged—appointments, as always, were made by the Queen on the advice of the prime minister, advised in turn by the Secretary of State responsible for environmental affairs (and, latterly, the equivalents in the devolved administrations). Members continued to be appointed in an individual, rather than a representative, capacity (except in the sense of 'balance' on such grounds as disciplinary background, gender, and geographical location). What was transformed, however, when Nolan 'curtailed some of [the old] methods of doing business',[31] was the way in which candidates were selected for approval at ministerial level. An opaque system involving consultation within tightly defined policy communities gave way, over a period of several years, to more open and transparent practices for choosing potential members and chairs of the Commission.

For a long period, however, the appointments process had conformed to the 'old methods', exploiting 'the well-established tradition of informal interdependence in Britain between scientific and political elites' (Weale 1992: 14). When it came to selecting new members, the Chair and Secretary of the Commission, and senior officials in the environment department, would be closely involved. Existing members might themselves suggest candidates, especially when retiring (in effect, a system of reproduction), and special advisors or sometimes witnesses who had impressed the Commission could also be in the frame. The 'trawl' for names (this word recurred frequently in interviews) extended to cognate bodies such as the Royal Society and relevant committees of the House of Lords, as well as the famous Whitehall list of 'the Great and the Good'.[32] Other departments would be consulted, with some taking more active interest than others: the Ministry of Agriculture, Fisheries and Food, for example, took the process '*immensely* seriously and made sure that they had someone on it. The Scots and Welsh were also fairly serious'.[33] In these various ways, names would come up through the grapevine, a list would be compiled, and a final selection made for the Secretary of State to approve and put to the prime minister. It was, of course, prudent to sound out prospective members (sometimes to their considerable surprise) before the process was too far advanced, because appointments were 'a terrible palaver...and you don't want to go round that procedure again'.[34] The process involved 'quite a lot of horse-trading', as recalled by a Secretary from the 1980s,[35] and also as remembered by a former Chair:

> [I]n the old days, when there was turnover, I would be invited to lunch by the Permanent Secretary...we would look through the list of the Great and the Good and see who might be suitable, and we would say, 'so and so I know personally, is a good committee man and willing to listen to reason and willing to work hard', and they'd say, 'great, we'll ask him'. And they would say 'how about him?', and you would say, 'very contentious, cantankerous, bigoted, pig-headed', and they'd say, 'well, perhaps someone else'. So gradually names would emerge.[36]

Procedures for appointing chairs were similar in many respects, though the field was smaller and the Commission itself had less of a role. A senior civil servant who had been involved in the process explained that: 'The chairmanship is a very key appointment of course...and we in the DoE and in...wider Whitehall devote a great deal of time to that kind of appointment...well in advance.'[37]

The post-Nolan system of appointments was very different, at least in outward appearance, partly because the reforms were implemented with a degree of rigour on this occasion. Vacancies on the Commission were now advertised, applicants 'sifted' for desirable qualities, and shortlisted candidates interviewed by a formal panel.[38] The names of successful applicants were then forwarded for ministerial and prime ministerial approval, as before. Interpretation of the terms for which members were appointed was also tightened. Traditionally, most members had stayed beyond their initial three years, having been 'sounded out' by the Secretary about their willingness to continue (though the way in which the question was put might depend on 'how useful they seem[ed] to be').[39] Not surprisingly, since most former members reported a positive experience of the Commission (as will be shown in the following section), some went on to serve for three or even four terms—'everybody was on elastic', as one senior civil servant recalled.[40] After Nolan, members could be re-appointed for a second three-year term without much fuss, but anyone wishing to continue beyond that had to apply for an openly advertised position in competition with other candidates.[41] Inevitably, this resulted in some disappointments.

It is worth dwelling on appointments for a little longer, since the process presents one of the most obvious opportunities for political 'meddling' in the functions of an advisory body. If such bodies have strategic and symbolic purpose, we might expect members to be selected for reasons beyond their expertise—more so when the body has some permanence and its enquiries are not easily directed. Interestingly, in the case of the Commission, there is limited evidence of such interference. One former Chair spoke of resistance to departmental nominees who were 'being put there for the wrong reasons',[42] and a former Secretary alluded to certain members being 'foisted upon' the Commission,[43] though neither implied that these occurrences were frequent. The same Chair remembered Margaret Thatcher, as prime minister, taking an unusually keen interest in appointments, and 'blackballing' one or two: '[i]t was all done with great delicacy but I . . . was persistent on two names and was just told they were not politically acceptable'. There were sectoral interests, too, both within Whitehall, as noted above, and to some extent outside; for example, a prominent trade unionist, a member in the 1980s, said that he had been reluctant to resign (though wishing to do so because of his professional commitments) without knowing that a member with similar background would be appointed.[44] In fact, the next trade unionist to serve on the Commission (until 1994) was the last such appointment. Nor, after 1998 (when Lord Selborne's term of office came to an end), was there a practising farmer. Such changes may reflect the perceived significance of particular constituencies, as well as reforms to the system of appointments.

In spite of the 'horse-trading', those who had been directly involved in the process insisted that expertise and authority were the key criteria in appointing members of the Commission: 'the important thing', said one former Deputy Secretary from the department, 'is that they should be known to carry weight'.[45] The demands of specific studies were a further consideration, as was some notion of 'balance', as noted above. High-level interference was unusual: once the names were ready to go forward to the prime minister, they were normally regarded as 'agreed'.[46] It is possible, of course, that choices were influenced in ways that are no longer discernible; certainly, the old system had 'a self-perpetuating element about it',[47] and embodied tacit as well as explicit criteria of 'suitability'. These were never likely to be eliminated by the changes of the late 1990s, since selected individuals could still be encouraged to apply, and unsolicited applicants sifted out if deemed unsuitable. Even so, one former Secretary felt that the new procedures had produced 'some very good members who may not yet have been on a list of the Great and the Good, and would not have been thrown up by the traditional ways'.[48] At the same time, the reforms restricted the pool in other dimensions: individuals in senior positions, who might have felt honoured by an invitation to serve, were possibly reluctant to invest time, and risk rejection, in a more complex and competitive process of appointment: as an earlier Secretary saw it, there was a danger in the new arrangements that 'you might lose people'.[49] By the end, for good or ill, the Commission included a smaller constituent of 'the Great and the Good', as conventionally conceived, than it had had in its earlier incarnations.

Overall, we might conclude that while there was political and sectoral interest in appointments, there was never obviously a crude instrumentalism at work in shaping the Royal Commission. We should note in any case that a lack of overt interference does not in itself make for a body of disinterested, 'rational analysts', nor exclude the possibility of the strategic and selective use of its advice: these important issues will be developed in Chapters 4–7. It should be recognized, too, that the effectiveness of an advisory body depends critically on the conduct of its members once they have a seat at the table, however they might have come to be appointed. In this respect, the culture and practices of the Commission are of particular interest here, since, over the long life of this body, they were both a product of the membership and a profound influence on the individuals who were appointed to serve.

A 'Collegiate Body'

Whatever their shortcomings, the procedures outlined above produced a diverse membership, at least in terms of disciplinary and sectoral background. The Commission was a predominantly academic body, with its members drawn from an increasingly wide range of disciplines over time, even if all of its chairs were natural scientists and Fellows of the Royal Society (Appendix 5).[50] In addition, there were always members with experience in industry and, at various times, in agriculture, local government, the statutory agencies, and the trade unions. This composition (shown for several points in Appendix 5) was distinctive. The body retained the breadth that might have been expected of a royal commission, and was often described approvingly by its members as 'a committee of experts' rather than a traditional 'expert committee'. At the same time, it maintained something of the aura of a 'scientific body', in a field in which it was deemed necessary for policy to be grounded in 'sound science'. The Commission was said to be located, certainly in the earlier years, in the 'hard science' section of the environment department's 'mental map',[51] and even at the end it was lamented for the loss of 'independent scientific advice'.[52] The significance of these characteristics will emerge more clearly in later chapters. For now, the benefits of diversity can be summed up neatly by a former member of the Secretariat:

> The Commission ... was broadly based and it always saw [as] an essential element of its strength and credibility the fact that it was able to take a sort of holistic view,

that it had not just environmental scientists...[but]...had credibility in a wide range of areas...I think that has been one of the strengths of the Commission and why it has lasted so long, that it has actually looked right across the board and its conclusions have...been more credible and robust as a result.[53]

Once appointed, most members seem quickly to have been absorbed into the general spirit of the Commission.[54] A striking feature of the interviews for this study was the evident pleasure that many had taken in their service, reflected in comments such as: 'absolutely fascinating', 'one of the most interesting things I've done in my professional life', and 'a marvellous ten years'; one said simply, 'I am proud and grateful to have been a member'.[55] Such gratitude was in spite of a heavy workload: 'two days a month is a hell of a commitment actually, plus all the reading'.[56] The generally positive experience was attributed by interviewees to a strong sense of identity and common cause, to the high standard of debate, and to the fact that, for the most part, they liked and respected their colleagues. Members who had served at widely differing times spoke of 'comradeship', a 'family atmosphere', and a sense of 'loyalty'.[57] The following comment encapsulates a widely held view:

[O]ne of the most fascinating things about the Commission was the sort of *esprit de corps* that was developed by the members...the standard of debate and argument was very high...there was considerable respect for individuals' abilities and experience in really quite diverse areas...the non-scientists...were able to hold their own and *their* views and their ideas represented a particularly important input...We could have some quite solid...arguments...and you might have thought in the morning that we were going to murder each other, and we'd have lunch and then we'd get somebody like the gang who were obstructing the implementation of Part II of the CoPA [*Control of Pollution Act* 1974] in from the government and...suddenly we'd all come together and we'd pounce on them.[58]

It is interesting in this context that members were unpaid (receiving only expenses) until 1992; after that there was, in addition, a per diem fee for attendance at meetings, and (from 2009) some preparatory work was also remunerated. At the time of the second FMPR (Defra 2007b), per diem payments were a little over £200 for members and £250 for the Chair. Some members refused to accept the fee when it was first introduced. One, from an earlier period, said, 'it never occurred to me that one needed to be paid, I felt it an honour to be invited to be a member'.[59]

There were some dissenting voices amongst the camaraderie and mutual respect. One member of the Secretariat felt that Commissioners had often been 'disappointing intellectually...they didn't have the breadth of intellectual grasp'; but they did ride 'hobbyhorses' and some 'would just say exactly the same things at every meeting'.[60] Some of the non-academics found the proceedings exasperating, and one reported that he 'usually emerged from the

Commission meetings...[feeling] deeply frustrated'.[61] One of the earlier members (an academic) was more direct, dismissing his colleagues as 'senile old coots'.[62] Inevitably, in such a body, there were intellectual, ideological, and occasionally personal tensions, with one of the more noticeable fault lines—or, as a former Secretary put it, 'a severe wavelength problem'— recurring at different times between economists and other members.[63] These tensions will be explored later in the context of specific studies. For now, it is important to emphasize two points. First, the dissenters were in a minority, and the overwhelming experience of those involved with the Commission (shared by the author) was a positive one, even if it has become somewhat idealized over time. Second, the Commission's composition was of great importance to the efficacy of its advice: as more detailed analysis will demonstrate, breadth of membership and mutual respect within a 'committee of experts' contributed to the robustness of the Commission's reports, and helped to underpin its claims to authority, independence, and legitimacy.

Working Practices

In the period 1970–2011, the Royal Commission produced thirty-three reports, twenty-nine of which were published as Command Papers (formally, 'presented to Parliament...by Command of Her Majesty') and four (after 2000) as special reports.[64] This amounts to a tangible product every fifteen months or so on average, though timescales for production varied from a few months to several years. Studies generally overlapped, so that at any one time the Commission would be finishing one report while in the scoping phase of the next, or even working on two studies in parallel—sometimes two substantial ones (as with nuclear power and air pollution in the mid-1970s), but more often one full-length investigation and one that was intended to be shorter, or more focused. While the reports in hand were always 'paramount',[65] the Commission also spent a significant amount of time on other matters, including follow-up work, correspondence, responding to consultations, and issuing statements (typically as press releases) on matters that it felt to be important.[66]

A pattern of working was established early on, the essentials of which changed remarkably little over the years. For most of its life, the Commission held monthly meetings over two days, mostly in London but with some, after devolution, in Scotland, Wales, and Northern Ireland.[67] There were retreats, too, once or twice a year, providing an opportunity for concentrated work on reports: a former Chair recalled that the Commission didn't do much on these occasions 'except read and eat and drink and argue about [the draft]'[68] (in the author's experience, there were bracing walks, too). More generally, the social dimension of meetings, with even the London sessions involving a dinner and an overnight stay for most members, was vital in generating trust and friendship, and allowing ideas to be tested in a relatively informal setting. As one member put it: 'There was a strong feeling of unity and identity and common purpose... we always had an excellent lunch... nice dinner in the evening. I think there was a strong sense of belonging'.[69] This is the 'invisible wiring' of an advisory body,[70] which contributes significantly to its functioning and internal cohesion. But social occasions were also valuable in other ways: in particular, they could be used to good effect in building networks and in 'anchoring' the Commission's ideas.[71] To this end, from around 1990 onwards, influential members of environmental policy communities would be invited for dinner, providing opportunities for off-the-record discussion and, in the words of one member, as 'a deliberate tactic to expose people' to the Royal Commission and its work.[72]

Once a topic was selected for study (of which more below), the Commission's normal practice was to determine the scope of the investigation,[73] take written and oral evidence, and draft the report, with these elements becoming increasingly intertwined as the study progressed. Evidence might be supplemented by specially commissioned contributions from consultants, and in some studies specialist advisors were appointed to assist the Commission in its analysis. Most studies involved visits, which offered a 'reality check', acted as a 'team-building activity', and sometimes had significant effect.[74] Examples of lasting influence include the impression of poor housekeeping at the Sellafield nuclear reprocessing plant during the study of nuclear power (RCEP 1976b) and encounters with 'cowboy' landfill operators (as described by one member[75]) during the waste study (RCEP 1985) (see Chapter 4). A former Chair remembered a woman bursting into tears during a visit for the agriculture study (RCEP 1979), as she recounted the problems of living downwind of a pig farm: '[visits] brought... home to us the human dimension of what we were talking about'.[76] The same could be said of the visits during the pesticides study (RCEP 2005) to residents who were convinced that their health had been damaged by pesticide spraying. For the author, the most lasting images came from the marine study, during which the Commission sailed across the Irish Sea in a trawler (RCEP 2004).[77]

In relation to the Commission's role and influence, three further elements of its studies are of particular interest here: the selection of the topics, relating to wider aspects of the policy agenda; the role of evidence, which can tell us something about the flow of information and ideas; and the processes through which the reports emerged, which hold some clues to their ultimate authority and impact. These elements will be considered in turn, though in practice, of course, they were closely interconnected.

Choosing Topics

Although the Royal Warrant granted ministers the power to refer matters to the Commission, only the studies of air pollution control (RCEP 1976a), marine oil pollution (RCEP 1981), and bystander exposure to pesticides (RCEP 2005) were directly requested in this way.[78] For the most part, the Commission's agenda was its own to define and its remit was interpreted with considerable flexibility. Studies ranged well beyond any narrow conception of pollution, and an initial focus on 'quasi-domestic issues'[79] broadened over the years to include topics with ramifications at the European and even the global level. In the 1990s, the Commission formally set down and adopted seven criteria to guide its choice of topics (Table 3.1), while accepting that not all would necessarily be satisfied in a given study (RCEP 1995a, 1997a).

The first stage in choosing a topic was normally to consider a shortlist of potentially suitable enquiries, typically suggested by members of the Commission and the Secretariat. The emergence of a favoured topic was described by one member as 'a wondrous crystallization process', in which the subject in question 'sort of hovers for a while, then takes form, and then fleshes itself out'.[80] Inevitably, the Commission was influenced by a wider policy agenda at this stage. Most of its studies were on issues that had been 'floating around', to borrow Kingdon's (2003) term, or had already been established as matters for political attention. Nor did the Commission 'go it alone' in making its choices: there was always consultation within Whitehall and, latterly, with a wider range of interested parties. One former Chair said that whilst he had never felt 'under any pressure to address any issue', he had been 'anxious to know [the environment department's] response, because if you address an issue that they're not interested in...you're very hard placed as a Royal Commission to bang a table enough to be successful'.[81] Members, similarly, resisted anything that looked like political direction, but recognized that 'you do have to

Table 3.1. Criteria Guiding the Commission in its Choice of Subjects for Study

a	The topics chosen should be what the Commission's First Report called 'priorities for enquiry': issues which require detailed and rigorous analysis before satisfactory policies can be adopted.
b	They should be of major importance for the long-term protection of the environment, measured against the objective of sustainable development.
c	They are likely to require action by the Government in the medium term, on a timescale to which the Commission's conclusions and recommendations can make a significant contribution.
d	They should raise wide issues, both intellectually (in the sense of spanning several disciplines) and organisationally (in the sense of not falling within the terms of reference of any other single body).
e	They are likely to involve general issues of principle.
f	They should not normally duplicate other studies already in progress or planned in the near future.
g	There should be a reasonable prospect that worthwhile conclusions can be produced within two years with the resources likely to be available to the Commission.

Source: RCEP (1995a: 2).

live in a practical world and take note of what is happening at the top'.[82] Whitehall officials offered views but did so relatively unobtrusively;[83] and the Commission was respectful, but by no means always cooperative. In the end, according to one member with long service, different perspectives were considered but the final choice was always made 'in an interplay between the Commission, the Chairman, and its Secretariat...government really takes a back seat'.[84] This was also the author's own experience. Not surprisingly, therefore, some studies were warmly welcomed in Whitehall while others were a source of annoyance, with different departments not always taking a unified or harmonious view. In the 1980s, for example, the environment department found the choice of lead pollution helpful, but was irritated by the Commission's decision to investigate freshwater quality in the run up to water privatization. The selection of agriculture in the 1970s displeased the Ministry of Agriculture, Fisheries and Food, which indicated that 'too detailed and too intense an enquiry might not be welcomed', and was 'obstructive' when the study commenced.[85] The Department of Transport was said to be 'furious' when the Commission turned its attention to that sector in the early 1990s.[86]

The Commission's choice of topics was also influenced by the specific interests and concerns of members at particular times and by a desire on the part of successive chairs to 'do something different'. The eventual choice of the contentious subject of lead pollution for the ninth report, for example, was made in part because the new Chair, Professor (later Sir) Richard Southwood, was anxious to raise the Commission's profile after a 'rather pedestrian'[87] report on oil pollution of the sea (RCEP 1981), and at a time when environmental advisory bodies were thought to be at risk. Thus, he was looking for 'a timely issue...something which is important and which is going to give [the Commission] a high profile and, we hope, a measure of success'.[88] Lead offered precisely such an opportunity, as the minutes of a meeting at the time make clear:

> It was futile for the Commission to try to change attitudes merely by statements of philosophical principles. The most effective way of bringing pressure to bear on government was by selecting the most topical and emotive topics; and the fact that a topic was being looked at elsewhere—or that the Commission had no immediate scientific answer to it—was not a sufficient reason for leaving it on one side.[89]

The topic of energy and climate change was another example, chosen under the new Chairmanship of Sir Tom Blundell in 1998, when the Commission wanted to find an arresting angle on an issue that had been widely discussed, and was determined that 'the Commission's report must not look like simply one more report on energy'.[90]

Overall, then, the Commission tended to select issues that were already in the air but which merited more rigorous investigation; in this sense, it amplified rather than set the environmental policy agenda. But, as Chapters 4–6 will show, it often treated its subjects in original and unpredictable ways and, in the process, developed new thinking across a wide range of issues and concerns. In choosing its studies, the Commission was influenced by others but exercised a considerable degree of autonomy, and its remit was moulded over time to fit an expanding environmental agenda as well as the interests and concerns of its members. Many commentators, from outside as well as within the Commission, felt that its hallmarks were to subject complex issues to rigorous, interdisciplinary treatment and to take a long-term view; they noted also that its constitution as a standing royal commission meant it was better equipped than other advisory bodies to adopt this distinctive approach. Finding subjects that lent themselves to such treatment, as well as meeting the Commission's self-imposed criteria, was, however, increasingly challenging towards the end of the Commission's life, as its field of interest became crowded and the nature of the dominant issues changed; this, too, was among the factors that weighed against it at the time of its dissolution.

Taking Evidence

Once the choice of topic was finalized, the process of gathering evidence could begin. The procedures changed very little over time. A call for written evidence was published,[91] and sent directly to bodies and individuals thought to have a particular interest in the study. Written submissions, often numerous and voluminous, were then sorted and summarized by the Secretariat and scrutinized by the Commission at its meetings. Selected witnesses were invited to give oral evidence—a process whose symbolic significance may have been as important as its role in gathering information.[92] While some were asked because the Commission had found their written submissions particularly interesting or provocative, it was seen as obligatory to invite 'the main players' in any given field.[93] In practice, there was no compulsion, but the Commission enjoyed the status of a prestigious body before which the majority of those approached would agree to appear.[94]

One earlier observer of committees taking evidence argued that many organizations are equipped to prepare submissions in a given sphere, and that '[t]he stronger they are, the less original their ideas are likely to be'; hearings, therefore, are 'often a waste of time' (Donnison 1980: 12). Certainly the Commission's evidence sessions were sometimes superficial, or predictable ('you could have written down on paper beforehand exactly what was going to be said'[95]), or in other ways frustrating for members and (especially) for the Secretariat. This sense of dissatisfaction is illustrated by the following extracts, the first from a former Secretary and the second from an industrialist and lawyer, who was a member in the 1980s and early 1990s:

> I don't think the Commission tended to handle the oral evidence sessions terribly well...we in the Secretariat...tried to define the agenda for the session beforehand, but we weren't very successful in getting guidance from Commission members on what they wanted the agenda to cover.... when it then came down to the question sessions, members...tended not to feel any ownership of the agenda and so subjects that may have been on the agenda...never got covered, other subjects which happened to come out of a chance remark during the course of the evidence session would be pursued to a far greater degree than they warranted.[96]

> People were never really...put on the spot, because it was a bit too cosy... Questions could be asked—and quite tough questions—but having asked the question, you say, well thanks very much and it was written down. It wasn't really followed right the way through...I think the whole way in which it was organized [was] a bit avuncular and rather bland...you can't really test how good or bad the evidence is without...really putting people under pressure now and again.[97]

Evidence sessions were not normally confrontational; indeed, the same member complained that they 'often degenerated into a philosophical debate between the Commission and witnesses rather than a cross-examination'.[98] For the most part, those presenting oral evidence were treated as 'guests not... victims', and bad-tempered or aggressive sessions were the exception rather than the rule.[99]

In spite of these criticisms, the best oral evidence could be informative, persuasive, and influential, especially when, as one member put it, 'someone in [a] group...really got through to us'.[100] Witnesses proffering in-depth knowledge, new information, clearly articulated arguments, or novel perspectives were most likely to 'get through', while those adopting a bland, predictable, or partisan stance often failed to impress—though even in these cases, according to the same member, there was always something in the evidence 'that relate[d] to the balanced view of the situation'.[101] In this sense, hearing oral evidence (contrary to Donnison's view, cited above) was rarely a waste of time. What does seem to have happened is that the evidence was subtly, perhaps unconsciously, screened, so that the routine and predictable, as well as views perceived to be extreme or unreasonable, were filtered out or afforded less significance. The filtering processes involved in taking evidence also meant that the Commission, while it was sometimes a conduit, was never uncritically so.

One further, intriguing feature of the process of gathering evidence merits comment here: it is that the process can sometimes affect the witnesses themselves as much as the receiving body. As Donnison (1980: 12) recognized in his discussion of committees of enquiry, 'if the committee is well briefed and led, opportunities can be made *to set fresh thinking going* that will prepare the ground for action on the committee's recommendations' (emphasis added). Similarly, preparing and presenting evidence to the Royal Commission sometimes encouraged those involved to consider their own position afresh. In contemporary terms, this might be thought of as a form of 'anchoring', and it was an effect that Lord Flowers had in mind when he spoke of 'doing good by stealth' (Southwood 1985: 347). As an important dimension of influence, this two-way function of evidence will be explored in more detail in Chapter 6.

Deliberation and Drafting

The Commission was a discursive and argumentative institution, debating issues at length and in depth during the preparation of reports. One member from the 1980s recalled that 'there was a remarkable amount of discussion, much more than on any other group I've served on',[102] and these exchanges were based on a great deal of preparatory work: 'We would all be sent the draft [of a report] and incubate this...in our homes and make copious notes'.[103] And while there was often some division of labour, with 'cells of members'[104] being particularly active on certain issues, report preparation was essentially a collective endeavour. Successive drafts, normally prepared by the Secretariat, but often with substantial input from members, were scrutinized and debated at Commission meetings until a final version was agreed. That this could be a prolonged and exasperating process is reflected in recollections of 'interminable redrafting', with 'substantial modifications' right up to the end; one member described dealing with drafts in plenary sessions as 'a nightmare'.[105] Certainly, members and Secretariat alike would have recognized David Donnison's (1980: 14–15) description of this stage of the proceedings, in his essay on committees of enquiry: 'As drafting begins in earnest, members of the committee become muddled, contradictory and sometimes ill-tempered. They complain about drafts, and then about revisions made to satisfy them. They always grumble about points of style...the committee's promised deadlines disintegrate'. In the Commission's case, what seems remarkable at first sight is that the iterative and frustrating process of drafting and deliberation could produce reports that came to be regarded as authoritative and influential; yet, as we shall see, the very practices that made for slow progress contributed to the robustness of the reports (Chapter 7). It is significant, too, that there was only one minority report in the history of the Commission (to be discussed in Chapter 4), and even this did not take issue with the fundamental conclusions of the study (RCEP 1972a). Since reports were often far from bland, successive chairs must have been skilful at 'negotiating floating icebergs and keeping everybody calm'.[106]

When a report was close to completion, it was the normal practice to send sections of the draft to external experts for peer review, while the full draft (without the recommendations) routinely went to the environment department (and others, as appropriate) for 'factual checking'. Members were on their guard against interference at this point ('in no sense did we feel that they had any veto on what we said'[107]), but the 'factual checking' was nevertheless a process of testing the water, and on some occasions the tone of the draft was modified at this stage to increase the chances of its recommendations being accepted. The report on incineration (RCEP 1993a)—discussed in more detail in Chapter 4—was an interesting case in point.

The Commission preferred to conduct both its own deliberations and the taking of oral evidence behind closed doors. According to one member who served in the 1970s, it resisted a proposal from Tony Benn, then Secretary of State for Energy, that it should open its meetings to the public:

> [W]e all agreed because...we found that [members] who were high up in... industry were quite willing to criticize the policies of their industries and we were able to tackle problems *in a genuinely scientific way*...if the meetings had been held in public, none of them would ever have said anything other than fitted in with the policy of their organization...[emphasis added].[108]

Such a position is not unusual in the world of policy advice. This member's reflections bear a striking resemblance to the rationale used by the *Gezondheidsraad* (the Health Council of the Netherlands) for holding its committee meetings in private. In their research on this body, Bijker, Bal, and Hendriks (2009: 84) found that the confidentiality of the committee process helped in the construction of 'authentic experts', enabling members to speak their minds with 'an attitude of disinterestedness'. Committees of the US National Academy of Sciences have similarly maintained confidentiality, at least for meetings at which they are deliberating their findings (US National Academies 2005).

Even in the post-Nolan climate, the Commission continued to hold its routine meetings in private. In 1999, in response to a proposal from the team conducting the first FMPR that it might open up the meetings at which it heard oral evidence, it delivered a firm rebuttal: 'Holding sessions of oral evidence in public...would not only involve significant additional complication and expense, but destroy their value.'[109] The question was revisited in 2001, when the Chair himself (Sir Tom Blundell) was of the view that it would be good for the Commission, and would eventually become necessary, to hold oral evidence sessions in public. But other members 'regarded such sessions as part of the process of forming the Commission's own views' and continued to resist any change.[110] However, the Commission did make agendas, minutes, written submissions, and transcripts of oral evidence available on its website from the late 1990s onwards.[111] It should be noted, too, that on a small number of occasions (for example during the transport and energy studies in the 1990s), public meetings were organized during the course of the Commission's investigations. Later, the Commission began to arrange seminars with invited participants to assist with the scoping of its studies and—in the first exercise of its kind—the study on bystander exposure to pesticide spraying was launched (in 2004) with a combined seminar and public meeting, which included an open discussion.

The Reports in Brief

As noted above, the Commission produced thirty-three reports (listed, with an indication of content, in Appendix 1). It was known for its rigorous approach to specific aspects of pollution (for example, in the third, eighth, and ninth reports, and the special report on aviation), for highlighting areas that it considered to have been neglected (in the sixth, eleventh, and nineteenth), and for scrutinizing the environmental implications of key policy sectors, such as agriculture, transport, and energy (in the seventh, eighteenth, twentieth, and twenty-second). Some reports (notably the sixth, thirteenth, seventeenth, and twenty-seventh) assessed the risks of established or emergent technologies. Others were cross-cutting, presenting overviews of pollution problems and priorities (for example, the first, fourth, and tenth), proposing new institutional arrangements (the fifth, twenty-third, and twenty-eighth), and developing the philosophy upon which (the Commission felt) environmental regulation should be based (notably in the fifth, twelfth, and twenty-first reports). An assessment of influence must await more detailed treatment in Chapter 6, but the present chapter would be incomplete without an acknowledgement that many—indeed, most—of the Commission's reports had at least some visible impact on policy, legislation, or regulations. Moreover, basic principles consistently advocated by the Commission, such as public access to environmental information (especially in the second and tenth reports), integrated pollution control (in the fifth and twelfth), and the 'duty of care' in waste management (in the eleventh), became widely accepted as part of the legislative basis for environmental protection. Why some reports led to changes in the short term, others over much longer periods, and a few not at all, is a key issue be explored in the rest of this book, drawing upon and developing the theoretical perspectives on knowledge and policy outlined in Chapter 1.

Formally, as we have seen, the Commission's reports were presented to the Queen and laid before Parliament as Command Papers (with a few exceptions, as discussed above), and the arguments and recommendations within them were addressed primarily to government, latterly including the devolved administrations.[112] In practice, however, the Commission's audiences were diverse. Some reports found a receptive international readership, especially in Brussels (where, as noted in Chapter 2, several of the later reports were launched, following the standard press conference in the UK). At the other end of the scale, topics such as waste, transport, and environmental planning had particular significance for local authorities. Beyond government, reports often made an impression on the relevant groups within policy communities, including professional associations, journalists, industry, statutory bodies, non-governmental organizations, and academia. Among a wider public, however, the Commission was never well known: occasionally it emerged into the limelight, when certain reports attracted the attention of the media, but it was more generally the case that 'the words RCEP [didn't] register, even amongst educated people.'[113] This was in spite of the aspiration of the first Chair of the Commission, Eric Ashby, that it should report in '"Radio 4 or even Radio 2 language"...to provide the press and parliament with ammunition which will prod governments into motion' (Tinker 1972: 579).

Views differed within the Commission, and over time, on the importance of wider recognition and dissemination. Earlier on, at least, some members were of the view it was 'not the Commission's duty to popularize its own reports', and one admitted that the reports had gone 'out onto the sea of consciousness' as far as he was concerned.[114] But there were always members who found such views 'naïve',[115] and in later years the Commission did become more attentive to its public profile. After the turn of the century, it became standard practice to produce short, digestible versions of the main reports for wider circulation; so, for example, summaries of the twenty-second and twenty-third reports (on energy and environmental planning respectively) were sent to secondary schools, colleges, and libraries throughout the UK, and the latter was distributed with the newsletter of one of the UK's leading environmental organizations (see Chapter 5). This effort reflected not only a new determination to be visible, but also the increasing length and complexity of the reports themselves. The earliest documents had been slim and elegant (with the main text of the second report, for example, filling only eight A5 pages), whereas later outputs often extended to several hundred pages of A4.[116] Whilst specialists welcomed the rigorous and comprehensive treatment of an issue, and many audiences would focus only on specific parts of a report, the summaries reflected an acknowledgement that weighty tomes were inaccessible not only for lay readers but also for pressurized professionals and decision-makers alike.[117]

Survival and Demise

One of the distinctive characteristics of the Royal Commission was its survival for more than four decades, through changes of administration, shifts in political ideology, lean periods for environmental policy, and re-ordering of policy priorities. This is all the more remarkable given that most other standing royal commissions had disappeared by the early 2000s, sometimes with barely a ripple to show where they had been. One former Chair attributed the Commission's tenacity to the skills of the late Lord Ashby, whom he felt had 'carved out a role for the Royal Commission which, without his wise chairmanship, might have disappeared like so many other[s]'.[118] The reasons for its long survival, however, are more complex than this, and will be explored in greater detail in Chapter 7.

Despite its tenacity, the Commission was never immune from critical scrutiny, nor from pressures to improve its performance and reduce costs; from time to time, its utility was questioned and it was not always popular with governments or with those whose interests were likely to be affected by its recommendations. It tended to be most at risk during the culls of quangos that governments engaged in from time to time, especially when the zeal to reduce expenditure coincided with questions about the Commission's efficacy or irritation with one of its reports; it was just such a combination that contributed, in 2010, to the Commission's disbanding, though previously it had managed to survive in similar political circumstances.

The first obvious threat to the Commission's existence was in the late 1970s, when a number of environmental advisory bodies were abolished by the incoming Thatcher government. As noted in Chapter 2, the Commission escaped a similar fate, in part through a form of boundary work: it was seen as 'too influential to axe, being oriented more towards the scientific establishment than the environmental lobby'; and, having survived, it 'assumed enhanced significance as an authoritative source of independent judgement and policy analysis' (Lowe and Flynn 1989: 264). Some ten years later, quangos were once again in ministers' sights, though on this occasion there were also concerns about the Commission's productivity, and several interviewees had been conscious of 'a mood in Whitehall...to get rid of it'.[119] The mood was undoubtedly influenced by the long-running study of freshwater quality, which emphasized both the slowness of the Commission's proceedings and its propensity, on some occasions, to be inconvenient. According to one Deputy Secretary, the report (RCEP 1992) 'took forever, and we got a bit restless',[120] while another official recalled the irritation at what was seen as an 'insensitive and unhelpful' choice of topic in the context of water privatization: 'if they could have killed [the Commission] there and then I believe they would have done.... of course, in the political context that was impossible but it did put pressure on the Commission to be more productive'.[121]

Such pressures were apparent during 1989, when the Commission's future role, including a range of possible new activities, was under discussion with the department—and, of course, at the Commission's monthly meetings.[122]

After extended consideration, the Commission agreed to undertake additional, 'more tightly focused' studies and to maintain 'a closer continuing interest in the subjects of its recent reports' (RCEP 1990);[123] in return, there would be a modest expansion of the Secretariat. Other proposals from the department, including the idea that the Commission might comment on a wide variety of environmental topics as they arose, were resisted.[124] One further effect of the pressure was to raise questions within the Commission about its international profile. In setting out ideas that might be put to the department, for example, the Secretary wondered whether 'the Commission may not want to tie itself too narrowly to UK interests but may want to look at subjects of interest internationally, and also to develop links with similar bodies elsewhere, particularly in Europe'.[125] A little later, the 1991 audit of the Commission (mentioned earlier) provided another channel for departmental anxieties about productivity and programme planning; this review did lead to some modest changes, but afterwards the issue seems to have died away.[126]

Generally, then, the Commission emerged relatively unscathed from an episode of rather intensive scrutiny in the late 1980s and early 1990s, shielded by its own standing and authority, by the rising profile of the environment (which made it 'politically counter-productive to bash [the Commission] on the head'[127]), and, according to one former member, by a fear of 'the uproar that there would be in the [House of] Lords' if it were to be abolished.[128] This last point is particularly interesting in the light of the significance of policy networks, which will be explored in more detail in Chapter 7. Suffice it to note here that for three decades the Commission was well connected with the Upper House of the British Parliament and had a productive relationship with several of its influential select committees.

The next significant point of vulnerability came at the time of the second FMPR, which took place in 2006–7 (Defra 2007b). On this occasion, there was a feeling within the Commission that the review might have been set up with hostile intent. While the first FMPR (DETR 2000a) had been conducted by the department's In House Policy Consultancy, the second was undertaken by external management consultants (PricewaterhouseCoopers), at a time when the Commission's relations with the department were once again at a low ebb. The review followed publication of a special report on bystander exposure to pesticide spraying (RCEP 2005), which the department had considered 'unscientific' (see Chapter 4), and coincided with a study of the urban environment in which there had been tetchy oral evidence sessions with civil servants from both of the main departments involved—environment and Communities and Local Government. Inevitably, the new form of accountability, which relied heavily on consultation, exposed the Commission more directly to both its friends and its critics—and such consultation had become more extensive by the time of the second FMPR.[129]

In the event, the 2007 review concluded that the Commission had 'historically exhibited independence, scientific rigour and depth' in giving advice, and that its important functions were unlikely to be performed by any other body (Defra 2007b: 3). But the consultants were clearly of the view that it was 'in need of modernisation' (ibid.) and the use of the term 'historically', with its hint that the Commission might have lost its edge, may not have been accidental. The review recommended a number of changes, including diversification of the membership—'fewer academics' (ibid.: 66)—and speedier production of reports, but also, and more positively, an increase in the Commission's budget. At a time of extreme financial stringency there followed a wrangle between the Commission and the department, largely about resources, which emphasized again the particular nature of the Commission's 'independence' and (according to one senior civil servant) caused its abolition seriously to be contemplated at that point.[130] In the end it was decided that the Commission should continue, with modernization being funded from savings elsewhere within its budget (Defra 2008). As part of the compromise, the Commission finally agreed to move into departmental premises (which had symbolic significance, even if the particular building housed no departmental functions). It also agreed to work with the department to 'seek applicants with experience in business, policy-making and regulation',[131] to restructure the Secretariat, and to make changes to its working practices, including the more efficacious production of reports. Otherwise, it seemed once again to emerge in much the same form and with its budget intact.[132] It is difficult not to conclude, however, that the Commission, which had increasingly been challenged by audit and review, was somewhat weakened at this point, though it went on to complete three more major studies and another special report before it was finally abolished.

The denouement came after the general election of May 2010, when the new Conservative–Liberal Democrat coalition government set about reducing the number of quangos in a particularly determined way. In a Parliamentary written statement in July of that year, the Secretary of State for Environment, Food and Rural Affairs set out plans to abolish thirty 'Defra bodies', including the Royal Commission:

> When the RCEP was set up in 1970, there was very little awareness of environmental issues, with few organisations capable of offering relevant advice. The situation now is very different, and the Government have many such sources of expert, independent advice and challenge. Protecting the environment remains a key Government aim, and DEFRA intends to draw on the full range of expertise available... I pay tribute to the work of the Royal Commission and its current chair, Sir John Lawton. Over the last 40 years the commissioners have made a significant contribution to raising the profile of environmental issues in the UK.[133]

Some felt that the decision to abolish the Commission, after its long and distinguished service, was peremptory and ill-advised.[134] There had been relatively little consultation, none with the Commission itself. When the news broke, there was a flurry of correspondence among certain members, former members (including the author), the Chair, the Secretary, and other interested parties about the implications of the Secretary of State's announcement—and, indeed, about its legitimacy, given that the Commission was not, in any substantive sense, a 'Defra body' but an independent commission appointed by Royal Warrant. It was clear that the Secretary of State did not herself have the power to abolish a royal commission; it was equally clear, however, that on this matter of public policy the Queen would act on the advice of her ministers; Her Majesty had been informed, and the devolved administrations consulted. The decision was given legal effect when the Queen issued a new Royal Warrant, terminating the Commission's service from the end of March 2011; its funding was withdrawn under the Comprehensive Spending Review and the Secretariat was largely re-absorbed into the department. The general view amongst the Commission and its supporters was that lobbying for its continuation would be futile, and the thrust of its valedictory seminar, held on 8 March 2011, was not to mourn the passing of this particular advisory body, but to consider 'the future of environmental evidence, advice and policy'.

Concluding Comment

Together with the overview of environmental policy in Chapter 2, the account of the Commission's history and practices in this chapter provides the context in which to address questions about its influence and the attributes that enabled it to have effect. To the questions identified in Chapter 1 we might now add another: why had it become possible, by 2011, for a government to sweep away this advisory body, which had survived disapproval and financial stringency on a number of previous occasions? Chapters 4 and 5 offer in-depth analysis of the Commission's investigations and recommendations, in different settings over the forty or so years of its existence. Chapter 6 then draws the findings together to offer a typology through which 'influence', in the context of policy advice, might most usefully be characterized and understood. Chapter 7, with a view to the future, considers the Commission's attributes, its long survival, and its ultimate abolition, and seeks to draw wider lessons about knowledge–policy interactions, relating the findings of this study to the conceptual framework set out in Chapter 1.

4

Risk, Precaution, and Governance

An Evolution of Ideas

with Tim Rayner

> We expect government to behave like a manager, not a scientific referee...
> This involves weighing up all the issues, both scientific and non-scientific, and arriving at reasonable judgements on the predictions which it would be prudent to adopt and on the action which should, in consequence, be taken.
> RCEP, *Tackling Pollution: Experience and Prospects* (1984: para. 2.32)

> The debate about how risks should be perceived, measured, communicated and managed throws up most of the issues that arise in the more general debate about the proper role(s) of science and scientists in society.
> John Ashworth, in Royal Society, *Science, Policy and Risk* (1997: 5)

> Setting an environmental standard is an exercise in practical judgement.
> RCEP, *Setting Environmental Standards* (1998b: para. 8.31)

Introduction

Anticipation and prevention of environmental damage, and precaution in the face of risk and uncertainties, have been among the hallmarks of ecological modernization. The precautionary principle—the most formal expression of the need for prudence in environmental affairs—involves all of these concepts to some degree, and in particular implies a willingness to take pre-emptive action rather than waiting for proof of harm. There is no universal definition of the principle, but its broad intent was encapsulated in the *Rio Declaration* of 1992: 'Where there are threats of serious or irreversible damage, lack of full scientific certainty shall not be used as a reason for postponing cost effective measures to prevent environmental degradation' (UN 1992: Principle 15).[1]

The UN wording implies a shift in the burden of proof within a given but incomplete body of scientific knowledge, qualified with reference to the costs of appropriate action. The principle can be applied reactively, when damage is already discernible but uncertainties about causality may persist. More controversially, it can be invoked in a *proactive* mode (Levidow and Tait 1992), when harm is postulated but has not yet been shown to occur. Some have gone further to argue for a more radical, *preventive* paradigm, demanding reconsideration of the very categories that might constitute 'harm' (Wynne 1992);[2] this thinking relates closely to changing conceptualizations of risk, and serves to emphasize that 'environmental problems' are not self-evident but have to be constructed through argumentative practices (Liberatore 1995). As we shall see, such practices have been important in the Royal Commission's treatment of risk and precaution, and in the presentation of its findings and recommendations.

The precautionary principle was greeted initially with suspicion in the UK, where it was seen in some quarters as a 'continental' philosophy, which threatened the British 'science-based policy approach' (Hajer 1995: 141); such reservations were shared by at least some members of the early Royal Commission. Nevertheless, as we have seen in Chapter 2, environmental policy in advanced economies evolved in a broadly precautionary direction during the latter part of the twentieth century. At the same time, and in ways that were often connected, the dominant discourses of risk and risk assessment were themselves being vigorously challenged; this was especially true of the so-called 'manufactured risk' (Giddens 1999: 4) associated with late modernity, and involving a high degree of human agency. Such issues were present in virtually all of the Commission's work, and in some studies became the focal point of its deliberations.

It is instructive, therefore, to trace the evolution of the Commission's thought on risk and precaution, showing how certain themes—such as the need to avoid irreparable environmental damage—were prominent from the outset, and how concepts and categories were constructed, refined, and sometimes undone through successive studies and reports. Of particular interest are the Commission's perspectives on risk governance, and on relations between science and politics, including its propensity to perform 'boundary work' in making a clear distinction between the two. Certain reports stand out as breaking new ground, while others contributed to a more gradual shifting of frames—though the progression was not always linear, as we shall see. Much can be learnt from the Commission's treatment of broadly comparable topics at different times. In reports on nuclear power (RCEP 1976b), genetically modified organisms (GMOs) (RCEP 1989), and nanotechnologies (RCEP

2008), it explored the environmental implications of science-based technologies on the threshold of major development or expansion. In others, it engaged with specific controversies—lead in petrol (RCEP 1983), incineration of waste (RCEP 1993a), and 'bystander' exposure to pesticides (RCEP 2005)— all of which involved public disquiet, official reassurance, and considerable scientific uncertainty. A third interesting group includes studies in which the Commission reflected on progress, or set out new ideas about the basis for environmental regulation (with important examples including RCEP 1984, 1988, 1998b).

The aim of this chapter is to identify landmarks, continuities, progressions, and what might be seen as regressions in the Commission's treatment of risk. This means that in a broadly chronological account the narrative will sometimes jump forward, following key concepts as they evolved. The Commission never operated in a vacuum, however, so it is helpful, first, to consider how broader perspectives on risk, in the scholarly literature and in policy discourse, changed over the period of its existence.[3]

Risk Paradigms

When the Royal Commission was first established, the risks associated with techno-scientific innovation were just beginning to command serious attention on policy and political agendas.[4] In the prevailing paradigm, such risks were seen as phenomena to be apprehended in the physical world, characterized and quantified through a process of risk assessment, then prioritized and managed in a rational and cost-efficient way. Scientists and technologists would determine the nature and magnitude of the risks, while welfare economists would help weigh them against the benefits of the activities in question: the role of such experts was to provide objective advice to decision-makers, much as envisaged in the linear–rational model outlined in Chapter 1. Public perspectives on risk, when they conflicted with those of experts, were dismissed as irrational, and there was a powerful instinct to 'correct' them through the relentless provision of 'facts'. Resistance to such correction was seen by many as an obstacle to rational policy—hence the growing interest, in the 1970s and 1980s, in analysis of 'public risk perception'.

Challenges to the dominant paradigm came from several directions. In anthropology, the 'cultural theory of risk', developed by Mary Douglas and Aaron Wildavsky, suggested that individuals' perspectives were socially and culturally conditioned by their worldviews, which might embrace, for example, the fragility of natural systems or the desirability of the risky activities in question (Douglas 1992; Douglas and Wildavsky 1982; Thompson, Ellis, and Wildavsky 1990). Such 'cultural biases' might be widely shared within social groups, or even by whole societies; the important point was that risk was a social construct, and information about specific risks would always be filtered by the recipient's worldview.[5] In a different vein, a new generation of psychometric studies sought explanations for public perceptions in the attributes of the risks themselves. Particularly influential was the work of Paul Slovic and his colleagues, who used factor analytic techniques to reduce many such attributes to two orthogonal dimensions ('dread risk' and 'unknown risk'), and showed that lay people's responses related closely to the positioning of different risks on these axes (Slovic 1987, 1993, 2000; Slovic, Fischhoff, and Lichtenstein 1980).[6] Both approaches—cultural and psychometric—offered important insights into the failure of probabilistic risk estimates, or simple, quantitative comparisons between different *kinds* of risk, to convince and reassure. Cultural theory showed how risks were interpreted within different frames, while the psychometric studies found 'wisdom as well as error in public attitudes and perceptions', and suggested that risk communication was destined to fail unless 'structured as a two-way process' (Slovic 1987: 285).

By the end of the 1980s the emergent view was that public perspectives on risk were complex, rather than simply irrational, and could certainly not be ignored in the development of appropriate policies. Nevertheless, a conviction lingered within scientific and policy communities that the essentially subjective nature of 'risk perception' was still to be set against some benchmark of real or objective risk; the notion that facts could (and should) be distinguished from values, judgements, and other influences on perception was, if anything, reinforced in a number of reports and reviews (see, for example, Royal Society 1983; US NRC 1983; for a discussion of these perspectives, see Expert Group on Science and Governance 2007). The corollary was that the 'scientific' process of risk assessment should remain separate from (and prior to) the politically infused task of risk management, in which different perceptions and judgements should be afforded a significant role.

This separation proved untenable, however, when applied to situations characterized not only by risks with known probabilities, or by uncertainties that might in principle be capable of resolution, but also by ambiguities, indeterminacies, and ignorance—they were subject, in other words, to different dimensions of incertitude.[7] In such circumstances, it could be shown that the assessment of risk was itself 'unavoidably influenced by the categories, presuppositions and models' of the experts involved (Shrader-Frechette 1995: 118; see also MacGillivray 2014), while lay perceptions might well be 'rationally based in judgements of the behaviour and trustworthiness of expert institutions' (Wynne 1996: 57). Nor could a model in which objective assessment 'informed' the policy process easily be upheld; indeed, environmental risks furnished much of the evidence that facts and values interpenetrate in the complex controversies of regulatory science. Some influential thinkers argued that anxiety about the risks inherent in late modernity was leading to critical scrutiny of the institutions of science and technology themselves, in a process of 'reflexive modernization' (Beck 1992; Beck, Giddens, and Lash 1991; Giddens 1990). Certainly it seemed that people worried as much about the social, ethical, and political dimensions of science-based technologies as about possible harms to human health and the environment (Grove-White 2001; Jasanoff 1995; Wilsdon and Willis 2004; Wilsdon, Wynne, and Stilgoe 2005).[8]

Collectively, these new understandings suggested that 'riskiness' should not be narrowly conceived, and pointed to a need for more open, deliberative, and participatory processes in risk governance (Jasanoff 2003b; Stirling 2003, 2007; Weale 2002). By the 1990s, more nuanced approaches to incertitude were having effect in both scientific and policy communities—reinforced, no doubt, by the visible failure of the technocratic approach to settle high-profile science–policy controversies (DETR, Environment Agency, and Institute for Environment and Health [IEH] 2000; DoE 1995a; Royal Society 1997).[9] In the UK, the environment department acknowledged that risk assessments must often and necessarily be subjective; that benefits and costs might not fall on the same people; and that risk judgements depended not only on the characteristics of the hazard, but also on 'broader psychological and sociological considerations' (DETR with Environment Agency and IEH 2000: 24; see also Defra and Cranfield University 2011; DoE 1995a). There were conscious efforts to open up debate about contentious risks and technologies, though governments did not always welcome (or know what to do with) the outcomes.[10]

By the early twenty-first century, these developments had resulted in a partial transition to a new risk paradigm, with the effect that different discourses co-existed, sometimes in the same assessments, and often in tension with one another; the merits of realist and constructivist perspectives on risk, and the appropriateness and practicality of wider engagement, continued to be vigorously debated. While the model of the neutral expert became increasingly untenable, the most typical response was to call for a clearer definition of boundaries, with experts being exhorted to make their judgements explicit and distinguish them from scientific facts (for example, OECD 2000; RCEP 1998b). The urge to separate the technical and the political remained potent, therefore, in spite of important insights from science and technology studies on the interpenetration of facts and values, and the co-production of knowledge and politics (Owens 2011b). At the same time, demands for wider engagement in risk governance steadily increased, still with some puzzlement about the end result: as the Royal Commission observed in 2008, 'enthusiasm to be seen to engage has sometimes run ahead of any real commitment or institutional capacity... to make intelligent and transparent use of the findings'

(RCEP 2008: para 4.103). The Commission was a keen observer of change, but it was also a driver which itself had substantial influence on the shifting discourses on risk and precaution. It is to these reciprocal processes, and their implications for environmental policy, that this chapter now turns.

Risk and the Royal Commission: Early Perspectives

The Commission's earliest pronouncements on risks to the environment and human health struck what might best be described as a cautious precautionary note. Its very first report (RCEP 1971), which generally adopted a 'doing well, must try harder' tone in an overview of environmental quality and pollution control, nevertheless warned that natural environments did not have an infinite capacity to absorb growing volumes of pollution. A year later, in its brief but pithy second report (RCEP 1972b), the Commission showed early signs of challenging the 'innocent until proved guilty' paradigm associated with British pragmatism. Setting out its views on the potential impacts of new products and associated wastes, it expressed particular concern about substances with properties such as persistence and fat solubility, and gently nudged at the burden of proof: 'While it would not be reasonable to regard substances with these properties as "guilty until proved innocent" it is reasonable to regard them as "under suspicion"' (ibid.: para. 13).[11] The Commission called for toxicological testing of such products prior to marketing (later to become a requirement in the European Union) and for 'sustained monitoring of their impact on the environment' (para. 13). More than thirty years later, testing and monitoring featured prominently in its twenty-fourth report, which dealt specifically with chemicals in products, including the tens of thousands of 'existing' substances that had not been subject to risk assessment prior to marketing in the EU.

Of the earlier reports, it was the third, dealing with pollution in estuaries and coastal waters (RCEP 1972a), that developed the most explicit position on precaution. Here the Commission took forward its thinking on assimilative capacities, reiterated concern about substances that could persist and accumulate in the environment, and fretted about irreversibility—'there could be points of no return in the deterioration of water' (ibid.: para. 10). On discharges of potentially dangerous substances to the aquatic environment, it argued that '[s]ome pollutants are rendered harmless by natural processes but others...are not'; the latter, it felt, should be abated at source, *'before* discharge into rivers or tidal waters' (ibid.: para. 21 [emphasis added]). This formulation lent support to the prevailing British position on utilizing assimilative capacities, but at the same time acknowledged the merits of the 'continental' discourse, favouring uniform, source-based controls, at least for persistent and bio-accumulative pollutants. The issue was soon to become a highly charged one, on which future Commission reports would have a great deal more to say.

The Commission recognized from the outset that the governance of environmental risk must involve ethical and political choices as well as scientific and technical judgements. If, as one of its secretaries suggested, it had been established 'to deal with environmental pollution not ethics',[12] it must quickly have realized that the two were not easily kept apart. Environmental issues raised fundamental questions about the boundaries of the moral community (who or what was to be protected?), about the ethical basis for decisions (welfare maximizing, or grounded in other imperatives?), and about the roles of knowledge and judgement in policy- and decision-making processes. Such questions were addressed more or less explicitly throughout the Commission's work, with significant implications for its framing of environmental risks and its policy recommendations. From an early stage, for example, it included both future humans and non-human nature within the scope of its deliberations. The third report (RCEP 1972a) emphasized society's long-term dependence on biological cycles, while what came to be known as the 'Flowers criterion' in the seminal report on nuclear power (discussed below) explicitly invoked a duty to future generations: 'it would be irresponsible and *morally wrong* to commit future generations to the consequences of fission power on a massive scale unless it has been demonstrated beyond reasonable doubt that at least one method exists for the safe isolation of [radioactive] wastes for the indefinite future' (RCEP 1976b: 81 [emphasis added]). Similarly, while the Commission often argued that the integrity of natural systems should be safeguarded as a matter of prudence, it recognized, too, that there were concerns to protect the environment 'even if this does not obviously benefit human beings' (RCEP 1972a: para. 15). The concept of 'responsible stewardship' runs through many of its reports (RCEP 1984: para. 1.13; also RCEP 1998b, 2002a, 2004), influencing its conceptions of harm and its judgements about the appropriate degree of precaution.

A frequently recurring theme in the reports—not always explicit—is that of the ethical underpinnings for the governance of environmental risk. There was always a tension in the Commission's work between the utilitarianism of a welfare economic approach and a sense that protecting human health and the environment from serious harm was a matter of moral obligation. The first report began with a clear statement of the former philosophy: the 'basic criterion' for deciding how much to spend on pollution abatement was that 'pollution should be reduced to the point where the costs of doing so are covered by the benefits from the reduction' (RCEP 1971: para. 20). Yet it was acknowledged that there could be 'no completely scientific and objective means of arriving at such decisions', in part because of the 'great practical and theoretical difficulties' in measuring all the costs and benefits (ibid.: para. 23). Beyond these methodological problems, the Commission noted that 'acceptance of these data as "true" measures of the relevant social costs *would involve ethical rather than scientific judgements*' (ibid. [emphasis added]).[13] It also emphasized that risks were unevenly spread, so that regulatory decisions would have important distributional implications. These were to become crucial considerations in a number of its later studies.

The need to consider costs and benefits was further emphasized in the third report, which dealt with estuaries and coastal waters (RCEP 1972a). In the only minority report in the Commission's history (included with the main report) two members called for extensive use of economic instruments in pollution control.[14] But the prevailing view in these early reports is that (whatever the means employed) the ends of regulatory policy could not be determined by an economic calculus alone. The fifth report, on the subject of air pollution, attempted to define what those ends should be:

> The aim of air pollution control should be to reduce and when necessary eliminate hazards to human health and safety, taking into account both the magnitude and certainty of risks, including the susceptibilities of critical groups, and the resulting costs to the community; to reduce damage to amenity, property, plant and animal life to a minimum compatible with the wider public interest (which will take into account such factors as economics, employment and trade); and to prevent irreversible damage to the natural environment.
>
> (RCEP 1976a: para. 41)

The tone is interesting. Costs would have to be considered, but there is an implied obligation on the part of governments and regulatory authorities to protect people and the environment from serious or irreversible harm; in such cases, the appropriate level of abatement could not be determined simply by balancing costs and benefits, even if these could be ascertained. This position would be argued and defended more explicitly in the Commission's later work. For now, it is worth jumping slightly ahead to the tenth report, another overview of the field, to cite its interesting distinction between ends and means: 'A more straightforward approach [than cost-benefit analysis]...is *to pre-determine*, not necessarily on economic grounds, a maximum "acceptable" level of risk, and then to formulate a policy which will secure this level at lowest social cost' (RCEP 1984: para 2.2 [emphasis added]).

But how, and by whom, should 'an acceptable level of risk' be pre-determined? In the British regulatory system of the 1970s, the answer was essentially technocratic: it fell largely to the pollution control agencies to interpret national legislation in local circumstances and thereby to determine the appropriate level of abatement (often in close consultation with industry). The early Commission, too, leant towards technocracy, lending support to British pragmatism and to the prevailing system (for air pollution) of 'best practicable means' (BPM). According to its fifth report, it was the controlling authority that would have to use 'informed judgement' in decisions about pollution, and this because many of the costs and benefits were 'inherently unquantifiable' (RCEP 1976a: para. 30). It did, however, inveigh against secrecy, and called for the machinery of control to be 'more open to public understanding and influence' (ibid.: para. 186; see also RCEP 1972b)—so, one might argue that the early Commission favoured a tempered, and more accountable, technocracy. Later, the respective roles of expertise, values, and judgements were to feature more explicitly in the Commission's reasoning, and in the late 1990s became the focus of a substantial report on setting environmental standards (RCEP 1998b). Meanwhile, the Commission's thinking about risk and 'risk perception' developed in interesting ways, with the sixth report, *Nuclear Power and the Environment*, marking something of a turning point. This was the first of the reports in which the Commission in effect conducted a technology assessment, at a time when, in its own words, the world was 'on the threshold of a huge commitment to fission power which, once fully entered into, may be effectively impossible to reverse for a century or more' (RCEP 1976b: para. 196).

Nuclear Power: Precaution, Ethics, and 'Deliberation in Public'

In some respects, the nuclear report showed the Commission in decidedly rational–objectivist mode. The debate about nuclear power was polarized, it noted, but 'not always well-informed' (RCEP 1976b: para. 9), and there was a tendency 'to dramatise the risks' in ways that could mislead people who had 'no basic understanding of the subject' (ibid.: para. 162). An important aim of the study was to 'present *the main facts* about radioactivity and nuclear power clearly and simply ... and thus enable the issues to be discussed objectively and in a proper perspective' (ibid.: para. 9 [emphasis added]). But the report was more memorable for its departures from the dominant discourse on risk in the 1970s. One was an explicit engagement with the social and political risks of a 'plutonium economy', to the extent that contemporary press coverage could characterize publication of the report as a 'moment of emancipation for ideas which were once considered subversive' (Hawkes 1976: 9). The other was a distinctive approach to the problem of radioactive waste.

Consciously stretching its remit, the Commission expressed strong reservations about the long-term implications of an expanded nuclear power programme, with both burner and breeder reactors, which would inevitably involve the movement of plutonium. It was especially concerned about the 'long-term dangers to the fabric and freedom of our society' (RCEP 1976b: para. 507) that the necessary security measures might present—thus being 'preventive' in extending the concept of harm into the social and political realm (Wynne 1992: 111; see also Flood and Grove-White 1976):

> We emphasize...that our concern...is not with the position at present, or even in the next decade, but with what it might become within the next fifty years. In speculating on developments on such a time scale, no one has a prerogative of vision. It appears to us, however, that the dangers of the creation of plutonium in large quantities in conditions of increasing world unrest are genuine and serious.
>
> (RCEP 1976b: para. 506)

The Commission concluded that Britain 'should not rely for energy supply on a process that produces such a hazardous substance as plutonium unless there is no reasonable alternative' (ibid.: para. 535); certainly, '[t]here should be no commitment to a large nuclear programme including fast reactors until the issues have been fully appreciated and weighed in the light of wide public understanding' (ibid.).

The Commission's most radical stance, however, was on the vexed issue of radioactive waste disposal. While for some in the industry, this was 'the biggest non-problem of the century',[15] the Commission thought otherwise. It was disturbed by the inadequacy of existing arrangements and unimpressed by the efforts to find a longer-term solution.[16] Some members were more confident than others about the technical feasibility of safe containment, but a collective view emerged that even if wastes could in practice be dealt with safely, this had to be *demonstrated*, and demonstrated in a way that could assuage public fears. As one former member (an industrialist) put it: 'disposal of nuclear waste...presents some problems but I think the solutions are as safe as walking into a supermarket. And at that time I think we would have been saying, "what are we worried about, it can be done safely", but we needed to have somebody show it before we could go ahead without worrying.'[17] The Chair, Sir Brian (later Lord) Flowers, although himself a part-time board member of the UK Atomic Energy Authority, was convinced of the necessity of such demonstration:

> I was seen by some...in the Atomic Energy Authority as a traitor...because I had allowed myself to see things through others' eyes. That's what's wrong with the nuclear industry—that it cannot do this, cannot take seriously what little it does see. Walter Marshall[18] used to say, 'I know these people have these views—I know very well—but they're wrong, so why should I take them into account? Their imaginings are not real', and I say, 'no, Walter, but their *worries* are real...and that is the problem you have to deal with'.[19]

The sixth report was the first in which the Commission made a strong case for 'a searching examination' of a technology on the verge of major expansion (RCEP 1976b: para. 163), though the sense of its recommendations was that such scrutiny should entail deliberation in public (essentially among experts), rather than public deliberation (with the inclusion of 'lay' voices). The report is remembered most clearly for the Flowers criterion, which, in adopting what

Wynne might call a preventive paradigm, helped to change the terms of the debate about civil nuclear power in the UK. In addition, as Chapter 6 will show, the Commission's scrutiny contributed to the opening up of policy communities and led to significant changes in the regulatory framework for radioactive waste. In many respects, the sixth report can be seen as one in which the framing of socio-technical risk was well ahead of its time.

The next landmark in the Commission's treatment of risk and precaution came early in the 1980s—a decade in which the changes associated with ecological modernization were beginning to make themselves felt. Arguably, its ninth report, on lead in the environment (RCEP 1983), made a significant contribution to this process in the UK. Certainly, it hastened the departure from the dominant techno-scientific discourse on risk.

Risk, Uncertainty, and Precaution: Towards an Ecologically Modern Perspective?

At the beginning of the 1980s, the question of whether lead additives in petrol could be a source of damage to human health had become highly contentious. One of the emblematic issues of the time, this controversy exemplified the turning of attention from the depredations of gross pollution towards less visible and more subtle effects. The harm caused by lead at high levels of exposure was well documented; what was at issue was whether human health was threatened by much lower concentrations in the environment (Millstone 1997). It was known that exposure could occur through a number of pathways, including drinking water and food, but the controversy centred primarily on airborne lead coming from additives in petrol and its possible impacts on children's intelligence and behaviour.[20] By 1982, when the Commission decided to make lead pollution the subject of a dedicated study, rival coalitions were already deeply entrenched, with one side demanding the urgent elimination of the lead additives and the other insisting that such a move would be disproportionate, since it could not be justified by the science. The ninth report would prove to be a decisive intervention in the debate.

The government's position in the early 1980s was delicate. The prevailing policy on ambient lead levels was one of containment, aiming for no increase overall and for reduction 'in those areas and circumstances where people are most exposed to risk' (DoE 1974: iii).[21] An expert committee set up by the Department of Health (the Lawther Committee) had found no scientific case for the removal of lead from petrol (DHSS 1980). In a decision clearly influenced by 'the balance of organized forces' (Kingdon 2003: 163), the government compromised, announcing in 1981 that the lead content of petrol would be reduced from 0.4 to 0.15 grams per litre (g/l), the lowest level compatible with engine technology at the time. Predictably, the anti-lead coalition dismissed the Lawther Report as complacent and the new policy as an inadequate 'half-measure' (Wilson 1983: ix). By now this coalition was spearheaded by a well-resourced and effective pressure group, the Campaign for Lead-free Air (CLEAR), which had secured commitment to lead-free petrol amongst the main opposition parties in Westminster.

The adversaries in the controversy, in their relatively loose alliances and use of distinctive storylines, might best be characterized as discourse coalitions. The 'traditional pragmatists'—including the government,[22] industry, and some scientists and journalists—emphasized uncertainties about cause and effect, and made much of the potential costs of abatement: a favourite line was that lead came from multiple sources and eliminating it from petrol would not be a cost-effective way of protecting human health. Their opponents—a coalition of health and environmental pressure groups, which also included scientists, journalists, and members of Parliament—offered an alternative, precautionary storyline, demanding 'an in-built prejudice towards environmental and health protection' (Wilson 1983: 178); waiting for definitive answers would, in their view, be tantamount to conducting an unethical experiment upon children. And if there were many sources and pathways for lead, this was all the more reason for action, since its removal from petrol was technically less difficult than a number of other possible measures for reducing human exposure.

Lead pollution had been flagged as a problem in the Commission's fourth report (RCEP 1974) but not immediately pursued. In 1982, in the early stages of a study intended to be one of its broader reviews, the Commission invited detailed submissions on a range of issues, including lead in the environment, and atmospheric and aquatic (estuarine, coastal, and marine) pollution. At this point, it had no intention (so soon after Lawther) of re-opening the issue of lead and human health. It quickly became apparent, however, that to deal with lead but not with health would have been like 'Hamlet without the prince',[23] and members were persuaded (in part by a submission from CLEAR) that there would be 'presentational difficulties' in producing a report which offered 'no substantive comments on the health debate'.[24] They were also aware that inclusion of such a sensitive issue in the context of a wide-ranging investigation 'would distort the balance of the report and could mean that findings in other areas failed to attract the attention they deserved'.[25] It was decided, therefore (at the September meeting in 1982), to defer the overview of pollution and concentrate entirely on lead—a decision reinforced by the determination of the new Chair, Professor (later Sir) Richard Southwood, to raise the Commission's profile.

Not all members were enthusiastic. Some were themselves aligned—or at least perceived to be aligned—with the 'traditional pragmatist' view, exposing the Commission to an unusual degree of scrutiny. CLEAR regarded Barbara Clayton, a professor of medicine who had served on the Lawther Committee, as 'a well known sceptic on the health effects of lead in petrol' and was similarly suspicious of Geoffrey Larminie, an environmental manager with British Petroleum (Wilson 1983: 112). Within the Commission, both Clayton and Donald Acheson (a clinical epidemiologist) expressed reservations about re-opening the question of the effects of lead on children's IQ.[26] At the outset of the study, the Chair was 'very conscious...that the Commission was divided on the lead issue...some of them already having committed themselves to certain lines. And so [the choice of study] was to some extent a high risk strategy.'[27]

Lead in Petrol: Framing the Problem

How, then, did the Commission come to its radical recommendation that lead additives should be phased out of petrol according to the 'best environmental timetable'? Certainly it did so at a rather late stage of its enquiry: at the end of 1982, alternative recommendations were still being considered, including one that endorsed the official position (a reduction to 0.15g/l).[28] One factor in its ultimate decision was the slowly accumulating weight of evidence: although the science was still far from conclusive, the limited amount of new work since Lawther had tended 'to indicate a more significant role for petrol lead in the determination of blood lead'.[29] More telling, however, was the Commission's *framing* of the issue, which differed from that of previous enquiries in a number of important respects.

Much was made of the fact that lead, a neurotoxin, was now present in significant concentrations throughout the environment as a result of human activity. The Commission's aversion to irreversible ecological disturbance, reinforced by Southwood's disciplinary background as an ecologist, made such contamination a cause for concern in itself, even if deleterious effects were not evident. This proactive—even 'preventive'—worldview, embracing a wider perspective on 'harm', was encapsulated by one of the members involved:

> [L]ead is a compound of known toxicity which builds up in the body and which is persistent, and the concentration of which in the world environment is gradually increasing. And that is not acceptable. That was our basic reason for saying we must cut back—*you just can't do that*. Now...that was a principle which became accepted by more and more members of the Commission—I wouldn't say...fully accepted, but it was an important element in the Commission's thought...that *permanently disturbing the composition of the environment is not acceptable* [emphasis added].[30]

On safety margins, a key argument crystallized early in 1983, when it was agreed that the blood lead concentration of the general population, averaging about 15µg/dl, was uncomfortably high in relation to the level at which clinical symptoms could sometimes occur (50µg/dl). As the minutes record:

The conclusion to be drawn from this was that the safety margin in the general population was inadequate and ought to be increased wherever practicable. This view was reached *without the Commission making any judgement of the medical evidence* of any effects of low concentrations of lead on neurological development or functioning [emphasis added].[31]

This formulation, though it would eventually be criticized,[32] was crucial in generating support for the recommendation to remove lead from petrol, even amongst those who had doubts about the scientific case; in other words, it allowed the Commission to arrive at a 'serviceable truth' (Jasanoff 1990: 250). One member summarized its significance thus:

> I thought that we cut through the dilemma of the medical evidence quite neatly, myself.... the decision... that levels of ambient lead were in many cases already near that which was quite openly poisonous... was a marvelous [one] because it actually wiped out all further disputation and it allowed [one of the medical members] to be alongside other people.[33]

The breadth of perspectives within the Commission was undoubtedly significant. One member, a philosopher, saw it as part of the role of the non-scientists to raise questions about the traditional scientific approach:

> [T]here were scientists, I think, who were constantly aware that the kinds of recommendations we were making could not stand up to the scrutiny of [their] peers... and the evidence was very shaky.... [M]y role was partly to say we can't expect [the] level of proof that would absolutely satisfy because we can't set up controls, we can't do any of the things that you do if you were conducting a proper scientific programme.[34]

In another sense, too, the breadth of membership and remit proved critical. In contrast to the Lawther Committee, the Commission could interrogate claims not only about health and environmental impacts but about technological developments and costs. As a member who had served on both bodies explained:

> [The Lawther Committee] was a Department of Health committee and its remit was tight, and it didn't allow us to tread on the ground that belonged properly to the Department of Transport.... When the Commission came to look at lead, of course our terms of reference are so broad that we can go where we like for evidence and involve who we like and of course we were able to come to a much better, much more sensible, much more meaningful report.[35]

Having cast its net more widely, the Commission became sceptical about claims that a move to unleaded petrol would impose substantial costs on industry, the economy, and the motorist. On the first of these, it deftly shifted the burden of proof to the manufacturers (RCEP 1983: para. 7.81). Perhaps its cleverest move, however, was to set its proposals against a background of technological advances in fuel economy. This allowed costs to the motorist, and a marginal penalty in national energy consumption, to be reframed as 'benefits foregone'. For an economist then serving on the Commission, this framing was uncomfortable,[36] but for some commentators it was the report's 'main, and remarkably straightforward, achievement' (ENDS 1983: 11).

Arguably, then, the Commission's precaution was *made feasible* by its finding that there were no insurmountable technical or economic barriers to the introduction of lead-free petrol, especially if it was phased in over time. The argument that lead additives were not necessary, together with the general desirability of reducing ambient levels of lead, meant that uncertainties about the health impacts could almost be set aside, and the boundaries around 'sound science' not transgressed too grievously. Things might have been different, as the Commission's former Secretary reflected, if members had been persuaded 'that the economic penalties of moving to unleaded petrol were insupportable—then I think they would have had to go into the medical evidence in great depth'.[37] As it was, the outcome was a dramatic shift in policy. The Commission's uncompromising recommendation on petrol was accepted by ministers within an hour of publication of the ninth report.

The reasons for this rapid take-up will be explored in more detail in Chapter 6, but three important observations can be made here. One is that the shift of policy would almost certainly, in any case, have come about in the longer term. As with the move to integrated pollution control discussed in Chapter 5, the case for unleaded petrol was gaining ground across the western world. The second point, relating to events that were more contingent, is that the ninth report was a clear case of authoritative advice allowing a government, with minimal embarrassment, to change its mind. Most importantly, perhaps, as a senior civil servant explained, the Commission had a strongly reinforcing effect on existing, ecomodernist trends:

> [T]here was a shift in the burden of proof... more towards the precautionary principle or more toward saying, if there is a significant risk in this *extremely* crucial area of mental health of children and their subsequent development... then it is prudent to act without cast iron absolute proof. There was therefore, if you like, a shift in attitude. And then there was the view, it is not *necessary* to... incur this risk, substitution is possible... a prudent responsible country will therefore get rid of it. I think that was the attitude that prevailed and you will discern a move toward... the precautionary principle.[38]

Sliding Between Paradigms

Although its immediate impacts were substantial, the ninth report was also notable for the precision with which it justified its recommendations. In this sense, it both advanced and circumscribed the principle of precaution. A little later, in the postponed overview of pollution (the tenth report), the Commission sought to describe the general circumstances in which a shift in the burden of proof might be justified, though with a stern warning that moving 'too far ahead of evidence...may be counter-productive to the cause of environmental protection' (RCEP 1984: para. 1.22). In its view:

> The main criterion...should be that the scientific evidence is inconclusive and likely to remain so for some time, but with a significant probability of serious damage—particularly of persistent or cumulative damage of an irremediable nature...Other criteria should include genuine public concern, which deserves attention irrespective of its basis, and the strength of the economic or technical case for the substance's continued use.
>
> (RCEP 1984: para. 2.31)

For the Commission, dangerous discharges to the aquatic environment clearly met these criteria. With Britain staunchly resisting pressures to impose stringent, technology-based controls on European 'black list' substances (see Chapter 2), the tenth report acknowledged the exigencies of regulatory science and, in a passage that would be widely cited, argued for a change in the government's position:

> Foresight and prudence...suggest that the United Kingdom should reappraise its stance on irretrievable discharges to the sea of toxic substances which are unarguably persistent and bioaccumulative, and which provide the justification for the 'black list'...For such substances, it is doubtful whether a threshold can be established between 'contamination' and 'pollution'.
>
> (RCEP 1984: para. 3.26)

Later, in a report on freshwater quality, the Commission would adopt an even firmer stance, pressing the government to move beyond its (by then) limited acceptance of the European regime and to place 'progressively less reliance... on the environment as a mechanism for processing anthropogenic waste' (RCEP 1992: para. 9.44). Following this thread in the Commission's work, we see a shift towards a more proactive, and sometimes preventive, position on precaution, well ahead of that in official discourse in the UK.

It is intriguing, therefore, that the Commission's approach to the high-profile issue of acid rain seemed more akin to traditional pragmatism: in the tenth report, it called only for further research and a pilot emissions abatement programme, though a more precautionary approach might well have been justified on the basis of its own criteria. Hajer (1993, 1995) helps to explain this paradox in his account of policy formation on acid rain: during the Commission's study, the Chair (Sir Richard Southwood) had unexpectedly been asked to chair the committee overseeing an international research programme on acidification.[39] This made it difficult for the Commission to take a more proactive line, as Southwood himself explained:

> If the Royal Commission had come out with some sort of view, just a year or two before the [international research programme] had been finished, they would have said, 'Southwood has prejudged the issue and we cannot rely on the opinion he gives at the end of the two years';... I felt that one had to stop, one had to wait for the experiment that I was conducting in theory to be completed before I wished to give recommendations.[40]

What is so interesting is that the Commission was capable, even within one report, of being both precautionary and pragmatic, emphasizing in its different stances the tendency for the precautionary principle to be reconstructed in the context of specific issues.

Public 'Perceptions'

As well as grappling with precaution, the Commission was engaging at this time with emergent ideas on risk perception, which in many ways accorded with its own acknowledgement of 'genuine public concern'; indeed, the tenth report (RCEP 1984) dedicated a whole chapter to the subject of 'attitudes and risks'. However, the narrative of this chapter veered somewhat awkwardly between realist and constructivist perspectives, as if the Commission couldn't quite make up its mind where it stood (it was 'certainly not a Brian Wynne version', as one official wryly observed[41]). Scientific assessment was distinguished from 'political considerations', and lay perspectives from those of experts 'using objective measures of risk estimation' (ibid.: paras 2.20 and 2.18)—but having constructed these boundaries, the Commission then set about breaching them, and came close to suggesting that experts might not be so objective after all. 'In a democracy', it proclaimed, 'it is an unhealthy sign when authority claims omniscience and dismisses grass-roots concern as "irrational"' (ibid.: para. 2.25); further, it saw that the tension between public and experts could be a creative one, serving to 'dispel complacency among officials and encourage... a less narrow approach by scientists' (para. 2.20). These themes, and the evident ambivalence, were carried through into the eleventh report (on waste), which more explicitly recognized the crucial dimension of trust:

> NIMBY [in this case, resistance to waste management facilities] cannot be dismissed as irrational: it is the expression of concern about a risk as perceived by the local inhabitants. But the fact that there is public concern does not necessarily mean that there is a real environmental hazard. A proper evaluation of the risk requires access to the relevant information and its interpretation, and the public will not be reassured by interpretations provided by the putative polluter, who has an interest. In the absence of public confidence in the role of bodies that are both authoritative and independent, interpretations by the press or pressure groups are often accepted, even if they go beyond what an informed expert would regard as justified by the evidence.
>
> (RCEP 1985: para. 1.15)

In a passage anticipating Wynne's (1996: 57) argument that public fears might be 'rationally based' in judgements of institutional behaviours, the Commission went on to suggest that not only secrecy, but past mistakes and 'simple incompetence' (para. 1.17), could fuel public disquiet. Oddly, it then failed to apply this logic to incineration, which it touched upon in the same report and returned to in a dedicated study in the 1990s, to be discussed later. In the meantime, it tackled an issue on which there was little past experience of any kind: that of the newly emergent field of genetic modification—or 'genetic engineering', in the less nuanced terminology of the day.

Precaution Without Deliberation: The Case of Genetically Modified Organisms

The Commission's thirteenth report dealt with the release of genetically modified organisms (GMOs) to the environment and, in advocating the case-by-case licensing of such releases, was generally thought to have taken a strongly precautionary stance. The decision to address the issue of GMOs (which did not, at that stage, have a particularly high public profile in the UK) was taken in 1986, shortly after establishment of the first official scheme to regulate planned releases; the Commission's aim was to influence further regulatory initiatives. Ultimately, the report (RCEP 1989), though it appeared barely in time, did achieve this objective. While not all of its recommendations were accepted, it provided the intellectual justification for Part VI of the *Environmental Protection Act* 1990, and might also have had an effect on nascent European legislation.[42] The impact in the UK will be considered further in Chapter 6; for now the report is of interest because of the approach taken by the Commission to a new and potentially risky technology.

In fact, the novelty of the GMO question put the Royal Commission in uncharted territory. Here was a set of biotechnologies on the threshold of major development, with empirical evidence about their health and environmental implications in short supply. Implicitly, the Commission recognized Collingridge's (1980) 'control dilemma': in the early stages of development, not enough is known about the risks and impacts of a technology to enable effective control; but by the time a knowledge base has been established, the technology is likely to have become embedded, and more resistant to effective regulation. GM was also a field in which the precautionary principle might be subject to novel interpretations—how, for example, should it be applied to living organisms released into the environment, which might act as 'environmental benefactors' but could possibly become 'self-reproducing pollutants' (Levidow and Tait 1992: 94; 1991)? In this context, the Commission was faced with the rival claims of two disciplinary groupings: molecular biologists, whose confidence in the precision of their techniques inclined them to be relatively relaxed about the predictability of GMOs; and ecologists, whose experience of introduced exotic species tended to make them more cautious. Collectively, the Commission leaned towards the latter perspective, unlike the US National Research Council, which was reporting at around the same time (US NRC 1989; for an interesting discussion, see Jasanoff 1995, 2005). One member of the Commission, a retired industrial chemist, took it upon himself to conjecture a range of possible negative impacts (RCEP 1989: para. 4.2), and the various scenarios were explored with the special advisors for the study and with those giving oral evidence.[43] As a result, members thought quite imaginatively about environmental risks, including 'secondary social consequences' (Jasanoff 2005: 57), such as the possibility that herbicide resistant plants might lead to the greater use of herbicides and potentially to environmental damage (RCEP 1989: para. 4.10).

Concerned, as always, about irreversibilities, and conscious that barriers to entry in the industry could be relatively low, the Commission saw the release of GMOs as a prime candidate for precautionary controls. It called for case-by-case examination of all proposed releases, justifying the additional burden by appealing to enlightened self-interest:

> Some may consider our proposals onerous but we believe them to be necessary for the protection of the environment. Moreover, the biggest brake on the environmental application of genetic engineering could result from an inadequately scrutinised release which caused serious damage to human health or to the environment and destroyed public confidence in both the science and the scientists.
> (RCEP 1989: para. 5.47)

This storyline of self-interest was referred to on several occasions during passage of the *Environmental Protection Bill* in 1990.[44] It would feature again in the Commission's report on nanotechnologies (RCEP 2008), considered in more detail in the following section.

As in the nuclear report, the Commission refused to confine its time horizons in the investigation of GMOs, arguing that new discoveries would 'open the door to more ambitious and more fundamental interventions in natural processes with the possible emergence of new risks to the environment' (RCEP 1989: para. 5.37). Thus it would not be sufficient to develop a system of controls 'on the basis of what is currently possible' (ibid.). Nevertheless, and in contrast with the nuclear report, wider social and ethical issues were dealt with only briefly, or explicitly avoided:

> Genetic engineering raises issues across a wide spectrum—ethical, social and political as well as environmental. Issues such as animal welfare, the possible loss of genetic diversity through the promotion of fewer crop varieties and the possibilities of military or terrorist use are touched upon very briefly in this Report.

> Other important issues, however, such as human gene therapy, human embryo research and the fundamental question of whether mankind should seek to create new forms of life, fall outside our remit and are not considered.
>
> (RCEP 1989: para. 11.2)

The reference to remit is interesting since on previous occasions the Commission had hardly been constrained by its terms of reference. But with GMOs, as the Chair recalled, 'there were a lot of moral issues, which we sidestepped, I must admit, quite deliberately'.[45] This reticence might have been due to time pressures—the Commission was keen for its findings to influence forthcoming legislation—though the minutes suggest that 'members were unwilling to be pressed into producing their Report over-rapidly'.[46] More importantly, perhaps, there was a sense in which the Commission was 'converted' to the benefits of GMOs during the course of its investigations. It had set out, as one member recalled, 'with an aversion to [GM], with a belief that it was not safe',[47] but came to believe that the technologies were potentially beneficial and that the risks could be managed. Whereas the Flowers Commission had felt that a complacent nuclear establishment would benefit from a 'shock to the system', its successor was dealing with a new industry and wanted to give it a fair wind. It saw firm regulation as conducive to this end, perhaps feeling that too much emphasis on 'moral issues' would be a distraction. Whatever the reasons, the Commission failed to support calls for a public biotechnology commission, which might have explored wider issues in a 'preventive' sense (Gottweis 1998), and its proposals for the committee that would scrutinize releases did not envisage representation of green groups or lay publics. Indeed, the report said little about public debate at all, though there was a brief section on 'public education'; even the customary call for openness was justified on the grounds that benefits might be jeopardized by 'public opposition motivated by fear of the unknown' (RCEP 1989: para. 8.22).

In the thirteenth report, then, the Commission was essentially concerned with consequences, even if quite broadly conceived, rather than with wider issues of social meaning or democratic control. Still, there can be little doubt that its stance on GMOs was at least a proactive one, or that it contributed to strongly precautionary legislation, welcomed in environmental policy communities (Jordan and O'Riordan 1995). Indeed, the approach was *too* precautionary for some: a few years later, the House of Lords Select Committee on Science and Technology (1993: paras 6.10, 1.5) accused the Commission of identifying 'conjectural' hazards in its report on GMOs, and of contributing to an 'excessively precautionary' regulatory regime in the UK. It is interesting, too, that in its defence the Commission chose to emphasize the scientific authority of its members and consultants, rather than making the case (as it had done often enough in its reports) for a proactive interpretation of precaution, which embodied wider considerations.[48] In a report on incineration produced in the same year, however, the Commission itself seemed to step back from such an interpretation, and even to overlook some of its own earlier arguments on risk and risk perception.

Regression?

As noted above, the Commission had addressed the subject of incineration in its report on waste (RCEP 1985), where—somewhat inconsistently with statements in the same report about public confidence and 'past mistakes'—it had argued that the available evidence did 'not justify the current level of public concern' about pollution from incinerators (ibid.: para. 1.19). This position was maintained in its later report focusing specifically on incineration (RCEP 1993a), which addressed the impacts of the technology primarily in relation to those of landfill.[49] Falling back on a conventional, quantitative account of risk (ibid.: paras 6.33–6), the seventeenth report maintained that well-operated incinerators were 'most unlikely' to be detrimental to human health (ibid.: para. 6.37) and suggested that public disquiet could be assuaged by openness, better liaison, and the implementation of best practice in plant operation. Within the relatively narrow frame adopted in this study, the Commission saw incineration as the more sophisticated, less polluting means of waste disposal, and argued for its expansion in any future waste management strategy.

The report met with an unusually negative reception among environmental activists, who had come to regard the Royal Commission as an ally. Critics felt that the treatment of health issues had been cursory, and didn't share the Commission's confidence that the new, integrated pollution control regime would lead to the safe and reliable operation of incinerators (see, for example, ENDS 1991, 1993; Greenpeace 1993). They were concerned, in other words, not only with uncertainties but with indeterminacies, which the Commission's analysis had noticeably failed to embrace. Nor had it acknowledged the fundamental ambiguity underlying divergent perspectives on incineration. For many, the deepest objection to incineration was that it would allow 'the continuation of dirty industries and wasteful practices by providing a "solution" to these problems': in this sense, the concept of harm extended 'to the total environmental impact from dirty processes and products which depend on incinerators for their survival'.[50] Whatever it said about health, a report that framed the problem as one of waste disposal, and (as the critics saw it) underplayed the potential for waste minimization and recycling (ENDS 1993), was never likely to win over the dedicated opponents of incineration. Their worldview was summed up by one commentator in the *Ecologist* (1994: 19), for whom modern, state-of-the-art incinerators provided 'a sophisticated answer to the wrong question'.[51] It is interesting in this context to note that the study relied heavily on a working group of six members, at a time when the Commission was also engrossed in its investigation of transport; as one civil servant recalled, 'the engineers wrote [the report]' and 'fell down' on public perceptions.[52] In fact, the sub-group included only one engineer (its chair), but the point, perhaps, was that none of the Commission's social scientists was involved. Although drafts 'were discussed...and approved by the whole Commission' (RCEP 1993a: para. 1.23), the relatively narrow framing of the incineration report, and its treatment of risk, would certainly have been influenced by the composition of the smaller group. This is significant if the Commission's diversity was one of its greatest strengths, as will be argued in Chapter 7.

Beyond 'Enlightened Technocracy'?

For the most part, the Commission's thinking about risk and precaution crystallized and evolved in the context of tangible issues such as those discussed above—water pollution, lead in petrol, incineration, or chemicals (or GMOs) in the environment. Periodically, however, it paused to address the principles underpinning environmental regulation more directly. The overviews in the first, fourth, and tenth reports reflected on such issues, and the twelfth report focused entirely on the concept of 'best practicable environmental option' (BPEO), which will be discussed in Chapter 5. In fact, the BPEO report was the first in which the Commission addressed the precautionary principle explicitly, having commissioned a paper on the West German *Vorsorgeprinzip* (the principle of foresight) from the Institute for European Environmental Policy (RCEP 1988: Appendix 3). *Vorsorgeprinzip* reinforced the Commission's proactive inclinations, sanctioning actions before damage occurred and 'against risks which are not (yet) identifiable' (von Moltke 1988: 61); in an interesting example of 'lesson drawing' (Rose 1991: 3), this German ecomodernist idea influenced the deliberations on BPEO, as did emergent psychometric research on risk perception.[53] Once again, in the report, an explicit policing of boundaries was combined with recognition of the limitations of science in informing public policy. Thus, those responsible for determining the BPEO 'should make a clear distinction between facts, scientific deductions, and conclusions which depend on value judgements'. But, at the same time, it was 'essential not to lose sight of the need, in some situations, to act on evidence that may be less than convincing to the scientist' (RCEP 1988: para. 2.30); decisions in the face of uncertainty would 'make heavy demands on judgement—the play of experience, intelligence and intuition on a new situation' (ibid.: para. 2.31).

The Commission's next foray of a philosophical kind came in the late 1990s, when it delved into the complex process of setting environmental standards—a study undertaken in part because it was concerned about the erosion of public trust in science.[54] In the standards report (RCEP 1998b), its twenty-first, it rounded out many of its previous arguments on equity, trust, and the need for transparency in risk regulation, but also introduced new and different perspectives. If it had reverted to a realist paradigm in its treatment of incineration, it recovered ground in the *Standards* study, and went further than before in presenting risk as a social construct, not amenable to 'simplistic comparisons between... statistical probabilities' (RCEP 1998b: para. 4.27). Thus:

> Attempts to persuade people that particular risks are acceptable by comparing them with other risks may backfire. They may appear to be trivialising the issues, or to be patronising in implicitly devaluing the perspectives and knowledge of those who are being asked to accept the risks. Such attempts will also be counter-productive if they are seen as glossing over political aspects of many of the risk conflicts in society and ignoring differences in the social meanings of risks.
> (ibid.: para 4.38)

Nor did expert assessments provide the benchmark that had previously been assumed: in effect, the Commission invoked Shrader-Frechette's (1995: 118) argument that *all* risks and risk assessments are 'perceived':

> Although estimates of risk are often presented as the objective outcome of a scientific assessment... they frequently go well beyond what could be justified in terms of rigorous use of the scientific evidence.... Inevitably the assumptions used are those of the practitioners making the assessment.... Other people, making different but equally valid assumptions, may produce substantially different estimates of risk.
> (RCEP 1998b: para. 4.22)

Yet even as the distance between objective (expert) and subjective (lay) assessment is diminished in this report, the boundary around 'science' is reinforced. Expert claims are seen as problematic when they go 'beyond' the science—the science itself still characterized as a realm apart—so the solution is to delineate the boundary more carefully. Thus 'assumptions and uncertainties implicit in the assessment' should be openly acknowledged (ibid.: para. 2.80) and a *'clear dividing line should be drawn* between analysis of scientific evidence and consideration of ethical and social issues which are outside the scope of a scientific assessment' (para. 2.69 [emphasis added]). Intriguingly, when the draft was under discussion, one of the issues considered by members was whether it 'shed light on where science stops and policy begins'.[55]

The Commission also had interesting points to make about precaution. Drawing on the work of Mary Douglas, it considered whether the precautionary principle embodied a 'cultural bias' towards particular conceptions of nature, and might lead, as some of the evidence for the study had suggested, to excessive environmental protection (for an example of such a critique, see Burnett 2009). It was inclined to dismiss this view, instead seeing the principle as a rational response to uncertainty, but one whose interpretation must be 'part of a political process' (RCEP 1998b: para. 4.46). Its own reading was explicitly proactive: while the conventional use of safety factors in toxicology (applied when there is already some evidence of effect) was 'in a sense precautionary', the 'true application' of the principle was 'in cases where there is reason to think that there may be an effect, but no evidence has yet been obtained for its existence or the evidence is inconclusive' (ibid.: para. 4.47).

Precaution would incur costs, of course (accounting for a good deal of resistance), and the more proactive the interpretation, the more costly it was likely to be. Inevitably, in the standards study, the Commission confronted again the tension between utilitarian approaches to regulation and those grounded in a duty to protect vulnerable people and environments. There were lengthy discussions, in particular, about the role of economic appraisal in the regulation of environmental risk and, unusually, deep differences among the members threatened agreement on the final draft.[56] In the end, economic appraisal was presented as 'an aid to making decisions which also take other factors into account' (ibid.: para. 5.49), but there was a clearer articulation than in any previous report of an alternative ethical framework:

> Those who take [a critical] view of economic appraisal argue that... certain moral obligations have a character and logic which is different in kind to the character and logic of preferences. A commitment to sustainability... arguably involves a duty to protect the environment for the sake of future generations. This is not a matter of balancing costs and benefits and discount rates, but a fundamental obligation and a constraint on other policies (ibid.: para. 5.46).

All of these arguments had significant implications for the setting of standards and the governance of risk. The twenty-first report attracted most attention, however, for its challenge to the risk assessment/risk management dichotomy, notably in its treatment of lay perspectives. Rejecting the idea that these could be relevant only in the management phase, the Commission proposed that people's values should be taken into account 'from the earliest stage in what have been hitherto relatively technocratic procedures' (ibid.: para. 8.37); this would include the framing of questions for scientific or technological assessment, which, according to the Commission, needed to be more 'socially intelligent' (ibid.: para. 7.22; see also Weale 2001). A dedicated chapter reviewed the means through which public values might be elicited and, in line with the shifting risk discourses of the 1990s, placed the emphasis firmly on more inclusive deliberation. This was seen as crucial because values were not necessarily fixed but could emerge 'out of debate, discussion and challenge, as [people] encounter new facts, insights and judgments contributed by others' (ibid.: para. 7.3). The framework for wider, 'upstream' engagement was applied in the Commission's subsequent work on chemicals (RCEP 2003) and pesticides (RCEP 2005),[57] but in these later reports the very possibility of separating an expert/objective phase of assessment from a political/subjective one was called into question: not only was the Commission concerned about the 'unacknowledged subjectivity' of experts (RCEP 2003: para. 2.144),[58] but it wanted to see 'a range of different perspectives' influencing the characterization of the risk and the design of the assessment process (ibid.: para. 6.20).

In its second report—including the prescient discussion of chemicals in products—the Commission had maintained a conventional distinction between hazard and risk, the former defined in terms of inherent properties and the latter including an assessment of likely exposure; thus its concern was with new products that were 'likely to get into the environment on a widespread scale' (RCEP 1972b: para. 16). Thirty years later it was more circumspect about this distinction, recognizing that the pathways of chemicals entering the environment could be very complex, and that uncertainties in these processes might have to be regarded 'as inherent...and not rectifiable' (RCEP 2003: para. 6.16). In this we have another example of its increasing commitment to a proactive interpretation of precaution: '[I]t makes sense to assume that the continuing use of large numbers of synthetic chemicals will lead to serious effects, which we cannot predict on the basis of our current or foreseeable understandings of these processes. A sensible approach to this uncertainty would be one of precaution—to reduce the hazard wherever we have the opportunity to do so' (RCEP 2003: para. 6.11).[59]

What is interesting about these later reports is their more forceful acknowledgement of what Jasanoff (2011: 28) calls the 'hybrid normative–cognitive character of policy-relevant science'. While the Commission still sought to delineate boundaries between science and politics, it became more conscious of their porosity and ambiguity. Intriguingly, therefore, from the standards report onwards, we see an advisory body whose own 'scientific' authority owed much to a form of boundary work reflecting critically on the very processes by which such demarcation is often achieved. These issues were illustrated starkly by a controversy that followed the publication of a special report on the risks to the health of 'bystanders' from the spraying of pesticides on agricultural land (RCEP 2005).

An Uncharacteristic 'Spat'

The pesticides study was one of only three that the Commission was asked by ministers to undertake. It arose out of a controversy about involuntary exposure to agricultural chemicals on the part of 'bystanders'—residents and other users of properties bordering arable land. In 2004, in a situation reminiscent—albeit on a more modest scale—of the lead-in-petrol controversy, the government was caught uncomfortably between a coalition of residents and pressure groups seeking additional controls and an agricultural lobby staunchly resisting them. Taking advice from its Advisory Committee on Pesticides (ACP),[60] the environment department had declined to introduce further protective measures, specifically the compulsory 'no-spray' buffer zones that protesters had been demanding. Predictably, assurances that the existing system of risk assessment provided robust protection failed to assuage the campaigners, and two public consultations served only to confirm the polarization of views. At that point, the Minister for Rural Affairs and Local Environmental Quality, having joined the Commission for dinner at one of its London meetings, asked it to examine the science on which the claims about public safety had been based. The Commission agreed but insisted, in addition, on addressing wider issues of risk communication and governance. Thus it embarked on a 'short' study (RCEP 2005), which, when published, generated a fascinating episode of boundary work, in which both the Commission and the ACP cast aspersions on the other's use and interpretation of 'the science'. The science, in this case, also became intertwined with the law.

The Commission accepted that firm conclusions about a causal link between pesticide spraying and the reported (and observable) ill-health could not be drawn (RCEP 2005) but it thought that some connection was plausible and that neither the limitations in the data nor alternative interpretations of the science had hitherto been sufficiently acknowledged. Thus it was 'surprised' at the degree of confidence expressed in the ACP's advice to ministers, and argued that the level of assurance given to the public was 'not robustly founded in scientific evidence' (ibid.: para. 6.14). Nor was it convinced that the risk assessment had been 'objective'; instead, it suggested that political and ethical judgements had been implicit in the ACP's advice. The report made recommendations on a number of issues, including the exposure assessment, monitoring of human health, and access to information. Most significantly, however, it recommended the introduction of 5-metre buffer zones as a precautionary measure, 'pending more research' (ibid.: para. 6.19)—an interesting formulation, given the intractable nature of the controversy and the unlikelihood of its being fully resolved by the science.

Unusually, a public 'spat' ensued. The ACP issued a critical commentary on the pesticides report, emphasizing what it saw as scientific shortcomings in the Commission's approach (ACP 2005). It was 'unconvinced by the scientific case' for 5-metre buffer zones, which it considered to be a 'disproportionate response to uncertainties' (paras 3.27 and 3.44); further, it claimed that its own view was shared by 'most other scientists' in the field of pesticide risk assessment (ibid.: para. 3.45). The Commission (again, unusually) responded, defending both its report and its scientific reputation. It pointed out that 'in complex situations [the scientific] evidence may well not be conclusive' (RCEP 2006: para. 24), and maintained that its difference with the ACP was primarily about the action that it was appropriate to take in such circumstances, when human health might be at stake. The government, having sought advice from two further expert committees on health-related aspects of the Commission's report,[61] agreed that further research should be conducted but did not accept the Commission's recommendation concerning buffer zones (Defra 2006b). The reasons for this will be explored in Chapter 6. What is interesting here is that neither the Commission nor the ACP—both authoritative bodies—was able to bring about the closure of this issue; at root, their disagreement concerned the interpretation of precaution, the ethics of risk governance, and the boundaries between science and politics (see also Fisk 2007).

Unresolved, the pesticide spray drift controversy moved into what, by UK standards, was an unusual legal phase (Warren 2009).[62] In 2008, a prominent campaigner successfully sought review of the government's approach in the High Court, following an earlier claim which had been adjourned pending the Commission's report. The dispute between advisory bodies notwithstanding, the High Court judge ordered the Secretary of State to reconsider his position, having been persuaded that there was sufficient evidence of harm to bystanders and that the model used in the risk assessment was inadequate.[63] However, the Court of Appeal overturned this ruling on grounds that tell us much about perceived authority in questions of scientific controversy. Here the judge argued that a case for review must rest on demonstration of 'manifest error',[64] and took the divergence of view between the Commission and the ACP as evidence that no such error existed. He observed that the respondent, although she had acquired considerable knowledge of pesticides, had no formal scientific or medical qualifications. He noted, in addition, that members of the Royal Commission had a wide range of backgrounds, and contrasted these to the ACP's expertise, 'heavily weighted' towards the relevant branches of science.[65] When the Supreme Court refused leave for a further appeal, the campaigner took her case to the European Court of Human Rights (Downs 2010), where it remains to be determined at the time of writing (2013).

Nanomaterials: Precaution and Adaptive Governance

We turn finally to the Commission's 2008 report on nanomaterials (RCEP 2008), the last of its assessments of emergent science-based technologies. It had decided on this study early in 2006, intending initially to examine the environmental implications of a wide range of novel materials. Finding that most of the available evidence and concern centred on nanotechnologies, however, it decided to focus on this burgeoning field as a particularly challenging case. Nanotechnologies entail the manipulation of matter at the nanoscale, at which materials often display properties quite different from those associated with their bulk form.[66] As with the biotechnologies considered by the Commission two decades earlier, nanotechnologies were novel and possibly risky but held out the promise of a multitude of applications and potential benefits. In the face of rapid development, concerns were being voiced about the health and environmental implications of nanomaterials, many of which seemed likely to become widely used, incorporated into products, and dispersed in biological systems and the environment. There was also unease about the ethics of potential applications, and about issues of power and control.

The Commission devoted considerable effort to a synthesis of such information as was available. It found that little was known about the effects of nanomaterials on living organisms (though there was cause for concern in certain cases[67]), and still less about their impacts in the wider environment. Adding considerable complexity to any assessment of risk, there were very large numbers of possible nanomaterials and their variants and applications. Here was another case of Collingridge's 'control dilemma', this time explicitly acknowledged in the opening chapter of the report (RCEP 2008). Not surprisingly, the Commission recommended a major research effort to identify potential impacts, noting the paucity of such research compared with investment in product development and applications. What is most striking about the report, however, is its treatment of risk and incertitude and its attention to the challenging issues of governance. The Commission acknowledged that science would not always be able to reduce uncertainties, in part because of the sheer scale of the task—described as a 'factorial experiment' in the report (ibid.: para. 4.53)[68]—but also because of indeterminacies and ignorance. Ambiguities were acknowledged too, in that concerns extended 'beyond issues of risk and risk management to questions about the direction, application and control of innovation' (ibid.: para. 4.85). Thus closing 'gaps' in knowledge and regulation would be 'necessary but insufficient' (ibid.: para. 4.84); the more substantive challenge was 'to find the means through which civil society can engage with the social, political and ethical dimensions of science-based technologies, and democratise their "licence to operate"' (ibid.: para. 4.58). There was little of the old defence of boundaries.

As in the study of GMOs, the Commission rejected calls for a moratorium, arguing that this would be a disproportionate response to concern. Instead, and in addition to a step change in research efforts and urgent attention to regulatory measures, it advocated 'an open and adaptive system of governance grounded in reflective and informed technical and social intelligence' (ibid.: para. 4.12). Such a system, pluralistic in nature, would help to avoid 'lock in' and maintain adaptive capacity. These arguments clearly built on the standards study (RCEP 1998b), but the 'nano' report was more nuanced, and more critical, in its approach to public engagement. It was here that the Commission made the comments (quoted earlier) about enthusiasm for engagement running ahead of institutional capacity, and it suggested, further, that public deliberation on all conceivable issues that might warrant it was as distant a prospect as that of scientific certainty. What it proposed was something more fundamentally preventive:

> A different approach to the governance of innovation—given the problems of deliberating emergent developments on a case-by-case basis—might be to begin with questions of principle, instead of working from technologies through to implications. A key task would be to consider which kinds of interventions in the human and non-human worlds, controlled by whom, might be deemed acceptable or problematic. Such principles could then act as a filter, directing attention to aspects of particular science-based innovations that seemed worthy of special scrutiny.
> (RCEP 2008: para. 4.102)

Like the nuclear and GMO reports, the 'nano' report was both forward-looking and precautionary—indeed, the Commission consciously invoked the nuclear report to argue again that 'concern should not only be with the position at present... but with what it might become within the next fifty years' (ibid.: para. 4.4). As with GMOs, it appealed to the self-interest of a nascent industry in calling for precaution and regulation (whereas it had seen the nuclear industry as more established but complacent). Intriguingly, it is the GMO report (which fell between the other two in terms of timeline) that seems out of line in its more minimalist treatment of ethical concerns, governance, and wider engagement. Interestingly, too, while the discussion of nanomaterials left far behind the notion that public fears need to be 'assuaged', it came closer to the nuclear report's prescription of exhaustive 'deliberation in public' than to the Commission's call in the standards study for 'public deliberation'.

From Enlightened Technocracy to Cautious Constructivism

This chapter has shown that throughout its existence the Commission absorbed, developed, refracted, and sometimes initiated ideas about environmental risk and appropriate policy responses. Its thinking on these issues was often influential, especially (though not exclusively) in its first two decades, when the field was relatively uncrowded. Questions of influence will be addressed in Chapter 6, but it is appropriate to conclude the present chapter by highlighting some of the most significant features of the Commission's treatment of risk and precaution.

A fine-grained analysis shows that the development of the Commission's thinking was neither linear nor straightforward. At any one time, its perspectives emerged from an interplay of many different factors: the general intellectual climate and ideas that were already in the air; the characteristics of the issue at hand, particularly its economic ramifications and the extent of public concern; the Commission's (or the Chair's) agenda; and the expertise and predilections of the members (and probably of the Secretariat as well). Sometimes a line of argument would become compelling as it was built upon in successive reports; the folly of over-reliance on the assimilative capacities of the environment is a case in point. But the Commission could also arrive at carefully constructed positions (on 'real risk' and public fears, for example), only for them to be undone (or ignored) in ostensibly similar cases in its later investigations. As well as all the other factors at play, it is clear that different cultural biases dominated within the membership at different times, so that deliberations ended up in different places. Nevertheless, across the broad sweep of the Commission's work, it is possible to tease out three particularly significant developments.

One was a shift away from the enlightened technocracy of the earlier reports and towards more inclusive perspectives on risk governance, articulated first in terms of the need to take account of 'public perceptions', but latterly with a recognition that different judgements and worldviews were never likely to converge on any single understanding of 'real risk'. The Commission's later reports are characterized by a qualified constructivism, most apparent from the standards study onwards. In turn, this inclined members to embrace more holistic framings of problems and potential solutions, and to think beyond risks and consequences, narrowly defined, to questions about the governance and social control of particular activities and technologies (though it is interesting that it was the Flowers Commission, in the study of nuclear power, that first thought more expansively in this way).

The second, and related, development lay in the Commission's changing interpretations of precaution. It inclined towards prudence from the outset—and long before 'the precautionary principle' passed into common terminology. No doubt this was a reflection, in part, of a particular worldview—on the vulnerability of environmental systems to irreversible damage—shared by many of the Commission's members over time. But the breadth of perspective within the Commission also seems to have inclined it towards precaution, perhaps by acting as a check on members with backgrounds in the 'harder' sciences who might have seen 'compromises in the methodological standards applied to science... as an abdication of professional norms' (Jasanoff 1987: 388–9).[69] A reading of the various reports shows that reactive, proactive, and sometimes preventive dimensions of precaution were typically intertwined in the Commission's work, with each becoming dominant at different times (though not always in steady progression; arguably, the Commission took a preventive stance on nuclear power, a proactive one on GMOs, and a reactive one on incineration). In the broadest terms, however, it could be said to have developed an increasingly proactive interpretation of precaution over time, and latterly to have moved beyond this to 'reconceptualiz[e]...the relationship between social commitments, moral identities and "natural" knowledge' (Wynne 1992: 123–4). This essentially preventive approach connected closely with its later, more constructivist perspectives on risk.

A third significant change, linked to those discussed above, can be seen in the Commission's performance of boundary work. The Commission was never a *scientistic* body—it was always aware of the limitations of environmental science and the complexities of the policy–political process—but for many years it did construct boundaries between these spheres and defended them quite assiduously. The reports of the 1970s and 1980s tend to associate science and expertise with objectivity—hence the comment in the tenth report, noted earlier, about experts using objective methods of risk assessment. Later, there was recognition (first articulated in the standards study but hinted at in earlier reports) that scientific assessment might itself involve pre-suppositions and value judgements, which, the Commission urged, should be made explicit and separated, as far as possible, from 'the science'. It was only in its last decade that it came to see the boundaries as porous and negotiable in the context of risk governance, and in effect to acknowledge the possibility of co-production.

What does all this mean in terms of the ability of an advisory body to give effective advice? It might be argued that the Commission's progression from enlightened realism to cautious constructivism, and towards proactive and preventive interpretations of precaution, was simply a microcosm of more ubiquitous developments in the field of risk and environmental governance, as well as a reflection of its own changing composition. Certainly, the Commission was moving with the flow of ecological modernization and with the changing risk paradigms outlined at the beginning of this chapter: both its members and those who gave evidence would have acted as conduits and there is clear indication in the reports (and their references) that the Commission was influenced by new thinking both in research and in policy formation. But to see its work simply as a reflection of these trends would almost certainly lead to an underestimation of its influence. On some occasions, as we have seen, it was itself in the vanguard of new thinking and managed to change the framing of problems and policies with significant effect. On many others, its authority (and its networks) enabled it to amplify and direct the flow of ideas, bringing them more rapidly and directly to the centre of the policy-making process: in other words, it not only went with the flow, but modified its speed and direction. These impacts will be considered again in Chapter 6. But now it is time to turn, in Chapter 5, to another aspect of environmental affairs to which the Commission paid considerable attention—the need to take an integrated and holistic view of environmental processes and policies.

5

The Quest for Integration

> The environment is a whole and is not divided according to the neat categories of public policy.
>
> Albert Weale, *The New Politics of Pollution* (1992: 93)
>
> Nothing less than a comprehensive policy for the environment will suffice.
>
> RCEP, *First Report* (1971: para. 11)
>
> [T]he time is never right for reform in the eyes of those who have a stake in the status quo.
>
> Royal Commission meeting paper (1975: 8)[1]

Introduction

The Royal Commission's holistic approach to the environment, expressed in the quotation from its very first report above, was one of the distinctive characteristics of its work. Even when dealing with specific pollutants, as in the case of lead, it often tended to see the problem 'in the round'. On notable occasions, however, it went further than this, stepping back to take a critical look at the conceptual and institutional frameworks for environmental policy. Unsurprisingly, since frames are often deeply entrenched, the Commission tended to encounter resistance when advocating significant change. Two studies, each of which examined the underpinnings of policy and regulation, form the basis for this chapter. The fifth and twenty-third reports (RCEP 1976a, 2002a) were separated by a quarter of a century, but in both of them the Commission called for greater integration of a fragmented regulatory system, and in doing so promoted another of the central tenets of ecological modernization.

The fifth report (RCEP 1976a) became something of a cause célèbre. In it the Commission urged the adoption of a new system of integrated pollution control and introduced the concept of 'best practicable environmental option' (BPEO), which it later elaborated upon in its twelfth report (RCEP 1988). The important changes that it envisaged in the mid-1970s were resisted at first but were gradually implemented after a time lag of more than a decade; it is particularly interesting that they retained their Royal Commission 'branding', even after this considerable lapse of time. The second case involves a later—and arguably more ambitious—attempt to move towards an integrated system of land use and environmental planning (RCEP 2002a). These proposals were not so much resisted as brushed aside at a time when the government was bent on its own reforms to the planning system, as noted in Chapter 2. If planning policy seemed to move closer to the Commission's vision a few years later, such convergence occurred with little explicit reference to the Commission's work, and was, in any case, short-lived. The accounts of these two cases necessarily differ in their level of detail, given the disparity in timescales over which the impacts of the two studies can be considered. In a comparative context, however, they tell us something of the Commission's experience when it tried consciously to change the policy frame, and point to interesting differences in the treatment of the Commission's reports in its earlier and its later years.

New Institutions for Pollution Control

> How sweet is Perfect Purity think some,
> Others, Muck must accrue when Work is done.
> Ah! Let's use the Cash in Hand and spread the word
> That Practicable Environmental Options are more fun.
>
> Gordon Fogg, *The Rubáiyát of a Royal Commissioner* (1985)[2]

As Chapter 2 has shown, the early 1970s was a period of restiveness about pollution and its regulation, with critics in the UK mounting high profile campaigns and governments feeling increasing pressure to respond. At issue was not only the 'grievous damage' (Bugler 1972: 32) that exercised many observers, but also a lack of transparency and accountability on the part of the regulatory authorities. The Alkali Inspectorate, responsible for regulating industrial processes that could lead to serious air pollution, had become a particular target. There was concern, too, within industry as well as in Whitehall, about the sheer complexity of arrangements for air pollution control. Faced with these pressures, the Secretary of State for the Environment, Anthony Crosland, invited the Royal Commission to review the existing system—the first of the ministerial requests for particular studies. As the minutes of the Commission's June meeting in 1974 record: 'In recent months there had been increasing public concern about the role and attitude of the Alkali Inspectorate and the Secretary of State had concluded that the present review should be undertaken.'[3]

The origins of the air pollution study are therefore consistent with the 'traditional view' of royal commissions as bodies that can be used to placate, though Crosland's move could just as well be interpreted as a genuine quest for advice; quite possibly it involved a mixture of the two objectives. Whatever the motive for turning to the Commission, ministers and officials in the environment department 'wanted to keep their finger on the pulse of what was happening'[4]—rather too much for some members, as we have seen in Chapter 3. After a certain amount of negotiation, two civil servants from the department were seconded to the Secretariat and, exceptionally, six 'Associate Members' were appointed for the duration of the study, on the grounds that the Commission needed additional resources whilst engaged simultaneously in preparing a report on nuclear power.

Instead of restricting itself to matters of air pollution, however, the Commission soon found it necessary to look at the system of pollution control more comprehensively. It had long been interested in the potential for cross-media transfers of pollutants, and had included 'pollution of one medium at the expense of another' as one of the issues on which it would like to receive evidence for the new study.[5] A visit to Bankside Power Station, where members were persuaded that a tightening of air pollution controls had led to an increase in discharges to the River Thames, reinforced this concern. In its subsequent deliberations, the Commission became convinced of the need for a more integrated system of pollution control—'it was sheer common sense'[6]—and an apparently inexorable logic then led to the concept of a unified pollution inspectorate, combining those regulatory bodies that had traditionally focused their activities on different media. It was one of the Associate Members, Jon Tinker,[7] who christened the proposed body 'Her Majesty's Pollution Inspectorate' (HMPI) (or 'Humpy', as it was affectionately known). In a paper prepared for the Commission's May meeting in 1975, he argued that: 'The concept behind the HMPI scheme is, in my view, so simple and so logical that it will rapidly come to be seen as self-evident, and receive widespread support.'[8] This view was to prove optimistic, as we shall see.

The BPEO concept emerged from the Commission's deliberations about the role of the new inspectorate, as a natural extension of the long-standing concept of 'best practicable means' (BPM) in air pollution control (see Chapter 2). A former member recalled the moment when Brian Flowers, in the chair, observed that there had been much discussion of BPM, and 'somebody said, "this is not best practicable means, it's best environmental option"'.[9] In applying the new concept, a unified inspectorate would consider the effects of any discharges on air, water, and land, seeking the best practicable environmental option for control. This extension of BPM was attractive to the Commission, which, while pressing its case for a more open and accountable system of regulation, was not sympathetic to the general critique of British pragmatism, and defended the BPM approach as 'inherently superior' to a system of emission standards set at national level (RCEP 1976a: para. 16). What was novel was the proposal that a new, integrated regulatory body should approach its task by considering the environment as a whole. Blending tradition with a more radical approach, the Commission envisaged that HMPI 'would seek the optimum environmental improvement within the concept of "best practicable means", employing the knowledge... and many of the present techniques of the Alkali Inspectorate... In effect, we have in mind an expansion of the concept of "best practicable means" into an overall "best practicable environmental option"' (RCEP 1976a: para. 271).

The fifth report made ninety-four recommendations, many of them quite specific and focusing on the original air pollution remit. Amongst other things, it dealt with enforcement of regulations, domestic smoke control (still a significant issue at that time), and the need for effective monitoring and more research. It also recommended the extension of BPM powers to local authorities (the bodies responsible for less serious problems of air pollution) and the establishment of air quality guidelines—though not the uniform standards being advocated in Europe, which the Commission considered to be 'unenforceable' (ibid.: para. 183). The report's most prominent contribution, however, and the one for which it has been remembered, was a new and more holistic way of *thinking* about pollution and the necessary regulatory arrangements.

In taking this stance, the Commission went 'way off the remit that [the environment department] had given to them'.[10] It did so against the advice of its Secretariat, who (from their own experience in Whitehall) feared that far-reaching recommendations impinging on the machinery of government would delay much-needed reforms to air pollution control, perhaps for many years. One member recalled that 'the Secretariat said... very politely, I mean they weren't trying to bulldoze us... "look, you haven't got a hope in hell's chance of getting anybody to take any notice of this, they've only just changed... Forget it"'.[11] The environment department, for its part, had been looking for modest reforms that might simplify the regime for air pollution control while mollifying some of its critics; ministers and officials neither expected nor wanted 'the HMPI complication to be thrust upon them'.[12]

An 'Elephantine Gestation Period'

Not surprisingly, there was no rapid acceptance of the Commission's central recommendations.[13] It took nearly seven years, during which there was a change of government, even to extract a formal response to the fifth report, and when this finally came it rejected the key proposal for a unified pollution inspectorate. The long delay, and subsequent policy reversals, have been analysed in detail elsewhere (Owens 1989, 1990; Smith 1997, 2000; Weale, O'Riordan, and Kramme 1991), but it is worth noting here how conceptual issues—the Commission's *ideas* about pollution and pollution control— became intertwined with more prosaic questions about the institutional (and geographical) location of the relevant regulatory authorities. In 1975, the Alkali Inspectorate had been moved from the environment department to the Health and Safety Executive (HSE), which dealt with the working environment and came under the Department of Employment. The move had created tension between the departments and was deemed by the Royal Commission to be 'potentially damaging to the interests of the environment' (RCEP 1976a: para. 257). Ministers had agreed that, pending the air pollution report, the integrity of the Inspectorate would be maintained so that the option of returning it to the environment department would remain open. Meanwhile, senior staff of the Inspectorate continued to work alongside environment department officials in Whitehall, a physical proximity valued by the department for 'the many opportunities [it] provided for informal exchanges of view'.[14]

The Commission was clear that if there were to be an integrated pollution inspectorate, it belonged in a department of the environment. Establishment of HMPI therefore implied a return of the Alkali Inspectorate (which would form the core of the new body) from HSE, a move that in principle the department desired. Yet successive environment ministers were reluctant to press for the Inspectorate's return, focusing instead on operational arrangements (including co-location) to safeguard the integrity of the department's pollution control functions. The explanation for this muted support requires a digression into the 'political stream' (Kingdon 2003: 145). The Labour government that received the fifth report had only recently established the HSE, with trade union representation alongside industry, and 'in the decade of corporatism, the general climate was not right for disrupting the work of an emerging, and potentially successful, tripartite organization' (Weale, O'Riordan, and Kramme 1991: 149–50). For the Conservatives coming into office in 1979, weakening the HSE would have been a distraction from the wider goal of reforming trade union legislation (ibid.), and ministers had no wish to provoke a confrontation with the unions on what they saw as 'a relatively minor matter'.[15] Amidst these macro-political considerations there was also a need carefully to plan moves in the turf war over the location of the Alkali Inspectorate. If the environment department showed too much enthusiasm for HMPI, such a body might indeed be established. But it might be established within the HSE, in which case the department would not only fail to recapture the Inspectorate, but would also lose its other pollution control functions, 'which would surely be the worst of all worlds'.[16]

In any case, the department was inclined towards cautious incrementalism rather than fundamental change, and there were genuine intellectual doubts about the viability of the Commission's recommendations. In particular, while BPEO was accepted as an elegant idea, civil servants expressed reservations about the capacity to implement it in practice. One comment on the draft government response (as it stood at the end of 1980) encapsulates these concerns:

> We do not have...techniques which can measure total environmental damage as a precursor to minimising it. We cannot generally do this in one medium alone....I am concerned that the practice of a number of our key concepts is badly divergent for [sic] the theory and the existing draft response would add to their number. It would also promise something that we could not in the end deliver.[17]

There were doubts, too, that the creation of a unified inspectorate would lead to real environmental improvements. The evidence for this was thought by some officials to be 'extremely thin',[18] and there was a feeling that even if modest benefits could be delivered, they would scarcely be worth the disruption. Thus a compromise emerged, in which the government would broadly endorse the cross-media approach, but seek its delivery through better consultation. The caution and pragmatism are captured in exchanges between officials in the department: one wrote, 'I do not see much of a role for general prescriptions.... as a general rule I suggest that we should operate informally, by getting parties round a table and exposing arguments to discussion',[19] while another, seeking to give 'general support' to [the BPEO] approach, was keen to present its adoption 'as a matter of sensible evolution rather than as a dramatic change'.[20]

At least initially, then, there was a sense that a holistic approach to pollution control, especially if it entailed major administrative reorganization, was a solution looking for a problem. The Commission could actively promote the idea but was unable, by itself, to bring about the convergence of policy, problem, and political streams that might lead to policy change (Kingdon 2003). As one civil servant recalled, 'it all seemed a bit too difficult to begin with'.[21] Another blamed the inertia on institutional structures and politics: 'the organizational structure militated against integration... the environment was very low down on the political agenda...Politicians don't do things unless they have to, or there is tremendous advantage to them to do it. I think those are the fundamental reasons why [the response] took so long.'[22]

The government finally responded to the fifth report through the standard medium of a pollution paper (DoE 1982), though some officials had hoped, as time went on, that a lower key communication might save embarrassment.[23] The process had taken nearly seven years, and delivered a polite 'no' to the Commission's primary recommendation:

> [T]he Government accept that the Royal Commission have identified a genuine problem in drawing attention to the potential transfer of pollution from one medium to another... The Government acknowledge also that the bpeo concept is one of considerable power and utility. However, they do not see it as calling for organisational change (though they do not rule out the possibility of such change being needed in future...). Rather, they accept the need for a more integrated approach to the control of industrial pollution, within the existing organisational framework.
>
> (Department of the Environment 1982: para. 8)

Despite this cautious response, the concept of BPEO was already having an effect within the department, even if officials felt perplexed as to how to interpret it practice.[24] It took several more years for the department to accept the case for a unified inspectorate. This required not only a cognitive shift—a belief that moving from a fragmented to a more integrated system would make a difference—but also an environment in which the organizational change would seem more palatable. It is worth examining these developments in more detail, for they provide an excellent illustration of 'the interactive and multiplicative effects of the variables that contribute to policy change' (Baumgartner 2006: 40).

Changing the Frame

In the 1980s, as we have seen, the British approach to pollution control was under pressure. The Royal Commission was itself a force for change, taking an ecomodernist stance in many of its reports and continuing to press its proposals on integrated pollution control and BPEO. In its tenth and eleventh reports, it challenged the government's view that change was unnecessary, maintaining that '*most* of the present and future problems in environmental pollution' would involve cross-media issues (RCEP 1984: para. 6.35 [emphasis added]), and insisting that a unified inspectorate would be essential for 'the complex and demanding tasks' of implementation (RCEP 1985: para. 9.45). There were more subtle influences, too. The concept of BPEO, by this stage widely accepted, undermined the old tendency to look for lines of least resistance and resonated with European concerns about landfill. A senior civil servant summarized these developments in terms that anticipate

Baumgartner's (2006) description of pressures building prior to policy change, as well as Kingdon's (2003) account of the role of ideas in this process:

> [T]here was a growing awareness that the environment did actually matter and that the existing administrative methodologies were potentially counter-productive... There was a head of steam building up that people ought to be more sensitive about their land use [a reference to landfill]. Those are powerful intellectual ideas wherever they come from politically, *they change the climate in which you are thinking about things*... it all begins to come together—you must be familiar with that critical mass of ideas—and then you get an idea: its time has come.[25]

In the political stream, the appointment of William Waldegrave as environment minister in 1985 was a factor of considerable significance. Not only was Waldegrave sympathetic to the environmental cause (see Chapter 2), he also 'had his mind round' the issue of integrated pollution control.[26] In fact, Waldegrave had an even more ambitious vision: he favoured an independent environment agency, and saw the return of the air pollution inspectorate to the environment department as a crucial first step. A further important development (not unrelated to Waldegrave's ambition and cleverly in line with the government's desire to reduce bureaucracy) was an 'efficiency scrutiny' of central pollution control functions, conducted by the Cabinet Office (DoE and Department of Employment 1986). One outcome was that a new, combined inspectorate could now be presented as an efficiency gain—'the Treasury actively hoped to see staff saving out of it'[27]—rather than a measure that would add complexity and demand extra resources, as had previously been implied.[28]

It helped, too, that by the early 1980s, the case for a new inspectorate was supported by a range of actors who could mobilize around the storyline of 'integration', even if their motives and specific preferences were divergent. The trade unions, local authorities, and environmental groups were in favour, as was the Confederation of British Industry (CBI), which was 'moving from a position of resisting change to seeking to manage [it]' (Smith 2000: 106). Significantly, a reformist faction of industry actively supported the return of the air pollution inspectors to the environment department and, together with the National Society for Clean Air, lobbied the government on this issue (Smith 2000; Weale, O'Riordan, and Kramme 1991). These developments could be seen as symptomatic of wider processes of policy learning and ecological modernization, but the CBI's position was also a strategic one: it saw an expert, unified pollution inspectorate in the environment department as a bulwark for traditional British pragmatism within Europe (Owens 1990; Smith 2000). It also feared amalgamation of pollution control functions with the Factory Inspectorate (within HSE), as well as the loss of 'early pre-development discussions... which have been so successful in the planning of industrial plants and complexes'.[29]

The decision to set up 'Her Majesty's Inspectorate of Pollution' (HMIP) was announced on 7 August 1986, with the new organization due to begin operation in April 1987.[30] Although at first it was understaffed, insufficiently resourced, and could hardly be said to be 'integrated', 'Humpy' (with a slight change of name) was there. A former Deputy Secretary in the environment department recalled how different factors converged to produce this outcome at that particular time:

> [T]here was a succession of reports from the Royal Commission, there was William Waldegrave who was a very significant force in all this...and there was also Europe. I've mentioned the dislike of Europe, but Europe was and is an irresistible force in these sort of matters...presentationally and operationally it was quite a good thing to create HMIP: it enhanced the image, it did look as if it could save money in the course of efficiency, and it showed that we could deliver BATNEEC in a structured kind of way and even be a bit of an example to some countries [as] to how it should be done. *It was an idea whose hour had come.*[31]

The Royal Commission itself was in no doubt about the origin of the idea. It welcomed the new development and 'placed on record the fact that the Commission had played a significant role in encouraging the setting up of the Inspectorate'; it noted, too, that the establishment of HMIP gave added timeliness to a study that it was then undertaking, specifically on BPEO, which had been defined only vaguely in the fifth report.[32] Earlier that year members had agreed to produce a short report—an 'essay on the principles of BPEO'—as a contribution to European Year of the Environment.[33] They were also of the view that an 'authoritative exposition'[34] would be a corrective to what they saw as misuse of the concept, notably in a report from the environment department on options for radioactive waste management (DoE 1986a).[35] In the event, the exposition proved more elusive than expected, and the essay turned into one of the 'philosophical' reports that some members found to be such a 'nightmare' to draft (see Chapter 3). It is not surprising, perhaps, that recipients of the report—the Commission's twelfth (RCEP 1988)—were sometimes equally bemused; one civil servant thought that it 'clearly read as if they had struggled to find the answer',[36] and a senior scientist in the environment department claimed that 'nobody could understand what they meant...best practicable environmental option is just a jungle of words'.[37]

In fact, the emphasis in the twelfth report was on process and accountability. The definition of BPEO that the Commission settled upon was a broad one, not tying the concept to particular pollution streams or industrial processes:

> A BPEO is the outcome of a systematic consultative and decision making procedure which emphasises the protection and conservation of the environment across land, air and water. The BPEO procedure establishes, for a given set of objectives,

> the option that provides the most benefit or least damage to the environment as a whole, at acceptable cost, in the long term as well as the short term.
>
> (RCEP 1988: para. 5.3)

What is interesting about this attempt to pin down the concept, apart from the Commission's eternal optimism that 'short reports' might be completed within a year, is that it appeared to have less impact upon policy developments—notably the *Environmental Protection Act* 1990—than the original idea, set out in the fifth report. The *Act* itself can be seen as a product of the new-found enthusiasm for green issues in the late 1980s, building on the subtle but significant diffusion of ecomodernist ideas throughout most of that decade. After Thatcher's Royal Society speech in September 1988 (see Chapter 2), environment ministers were emboldened to pursue legislation, and policy entrepreneurs within the department seized the opportunity to provide the legislative framework for HMIP.[38] In a complex clause, the *Act* (Part I, Section 7[7]) required that for the relevant processes, the new inspectorate should

> include the objective of ensuring that the best available techniques not entailing excessive cost will be used for minimising the pollution which may be caused to the environment taken as a whole by the releases having regard to the best practicable environmental option available as respects the substances which may be released.

Practising Integration

The interpretation of BPEO in the *Environmental Protection Act* 1990 was narrower than the Royal Commission's, but even so it was perplexing for those who had to try to put it into practice. Interestingly, the Commission had never tackled the practicalities of BPEO, in spite of the view of one member, an industrialist, that it should try to show that firms would be able to carry out the necessary assessments without excessive cost.[39] It had felt that its task was 'to demonstrate BPEO as a philosophy and approach and leave it to others to adapt it to their circumstances, with HMIP as a primary target'.[40] One former official of HMIP echoed concerns expressed within the environment department during drafting of the response to the fifth report: 'we don't know enough about pathways... we just don't have that information about that many chemical species... What's the BPEO? I don't know!'[41] Attempts to operationalize the concept consumed a good deal of energy in the 1990s, causing much frustration and occasionally breaking out into public controversy.[42] Over time, however, it was the elegance of the idea put forward in the fifth report that proved enduring, rather than the detail of the twelfth (with the exception, perhaps, of the definition of BPEO in the latter, which is still cited twenty-five years later[43]). Certainly this was the view of officials who had been involved in drafting and implementing the legislation. One said that the twelfth report 'didn't take the seminal arguments of the fifth very much farther forward.... To be frank I skim read it and never looked at it again',[44] and another: 'I don't remember for the life of me what the twelfth one said... they go on about audit trails and paper trails—I've got no time for that... that is all peanuts—*the trick is to think of it in the first place*'.[45]

Once HMIP was up and running, the logic of integration pointed inexorably to a merger with the National Rivers Authority, the body that had been formed in 1989 to regulate the newly privatized water industry. There were logistical difficulties to overcome (see Weale 1996), but five years later the *Environment Act* 1995 combined the two bodies to create the independent Environment Agency that some had always favoured. There was movement in Europe, too, where integration of pollution control functions had been on the agenda since the Third and (more specifically) the Fourth Action Programmes on the Environment (Council of the European Communities 1983, 1987). With strong encouragement from the UK, a Directive on Integrated Pollution Prevention and Control (IPPC) was adopted in 1997—'one of the relatively infrequent occasions when British policies and practice were successfully implanted on the Continent' (Mills 1998: 10).

Legislation is never the end of the story, of course, and it would be interesting to scrutinize the implementation of integrated pollution control and assess how faithful it had been to the Royal Commission's ideas (for initial commentaries, see Skea and Smith 1998; Smith 2000; Weale 1992). The primary aim of this chapter, however, is to explore the ways in which expert advice might help to effect significant—perhaps 'third order' (Hall 1993)— policy change. And, at first sight, the Commission's fifth report would seem clearly to be implicated in such a process, even if its consequences unfolded over a period of several decades. The acceptance of a new logic and the creation of HMIP have been almost universally attributed to the Commission, and the Environment Agency has been described as its 'biggest child'.[46] Even the European IPPC Directive, although it differed in significant respects from the British legislation,[47] can (according to some observers) be traced back along a 'fairly direct line'[48] to the Commission's fifth report. The UK's experience informed the drafting of the European legislation;[49] and that experience in turn was seen to derive closely from the Royal Commission's ideas.

Learning to Integrate?

We should pause, however, to reflect on the nature and extent of the Commission's influence, even in such a seemingly straightforward case. One former member mused on this point: 'the fact that we have ultimately got IPC [integrated pollution control]...is that the Commission or is that just a general trend and changing of opinion to which the Commission contributed? I'm not entirely sure.'[50] Certainly the idea of integrated pollution control cannot be attributed to a single source, and the Royal Commission was not the only body promoting it; within Europe in the 1980s, for example, the Institute for European Environmental Policy (IEEP) was active as a policy entrepreneur on this issue[51] and, while in tune with the Royal Commission's ideas, was engaged with developments on a wider international scale (Haigh and Irwin 1990). As we have seen, there were diverse—even divergent—factors at work in the ultimate establishment of integrated pollution control in the UK. Much the same can be said about the British role in the later European legislation. By the mid-1990s, the UK government was keen to be in the vanguard of an ecomodernist approach to pollution control, but also wanted to pre-empt any leanings towards the continental style of regulation. Thus there were 'green and brown' reasons for supporting the IPPC Directive, and in both respects the home-grown concept of BPEO was thought to be rather helpful.[52] Taking a broader view, it might be argued that the logic of integration was so compelling that it was bound to influence legislation in the developed economies sooner or later; even a simple, 'single-loop' learning process might account for the changes that eventually took place, given that existing approaches to pollution control were failing to deliver their objectives.

It is impossible fully to disentangle the threads. Integrated pollution control was not unknown in the 1970s, but its practice was exceptional and as a concept it was barely developed. The Commission arrived at the idea independently in the course of its air pollution study, rounded it out and, in the words of an official of HMIP, gave it 'momentum' within UK environmental policy communities.[53] The elegant formulation of the concept of 'best practicable environmental option' provided a particularly powerful framing of the case for a cross-media approach, and the Commission's ability to keep ideas alive was helpful in the inhospitable political and economic climate of the 1970s and early 1980s.[54] The Secretary who had doubted the wisdom of advocating such radical policies in the first place conceded that, in the end, the fifth report *sowed the seed* for a better, more sensible pollution control organization in the future';[55] in this sense we might invoke a classic enlightenment function, with the Commission's ideas acting as a resource to be drawn upon as external conditions changed. And if it is unsurprising, in the bigger picture, that the logic of integration came to be widely recognized and accepted, then the Commission can at least be seen as a potent force in a more general and pervasive trend. Arguably, in this case, it helped instigate 'third order' policy change, though not all commentators agree that the shift to integrated pollution control constituted such a radical realignment of the regime—a point that will be returned to at the end of this chapter. But it is time now to turn to the second of the studies considered in this chapter, in which the Commission advocated substantial changes to the organization of land use and environmental planning. Intriguingly, the study was anticipated by the Secretary of the National Society for Clean Air, who, in commenting on the IPPC Directive, observed that there would be some who saw 'the linkage of environmental matters with planning, transport and economic development ... as essential to the sensible progress of environmental policy' (Mills 1998: 11).

Environmental Planning

Twenty-seven years after its initial advocacy of integrated pollution control, the Royal Commission took a further, substantial step in its quest for integration. Its twenty-third report (RCEP 2002a) looked beyond the system for pollution control (by this time substantially reformed) to develop a more ambitious vision of environmental planning, embracing a wide range of regulatory regimes and having an important spatial dimension. The environmental planning report, published in March 2002, was not received with enthusiasm by the UK government, and fared only slightly better in the devolved administrations (UK Government 2003c, a; Scottish Executive 2003; Welsh Assembly Government 2003). Indeed, in the short term it seemed almost to sink without trace, in spite of its relevance to important forthcoming legislation. Given that the planning report continued and extended the theme of integration, it is instructive to consider its fate alongside that of the fifth report, albeit over a more limited period of time.

The Commission decided to embark on a study of environmental planning early in 1999, after considering a number of possible alternatives, according to its usual practice. There were reservations about the topic. Some members thought it too close to the recently published report on environmental standards (RCEP 1998b), in that it would necessarily deal with 'broad social themes', and some were concerned that seeking to interpret the concept of sustainability in a planning context would be 'unrewarding'.[56] Nevertheless, after refinement, the proposal won general support, and in July 1999 the Commission announced its intention to investigate 'whether present arrangements for environmental planning are capable of achieving environmental policy objectives' (RCEP 1999: 1). It fully expected, in undertaking this study, to have to consider 'a broad range of relevant topics' (ibid.).

The study was certainly wide-ranging. Not only was responsibility for safeguarding the environment spread across many different public bodies, but the four constituent parts of the UK were able to develop their own policies in the field of environmental planning, broadly defined. Specialist agencies (dealing, for example, with pollution control or nature conservation) clearly had a central role, but the Commission was determined to engage, in addition, with the functions of the British land use ('Town and Country') planning system, through which, since 1947, the use and development of land had been substantially regulated in the UK.[57] Though not explicitly designed as an instrument of environmental protection, the planning system had been hailed by environmental groups as something of an 'unsung hero' in this respect (CPRE 2002: 2; see also Christie, Southgate, and Warburton 2002). The Commission agreed that the system had been 'a major force protecting the UK environment' and recognized, in addition, that procedures such as planning inquiries had provided an important public arena for the rehearsal of environmental controversies (RCEP 2002a: para. 1.5; see also Owens 2004).

This was new territory for the Commission. While it had addressed connections between planning and pollution control in a number of previous reports (including the fifth),[58] it had never attended specifically to the land use planning system; nor had it considered the role of the system in the delivery of environmental objectives, though arguably this had been enhanced by the requirements of environmental assessment (see Chapter 2). Once it embarked on the investigation, the Commission found the subject more challenging and less easy to circumscribe than it had expected, such that a report originally anticipated 'by the end of 2000' (RCEP 1999: 1) did not in fact appear until March 2002. The drafting was unusually difficult, with substantive changes being made to the text even after the 'final' version had gone to press.[59] Part of the problem was that the study differed in important respects from most of the Commission's previous investigations. It offered rather little in the way of 'scientific' content; its subject matter was diffuse and complex, as noted above; and some of the topics explored (such as agriculture and the environment, or public participation in planning) had been subjects of analysis, public controversy, and policy-making over many years. At times, the multi-faceted nature of the study seemed to endow it with almost hydra-like qualities.

The broad task that the Commission had set itself was 'to assess whether the various regimes... existing at different levels for setting and achieving environmental goals provide[d] an effective, accountable and transparent way of protecting the environment' in the UK (RCEP 2002a: para. 1.10). Its conclusion, in a nutshell, was that they did not. Instead, it found that the existing arrangements amounted to a fragmented and sometimes contradictory system, which took insufficient account of environmental limits and constraints. For any given geographical area, there was a bewildering array of plans and strategies, produced and reviewed through disparate procedures and over widely varying timescales. And while 'it might be possible in theory for such a plethora of plans to add up to a coherent and effective strategy' (RCEP 2002a: para. 4.15), in practice there was a serious lack of connectedness and coordination. Indeed, the Commission wondered whether devising new plans and additional layers of activity had become a way of 'denying the contradictions in a system that seeks both to accommodate growth and to protect and enhance the environment' (ibid.: para. 4.35). The report highlighted numerous other issues, including opportunities for civic engagement, on which it argued that 'policies, plans and programmes are more likely to be soundly based if they have been the subject of an open and accountable process of deliberation and careful judgement' (ibid.: para. 5.12).

Integrated Spatial Strategies

There were seventy-three recommendations in the report, covering diverse aspects of environmental planning. Its most far-reaching proposal, however, was for the introduction of a new kind of plan—'Integrated Spatial Strategies'—which would take account of all spatially related activities (including agriculture and forestry) and all spatially related aspects of environmental capacity.[60] The Commission envisaged that the new strategies would be 'four-dimensional', covering the atmosphere and groundwater as well as the land surface, and looking well into the future—at least twenty-five years ahead. They would become the dominant plans for the areas that they covered, with a firm statutory basis, a clearly designated lead body, and a requirement on all other public bodies to co-operate in their preparation.

The Commission baulked, however, at recommending what kind of geographical area an Integrated Spatial Strategy should cover. This was a tricky issue, not only because any spatial plan must deal with multiple phenomena whose boundaries are not co-terminous, but also because the question of the most appropriate level for strategic planning in the spatial hierarchy was becoming highly politicized. As we saw in Chapter 2, the Labour government at the time was bent on 'modernizing' the planning system, and proposals to this effect (DTLR 2001) were published in the form of a Green Paper while the Commission was engaged in its study. Amongst other changes, the government sought to delete the 'middle layer' of planning in England and to move the statutory, strategic planning function upwards to the regions. Critics (including several members of the Commission) were concerned about the 'remoteness' of the regional level and with what they saw as a democratic deficit in the bodies that would be charged with preparing the new plans; they worried, in addition, about the dominance of conventional growth objectives and inadequate treatment of environmental considerations (which had risen in prominence at the level of planning that would be removed).[61]

The Commission chose to sidestep the dilemma of scale by emphasizing 'the principles which should govern spatial planning, irrespective of the areas over which it is undertaken'.[62] Thus it argued that the area covered by an Integrated Spatial Strategy should 'be meaningful to the public, but also large enough to provide a valid basis for strategic planning. It should not be any smaller than the areas for which structure plans are prepared at present' (ibid.: 5). In fact, the Commission leaned towards the intermediate, sub-regional scale for its strategies, providing that such areas could be defined 'on the basis of functional coherence' (RCEP 2002a: para. 10.90). For the immediate future, however, it recommended that existing plans at the regional, county, and metropolitan levels should all be converted into Integrated Spatial Strategies.

The planning study, then, was one of those in which the Commission's investigation was contemporaneous, and overlapped in important respects, with the development of legislative proposals.[63] But whereas on previous occasions (during the study of genetically modified organisms in the late 1980s, for example) the relationship between the Commission and government had been constructive, this time it verged on being antagonistic: indeed, according to one commentary, the Commission took issue 'in an uncharacteristically combative way' with many of the proposals in the planning Green Paper (ENDS 2002: 22). A few days before the report was published, the Chair (Professor Sir Tom Blundell) engaged in a brief but tetchy exchange with the planning minister (Lord Falconer) on BBC Radio 4's *Today* Programme, in which Sir Tom claimed that the Green Paper focused on a business agenda at the expense of longer-term considerations and the minister seemed to accuse him of not having read the document.[64]

Attendance at the launch of the environmental planning report, on 21 March 2002, was disappointing, in part because of unfortunate timing (when many environment correspondents were at a North Sea conference in Norway) but also, undoubtedly, because Integrated Spatial Strategies did not make for compelling headlines.[65] Nevertheless, environmental and planning organizations were enthusiastic, with commentators agreeing that the report could—and indeed *should*—set a new agenda. For the editor of *Planning* (the weekly publication of the UK's professional association of planners) the Commission's message was 'what the government should have said in the planning green paper', even if the findings of the report were about 'as popular as a rattlesnake in a lucky dip' (Morris 2002: 11). The Royal Society for the Protection of Birds (RSPB), one of Britain's largest and most influential environmental organizations, issued copies of the summary of the report with its quarterly magazine, arguing that it was 'imperative that each administration [in the constituent parts of the UK] carefully digests and acts upon the Commission's recommendations' (Mitchell 2002: 3). The report also found favour in the

Journal of Environmental Law, where a review article deemed it 'a hugely provocative and stimulating document' (Layard 2002: 417).

The government responded to the report with one brief overarching paper (UK Government 2003c), accompanied by more extended 'daughter documents' for England, Scotland, and Wales (UK Government 2003a; Scottish Executive 2003; Welsh Assembly Government 2003). All claimed to be implementing many of the recommendations already, with the responses often simply re-stating existing policies. Some recommendations, such as the proposal that there should be third party rights of appeal against planning decisions, were roundly rejected. But the main thrust of the report—the Commission's integrative vision—was sidestepped, at least in England, by claiming that the new Regional Spatial Strategies envisaged by the government, combined with environmental and sustainability assessments, would do the job (UK Government 2003a: 29). There are interesting parallels with an earlier government's insistence, when faced with the Commission's proposals for integrated pollution control, that existing arrangements should be able to achieve these objectives perfectly well.[66]

The *Planning and Compulsory Purchase* Bill was wending its way through Parliament at the time of the government response. Although introduced in December 2002, well after publication of the environmental planning report, it seemed to pay little heed to the Commission's recommendations. Indeed, after this particular Bill gained Royal Assent in 2004, planning reform proceeded apace, with two major reports commissioned by the Treasury (Barker 2004, 2006), further legislation in the form of the *Planning Act* 2008, and—of particular interest here—the pursuit of additional, substantial changes to arrangements for regional planning. In 2007, a review of economic development and regeneration in England (the 'Sub-National Review') proposed that there should be a single, integrated strategy for each of the English regions, embracing social, economic, and environmental considerations (HM Treasury, BERR, and CLG 2007). In spite of a superficial similarity, however, the Sub-National Review had little in common with the Commission's vision of integration (and made no reference to the environmental planning report). Focusing on spatial and economic planning,[67] the Review paid scant attention to environmental capacities or even to the government's own objectives for sustainable development (SDC 2007; Townsend 2009). Instead, it reinforced the 'business agenda' by proposing that the Regional Development Agencies (RDAs)—business-led and non-elected bodies—should take lead responsibility for strategic planning and should be guided in this endeavour by 'a single over-arching growth objective' (HM Treasury, BERR, and CLG 2007: 9). It seemed, then, that the Commission's concept of holistic, long-term spatial strategies framed by environmental constraints had fallen on stony ground, with very little prospect of further nourishment.

There was nothing new, of course, in recommendations from the Royal Commission being sidelined or ignored in the first instance. But the brusque responses to the planning report, and the failure to engage with, let alone embrace, its most basic ideas, seemed indicative of a new impatience with the stance that the Commission had taken. Certainly, the report's proposals went against the grain of certain policy core beliefs (Jenkins-Smith and Sabatier 1994), but this had also happened before. What is interesting is that the responses of the UK government and the devolved administrations were produced more rapidly (and apparently with less concern for the Commission's dignity) than the tardy response to the fifth report two decades earlier. Perhaps, in the twenty-first century, it was less embarrassing to brush a Royal Commission aside, the more so when, as in the planning study, it was not seen to be deriving any special authority from some central core of scientific expertise.

A Diffuse Effect?

Still, it might be argued that the Commission exerted a degree of influence on the evolution of the planning system in the context of a wider debate. Shortly after the launch of the environmental planning report in 2002, the Chair sent a note to members to say that he was encouraged, in spite of the lacklustre event, that the report was having an early influence on government policy. He suggested that the Commission had added important issues, 'absent from the government's previous thinking', to the agenda, and that a recent speech by the planning minister had 'shown a real change in the way the government [was] presenting its intentions'.[68] The speech in question had been delivered on 18 March 2002 at the launch of another report on planning—this one produced by the RSPB (Christie, Southgate, and Warburton 2002). Acknowledging criticism of the planning Green Paper, the minister had announced his intention 'to put sustainability at the heart of the planning system' (Falconer 2002: 7) and promised that the system would be given a statutory purpose (one of the Commission's recommendations). On the same day, he had had a face-to-face discussion with Sir Tom Blundell (and a second Commission member), this time in a more conciliatory atmosphere than that of their Radio 4 encounter. While this meeting in itself is unlikely to have influenced the speech, it is clear that the Commission was adding its voice to a wider narrative of concern. Groups like the RSPB carried weight in planning and environmental policy communities, and could make good use of a report which resonated with their own objectives. The Commission made direct inputs to the policy process too: soon after publication of the report, for example, groups of members held meetings with officials at the environment department and with the Director of Planning and Chief Planner at the Department of Transport, Local Government and the Regions.[69]

The government pressed on with new legislation, as we have seen. But in the *Planning and Compulsory Purchase Act* 2004, the 'business agenda' was toned down, and subsequent guidance to planning authorities advised that policies and decisions should recognize 'the limits of the environment to accept further development without irreversible damage' (ODPM 2005: para. 19; see Cowell and Owens 2006).[70] Several years later, the bullish proposals in the Sub-National Review were also moderated, so that the regional planning system that eventually emerged—given effect in the *Local Democracy, Economic Development and Construction Act* 2009—went some way to addressing concerns about democratic deficit, short-termism, and the dearth of environmental considerations.[71] Interestingly, the strategies envisaged by this stage had a time horizon of between fifteen and twenty years (CLG and BERR 2009)—not so much less than the twenty-five years recommended by the Commission—and allowed for certain issues to be addressed at the level of sub-regions, identified on the basis of 'functional relationships between areas' (CLG and BIS 2009: para. 3.6). Seven years on from the Commission's planning report, it might have seemed, therefore, that something approximating its vision was being enacted, though by a less direct route than the Commission would have intended and certainly without its imprint, which had proved so durable in the case of integrated pollution control and BPEO.

Any convergence was short-lived, however, and what happened next nicely illustrates the contingencies of policy evolution. After the general election of 2010, the new Coalition government pursued an even more radical re-scaling and streamlining of the planning system, seeking (like its predecessor) to stimulate growth and competitiveness, but with the added ingredient of 'localism', its own particular agenda. Regional Strategies were revoked almost immediately, to 'put greater power in the hands of local people', and were formally abolished in the *Localism Act* of 2011.[72] The regional layer of planning was thereby effectively removed, though the *Act* did place a requirement on local authorities to co-operate on issues that transcended the local level. Later, most of the twenty-five National Planning Policy Statements, which had provided the context for local plans, were replaced by a single document of fifty-nine pages— the *National Planning Policy Framework* (CLG 2012)—in which one could find little or no trace of the Commission's environmental planning report.[73]

Conclusions: The Contingency of Influence

This chapter has considered the origins, evolution, and influence of two reports which promoted an integrative vision, one published in the Commission's first decade and one in its last. In both cases, the central recommendations, calling for better integration of the institutional arrangements for environmental protection in the UK, were unwelcome to the government of the day. The proposals in the fifth report sought radical changes to the system for pollution control at a time when the environment department was looking for much more modest reform; those in the twenty-third report were aimed at a closely interconnected system of land use and environmental planning, going deeply against the grain of government thinking on the use and development of land.

The ideas in the fifth report survived their initial, unpromising reception. Twenty-five years later, a system of integrated pollution control was in place in the UK and a Directive on Integrated Pollution Prevention and Control was coming into force in the European Union. The policy process had been a slow one: Her Majesty's Inspectorate of Pollution was established eleven years after the Commission's report; the legislative basis for integrated pollution control after fourteen; an independent Environment Agency after nearly twenty; and European legislation (influenced at least in part by British developments) a few years later. But in spite of the time lag, and differences between the original proposals and the system that was eventually enacted, the shift from fragmented towards more integrated pollution control in the UK remained strongly associated with the Royal Commission. In the eyes of informed commentators, it was the Commission that had endowed the idea with 'intellectual acceptability', and the fifth report was seen as one of its 'great achievement[s]'.[74]

At the time of writing, little more than a decade since publication of the environmental planning report, it might be argued that it is too soon to assess the ultimate effects of the Commission's arguments and recommendations in this area. Yet the circumstances, and the signals, suggest that the twenty-third report is unlikely to have the kind of delayed impact that could so readily be traced back to the fifth. While key proposals on integration were initially rejected in both cases, the earlier report had a marked influence on policy discourse in the years between its publication and its legislative results. The main effects of the planning report, in contrast, were both more immediate and more anonymous. The Commission's ideas about Integrated Spatial Strategies added weight to a broad-based, critical discourse seeking to influence planning legislation; and it was this wider critique, from a coalition of planning and environmental interests, which (if anything) helped to temper the streamlining inclinations of the government. Later, as we have seen, there was indeed some movement towards a form of integrated spatial planning at regional level, though once more the origins of this development were diverse, and in any case it didn't last. What is noticeable is that, aside from a modest amount of academic commentary, none of these policy developments relating to planning and the environment ever became closely associated with the Royal Commission's report.[75]

It is significant, perhaps, that the Commission itself, in contrast to its predecessors that pressed so hard for IPC and BPEO, seemed disinclined to become a 'torchbearer' for Integrated Spatial Strategies. There was a flurry of activity in the aftermath of publication,[76] but an intention, five years later, to submit critical comments on the Sub-National Review (HM Treasury, BERR, and CLG 2007) was never followed through.[77] Several later studies (on the urban environment, climate adaptation, and demographic change) referred to the environmental planning report and reiterated some of its concerns (RCEP 2007, 2010, 2011),[78] but none pressed the idea of Integrated Spatial Strategies, and for the last of these reports, at least, the whole context for spatial planning had substantially changed. The planning report seemed to be overlooked within wider policy communities, too (whereas the fifth report had remained very much alive): as we have seen, policy proposals and consultation documents in the planning field made little reference to the Commission's study, even when (in the context of Regional Strategies, for example) it might have furnished useful legitimation. Nor, at least in their submissions regarding the Sub-National Review, did the green groups draw explicitly upon the planning report, in spite of having received it so favourably only a few years before.[79]

What, then, might be concluded from looking at these two cases—the fifth and twenty-third reports—together? We might say, perhaps, that the Commission is more recognizable in both contexts as a cognitive agent than as a creature of government, but that its effectiveness in the former capacity could never be guaranteed. As noted earlier, the motivation for referring the air pollution issue to the Commission in 1974 may well have involved both rationalistic and strategic elements. As an attempt at pacification, the referral failed. As a quest for advice, and a means of eliciting proposals for 'first' and 'second order' policy change (Hall 1993), it was modestly successful; many of the less radical recommendations of the report were, in fact, accepted. The more fundamental changes ultimately associated with the fifth report were almost certainly unforeseen at the time but, in taking place over several decades, are consistent both with Hall's analysis of 'third order' change (ibid.) and with Weiss' (1977) concept of 'enlightenment'. Interestingly, several interviewees (and indeed the Commission) referred to the 'inherent logic' of integrated pollution control, implying that the learning involved in the shift towards integration had a strongly rationalistic component. But there was an important discursive dimension too, in that the fifth report (and the Commission's subsequent tactics) helped change the interpretive framework of policy, 'embedded in the very terminology through which policymakers communicate about their work' (Hall 1993: 279).[80]

In the case of environmental planning, while the sidelining of inconvenient recommendations was in some ways reminiscent of the fifth report, the resemblances were in fact superficial. The whole tenor of the response on planning was lower key, and nor did the government make use of the report to help rationalize its own, later, proposals. Perhaps the 'inherent logic' of Integrated Spatial Strategies was less compelling than that of integrated pollution control, and the planning report offered no equivalent of the conceptual elegance of BPEO; certainly, the diverse arrangements for planning and environmental protection presented a more complex and a more open system (if they could be said to constitute a system at all) than the pollution control regime of the 1970s. Furthermore, planning 'writ large' was a field in which the Commission stood out less clearly as an authoritative, expert body, and it is interesting that even those interest groups that had warmly welcomed the twenty-third report didn't feel the need to invoke it to lend authority to their own argumentation. If the Commission contributed through its planning study to policy learning, it did so in the diffuse and unattributable ways that James (2000: 163) has described as 'atmospheric'; in any case, learning was always going to be difficult when policy core beliefs about planning and growth were moving in a determinedly different direction. Most of all, perhaps, the fate of the twenty-third report demonstrates how difficult it can be to distinguish atmospheric influence from very little influence at all.

Exploring the Commission's impact and authority in these two cases amply illustrates the complexity of the factors contributing to policy change; we can see how cognitive and discursive factors, institutional inertia, political and ideological commitments, economic conditions, and random events combined to produce very different outcomes in each case. The resulting effects tell us something, too, about the authority of a high-level advisory body, and how this might vary in different contexts and at different times. These will be important considerations for the next chapter, which examines both the nature of influence and the circumstances in which advice is likely to have effect—or not, as the case may be. They feature too in Chapter 7, which seeks to identify those characteristics of the Commission that were most critical to its standing and influence, and (finally) reflects on the relationship of such characteristics to the larger project of 'good advice'.

6

The Circumstances of Influence

> Ideas, unless outward circumstances conspire with them, have in general no very rapid or immediate efficacy in human affairs; and the most favourable outward circumstances may pass by, or remain inoperative, for want of ideas suitable to the conjuncture.
>
> John Stuart Mill, *Dissertations and Discussions* (1859: 190)[1]

> Getting people to see new problems, or to see old problems in one way rather than another, is a major conceptual and political accomplishment.
>
> John Kingdon, *Agendas, Alternatives, and Public Policies* (2003: 121)*

> I do not despair; the history of pollution control is a long one.
>
> Lord Flowers, in debate on *The Environment*, House of Lords (1983)[2]

Introduction

A question frequently asked about advisory bodies is whether they have been influential. But the question is not easily answered. As preceding chapters have shown, influence can take many forms—direct or indirect, visible or subtle, immediate or long term, superficial or profound. Causal connections are elusive in complex environments and even when policies seem to flow directly from recommendations, the origins of ideas and initiatives can be difficult, if not impossible, to pin down. In searching for influence, therefore, as noted in Chapter 1, we must be wary of false positives—of crediting advisors with changes that have multiple (or entirely unrelated) causes. As one civil servant warned, we should tread carefully in assuming that '"the Royal Commission made this happen" because they happen to have recommended

---

* From Kingdon, John W., *Agendas, Alternatives, and Public Policies* (Longman Classics Edition), 2nd, ©2003. Printed and electronically reproduced by permission of Pearson Education, Inc., Upper Saddle River, New Jersey.

something that subsequently happened'.[3] But there is a risk of false negatives too. When advice has been proffered on diverse issues over extended periods of time, it is possible for significant impacts to be overlooked because the lines of influence have been diffuse, invisible, or obscure.

Even if these difficulties could be overcome, simply to document cases of influence (in terms of action or legislation, for example) would tell us too little about the nature of expert advice and its relevance in modern democracies. More fruitful lines of enquiry, as suggested in Chapter 1, embrace wider questions about knowledge, politics, and policy-making, and open for scrutiny those practices and characteristics of advisory bodies that endow them with authority—or not, as the case may be. No attempt is made here to document and measure the effectiveness of the Royal Commission in any exhaustive or mechanistic way; the emphasis, instead, is on 'the circumstances of influence' (Owens 2011a, 2012)—on the different ways in which the Commission's ideas and recommendations came to have effect and on the 'outward circumstances' (to borrow from J. S. Mill) in which they were most or least likely to do so. Of course, these circumstances cannot wholly be separated from the Commission's own attributes and actions, which helped to shape the environments in which its advice and recommendations were received; these important issues will be considered more fully in Chapter 7. It is helpful throughout to think in terms of networks rather than linear connections of cause and effect: in practice, the Commission was rarely a point source of influence from which internally generated ideas emanated to make contact with other objects.

In considering the 'circumstances of influence', the effects of the Commission's work can usefully be conceived of in terms of a spectrum or continuum, with direct, rapid responses to its recommendations at one end and subtle, long-term conditioning of the policy environment at the other. In between are time-lagged but traceable effects—'dormant seeds' (Owens 2011a: 84, 2012), 'atmospheric influence' (James 2000: 163), which is subtle and diffuse, changes in policy frames (which can be particularly difficult to attribute), and potent effects that might nevertheless be hard to see. Because different forms of influence shade into one another along the continuum, and sometimes co-exist, it is not straightforward to categorize the effects of particular reports or recommendations: a time-lagged, direct impact can be difficult to distinguish from an atmospheric one, for example. Of course, the single dimension is a simplification, and it is moot whether 'the dogs that didn't bark'—those outputs of the Commission that apparently sank without trace—should be characterized as occupying an 'ultraviolet' zone off the far end of the spectrum or a different dimension of (non-) influence altogether.

Mindful of these complexities, it is helpful, nevertheless, to organize this chapter around different forms of influence and the circumstances in which they can be identified. The discussion draws in part on the cases considered in Chapters 4 and 5 but, when appropriate, ranges more widely across the Commission's activities and reports. It begins, in the next section, with the positive responses and immediate actions that would seem, at first sight, to constitute the 'direct hits' of the world of policy advice (Owens 2005: 291).

Visible, Short-term Responses

Publication of a Royal Commission report, with the expectation of a formal government response, could be seen as a 'concentrated moment' of influence.[4] As a senior civil servant explained, the need to respond 'pushes things along.... [The government is] now in a position of either having to accept that...something needs to be done or to defend the position that nothing needs to be done'.[5] In such circumstances, the prompt acceptance of recommendations, followed without undue delay by visible changes in policies, practices, or institutions, might be seen as one of the least ambiguous indicators that an advisory body is having effect. Diane Stone (2004: 11), reflecting on the role of 'think tanks', suggests that direct effects of this kind are unusual: 'it is rare to find uncontested examples of a one-to-one correspondence between a...report and a policy adopted subsequently by government'. But in the case of the Commission, one-to-one correspondence, or at least a substantial visible impact, seems not to have been so rare. Examples of direct lineage between recommendations and action can be found across all four decades of the Commission's existence, sometimes associated with policy developments over remarkably short periods of time.

During the Commission's first ten years or so its advice seemed consistently to be effective—a reflection, perhaps, of the relative novelty of its field of investigation and the lack of coherent environmental policies at that time; there was a new environment department too, seeking to establish itself and looking for useful ideas. Reviewing (and formally responding to) the first four reports in 1975, the department pronounced that it had been possible 'to adopt a very large proportion of the [Commission's] recommendations in their entirety' (DoE 1975: 2). The third report, dealing with pollution in estuaries and coastal waters (RCEP 1972a), had been particularly influential in terms of legislation, its recommendations having been closely followed in Part II of the *Control of Pollution Act* 1974. The sixth report, on nuclear power and the environment, had an immediate impact in changing the regulatory oversight of nuclear waste (discussed later) and in legitimizing concerns about nuclear power that had previously been dismissed (Rough 2011: 198): for many, this report was a 'watershed in environmental politics' (ibid.), and the 'Flowers criterion' echoed through policy discourse for many decades to come.[6]

Even as the field matured, 'direct hits' remained a familiar feature of the Commission's interactions with government. In the 1980s, the most celebrated of these was the policy reversal on lead in petrol (DoE 1983b; RCEP 1983)—'a coup...for the Commission', according to its Secretary from that time.[7] Further examples include the rapid acceptance of the 'duty of care' principle in waste management, used by the Commission to frame its eleventh report (RCEP 1985; see also DoE 1986b), and the almost immediate adoption of a precautionary stance on GMOs, as advocated in its thirteenth (RCEP 1989); both ideas found a legislative basis in the *Environmental Protection Act* 1990 (along with provisions on integrated pollution control). A few years later, the Commission's recommendation 'that a levy be applied to all wastes deposited in landfill sites' (RCEP 1993a: para. 9.37) also seemed to have rapid effect when a 'landfill tax' was proposed in the budget of the following year and was introduced, as the UK's first environmental tax, in October 1996.[8]

Among the Commission's later reports, *Energy: the Changing Climate* (RCEP 2000), stands out for the visible and almost immediate impact of one of its most radical recommendations. This was a proposal for an ambitious, long-term target for the reduction of carbon dioxide emissions in the UK (by 60 per cent over half a century), supported in the report by four energy scenarios to indicate how such a target might be met. The government's initial response was to set up a review of energy policy objectives, looking fifty years into the future, as the Commission had urged. The energy review (PIU 2002) was then followed by the first comprehensive statement on energy policy for several decades, in which the government accepted the emissions reduction proposal, largely in the Commission's own words: 'We...accept the Royal Commission on Environmental Pollution's (RCEP's) recommendation that the UK should put itself on a path towards a reduction in carbon dioxide emissions of some 60% from current levels by about 2050' (DTI 2003: para. 1.10; cf. RCEP 2000: 199; see also UK Government 2003b).[9]

These examples of visible impact are by no means exhaustive, nor should we forget the quiet acceptance of many lower-profile recommendations, to which it is impossible to do justice here.[10] If the sequence of 'recommendation–acceptance–action' is indeed indicative of influence, then Richard Southwood's assessment that the Commission's impact in its first fifteen years had been 'moderately satisfactory' (Southwood 1985: 347) can reasonably be extended across the range of its subsequent reports. It is worth noting, too, that legitimizing *non*-action can be an important form of influence. The nuclear report in 1976 enabled the then Secretary of State for Energy (Tony Benn) to defer a controversial decision on commercial fast reactors (which were ultimately not pursued); and the energy report of 2000, in arguing that emissions targets could be achieved in a variety of different ways, helped keep new nuclear reactors in the background of UK energy policy (as a possibility but not an active one) for another seven or eight years (Rough 2011).

While recognizing influence, however, we should be careful about attributing causality. Even in what seem to be the most straightforward cases of direct effect, the Commission was rarely—perhaps never—the *only* agent at work. More typically, it added an authoritative voice to the debate at a crucial moment, sometimes acting as a catalyst in a complex and crowded field. What is distinctive about the 'direct hits' outlined above is that they occurred when recommendations resonated with the political temper of the times. So, for example, the Commission's earliest reports offered clear guidelines for governments that needed to 'do something' about gross pollution, and they were eagerly awaited.[11] We might say that in the new policy domain of 'the environment', the Commission fulfilled something close to a rational analytical role, though even here it was both responding and contributing to wider agendas and debates. There was a *tabula rasa* in the case of GMOs in the late 1980s, too, when civil servants drafting the legislation on releases to the environment were glad of reinforcement from the Commission (see Chapter 4): as a senior civil servant noted in oral evidence for the GMOs study, the environment department was 'looking to the Royal Commission for an authoritative Report' on this 'new area of activity'.[12] But others were also urging the government to take a strongly precautionary line; there was, as indicated by the former director of a prominent environmental organization, an advocacy coalition converging on a particular view: 'we developed a whole lobbying agenda on Part VI of the [Environmental Protection] Bill...A lot of [the issues] were prompted by things the Royal Commission was saying, and that all went into the briefings...So yes, when I say "we" I mean we in the broadest sense not just we the Green Alliance.'[13] The Commission was given the credit, nevertheless, during the Bill's passage through Parliament, when ministers made it clear that the proposals 'closely follow[ed] the recommendations of the RCEP's invaluable 13th report', and that 'the bulk of the legislation [on GMOs]...either was suggested by the Royal Commission or has been amended to meet its concerns'.[14]

In the case of lead in petrol, an established policy had become a matter of deep controversy (see Chapter 4), and the Commission's perspective was added to an already noisy debate. The ninth report was an important and timely intervention, however, because it enabled a beleaguered government to change its mind by invoking the authority of expert advice, rather than appearing to perform a 'U-turn' under pressure. The recollections of a civil servant and a former Commission member respectively are suggestive of a strategic model of relations between expertise and policy, as outlined in Chapter 1:

> [The government] couldn't just say 'no, there is no risk', and let the pressure groups go on; [it] couldn't just surrender to the pressure groups—industry would have crucified it. The only way you can do it is to turn to an external body and that is what [the Commission] was set up to do.... It's done a very good job, which is to take the heat off and give... the government a way out'.[15]
>
> CLEAR [the Campaign for Lead-free Air] softened [the government] up and the Commission gave them, as it were, a reputable reason for making a decision for which politically they were strongly inclined.[16]

So what was going on here? The Commission was influential, certainly, but its rapid impact could be explained, at least in part, by its offering the government 'a way out'. Yet in order to be convincing as a source of legitimacy, the Commission had to have some claim to independent authority in the first place (an issue to be taken up in Chapter 7), and the 'way out' was skilfully illuminated by its framing of ambient levels of lead as a problem and lead-free petrol as an important part of the solution. Neatly, the Commission presented the risk to human health, with all the attendant uncertainties, as one that it was simply unnecessary to take. In other respects, too, this was not the voice of science coming from some realm outside politics: if the lead-in-petrol issue had not already been politicized by a powerful anti-lead coalition, and if Richard Southwood had not been concerned with raising the Commission's profile at that point, the ninth report would not have focused exclusively on lead at all. So there were cognitive and discursive components, and aspects of co-production, in the Commission's interactions with policy–political processes at that time. On an even wider stage, all actors were 'locked into an international game' (Haigh 1986: 82) in which political, technical, and economic factors were beginning to point inexorably towards lead-free fuel. In short, this case involves elements of all four conceptualizations of the role of expertise outlined in Chapter 1; and it provides strong support for Kingdon's (2003: 73) assertion that 'when we try to track down the origins of an idea or proposal, we become involved in an infinite regress'.

Similar points can be made about the '60 per cent' recommendation in the energy report of 2000 (a detailed account can be found in Owens 2010). The figure itself was an interesting hybrid, derived from climate science, considerations of global fairness,[17] and the Commission's conscious efforts to produce something more arresting than just 'one more report on energy'.[18] Then, for a complex mix of reasons, the UK government was willing to add a long-term, strategic target to its more immediate actions and policies on climate change. The report came at a time of heightened awareness of potential climate impacts, with deep reductions in emissions (of the order put forward by the

Commission) being widely urged. Security of energy supply was also strongly emergent as a political concern, helping to justify intervention in an energy sector which had increasingly been governed by market principles. By 'stating the obvious that a radical shift in a non-carbon direction would be required', the energy report 'acted as a catalyst', helping to establish consensus about energy and climate change policies (Helm 2004: 370), and the '60 per cent' recommendation remained something of a benchmark as these policies converged (Chapter 2). The government's receptiveness may have been conditioned by external events, too: early in 2003, when the energy White Paper (DTI 2003) was being finalized, Prime Minister Tony Blair was mired in controversy about his alliance with US President Bush in the run up to the Iraq War. A bold carbon emissions target was perhaps a useful statement of distance on at least one important international issue.[19] But none of these factors, nor the retrospective wisdom that policy was in any case moving in that direction, should be taken to imply that the Commission did little more than produce a far-reaching recommendation at an auspicious political moment. Rather, the Commission consciously sought to make an intervention, and did so by cutting through complexity and broaching a difficult issue in a way that made it amenable to political action. It was these discursive dimensions of its role, at least as much as the authority of its analysis, that helped to shape the 'policy window' (Kingdon 2003: 165) and maximize the impact of its report.

A word should also be said here about the important contributions on waste management mentioned above—concerning the 'duty of care' and the landfill tax—both of which apparently had short-term, visible effects. On close examination, however, these 'direct hits' exemplify the problem of categorizing forms of influence, as well as the challenges of attribution. In the case of the 'duty of care', the idea was indeed accepted in principle with alacrity—five days after publication of the eleventh report—and was endorsed, with little delay, in the government's formal response (DoE 1986b: paras 3, 42).[20] Arguably, it was an idea whose time had come because of a widely perceived crisis in Britain's waste disposal arrangements (ENDS 1985d). But the Commission was not alone: policy inertia on waste had been roundly criticized by a chorus of authoritative voices;[21] the government was under pressure from Europe; and, just in advance of the Commission's eleventh report, an inter-departmental review of Special Waste Regulations had recommended something very similar to the 'duty of care' (Joint Review Committee 1985). In short, there was political pressure, a degree of urgency, and a clear convergence of view. The Commission's work—'the last in a series of high level appraisals' (ENDS 1985c: 13)—brought other proposals together into a coherent framework, providing 'a welcome addition to the reports which have influenced the Government's thinking in recent years' (DoE 1986b: para. 4). The eleventh report might thus be said to have had a direct but catalytic effect, time-lagged in the sense that the legislation itself took another four or five years. In this case, the *idea*—the 'duty of care'—was acceptable, even welcome, and the delay in implementation can be attributed to system inertia, political priorities, and competition for Parliamentary time.[22]

The landfill tax was even more obviously a case of the Commission giving a final push to an idea that had been in the policy stream for some time. The line of investigation in the incineration study had itself been influenced by government-commissioned research on economic instruments in waste management—one commentator noted, for example, that the report had 'join[ed] the list of Government-funded studies' emphasizing the role of such instruments[23]—and the Commission had employed consultants to analyse the economics of alternative disposal methods.[24] Its endorsement of a landfill tax was nevertheless helpful to the government, or at least to the environment department, as one official involved at the time explained:

> [T]he Commission were...pushing at an open door when they made that recommendation. They knew that we had done some work on it...But having...a body as eminent as the Royal Commission supporting [the tax] helped us to make the case...When it suits to rely on something the Royal Commission has said, then obviously we'll make use of it.[25]

So this apparently direct effect might be interpreted not only as catalytic, but also as an example of legitimation—here of the Commission enabling a particular departmental position within government to hold sway. As noted above, however, whether it 'suits' to use an advisory body in this way depends to a considerable degree on the extent to which it is trusted and on the prospect that its views will command respect. But it also depends on circumstances. In plenty of other cases, as we shall see, it did not 'suit' to accept the Commission's recommendations at a particular moment, and its influence was therefore attenuated, or much delayed.

Dormant Seeds

When the Commission's advice landed in less propitious political circumstances, it was unlikely to have immediate effect, especially if it ran counter to 'deep core' or 'policy core' values and beliefs (Jenkins-Smith and Sabatier 1994: 180). Over time, however, circumstances change, and even resilient core beliefs can be challenged. This helps to explain why some of the Commission's proposals, acting like dormant seeds, generated little activity in the first instance but came to fruition later, sometimes after a considerable delay. In the clearest example, it took between ten and twenty years for the

The Circumstances of Influence

Commission's ideas on integrated pollution control (IPC) to become substantially reflected in institutional and legislative arrangements. This complex and lengthy process was explored in Chapter 5, but several aspects are worth reiterating here. First, the initial proposals for cross-media pollution control and a unified inspectorate (RCEP 1976a) were unwelcome not only because of their novelty but also for institutional and political reasons, some of them unconnected with the environmental issues at stake. Second, the ideas were kept alive, not least by the Commission's own arguments and actions, and remained viable much later, when 'outward circumstances' had changed. Third, what we see in this case is a time-lagged effect, but one that is nevertheless direct and visible. Neither the intrinsic logic of the key concepts (as some of their advocates perceived it), nor the Commission's authority, were sufficient in the first instance to bring about change, but the fifth report helped set in train a more gradual process of policy evolution in the UK. When IPC and 'best practicable environmental option' (BPEO) were at last enshrined in legislation, they remained closely identified with the Royal Commission, in spite of the lapse of time and the fact that the trend towards integration was observable across the Western world. The Commission was not the sole agent of change, but nor was it a passive recipient of circumstance: rather, as discursive agent, persistent advocate, and policy entrepreneur, it played a significant role in the ultimate convergence of problem and policy streams.

In another example of dormancy, measures proposed in the Commission's seventh report, on agriculture and pollution (RCEP 1979), encountered stiff resistance at first but became easier to implement once the 'agricultural fundamentalism' (Wibberley 1981: 157) of the post-war decades began to wane. The report focused primarily on pesticides, nitrogen fertilizers, and farm wastes, and fell largely within the remit of the Ministry of Agriculture, Fisheries and Food (MAFF), which, as noted in Chapter 3, had been wary of the study in the first place. Reaching agreement within Whitehall on what to do with this report was far from easy.[26] The government was slow to respond and when it did so (four years after the report) it rejected many of the Commission's recommendations (DoE 1983a).

Yet gradually, over the next decade or so, the situation changed. According to a former Chair: '[MAFF] pretty well rubbished almost every part of the report, and then if you follow agricultural policy over the next five or six years, they've adopted almost everything'.[27] The proposals relating to pesticides, for example, had a delayed but nevertheless direct effect, though in this case the Commission had a relatively low profile. It had argued, in the agriculture report, that the existing, voluntary scheme for approval and use of pesticides (the Pesticides Safety Precautions Scheme—PSPS) was inadequate, and had urged the government to make provision, through reserve powers, for

133

a statutory system of control (RCEP 1979: paras 3.105–6). The response was tepid: the government thought such a move unnecessary but promised (in time-honoured fashion) to keep the PSPS 'under continual review' (DoE 1983a: para. 24). In the event, a statutory system was given effect two years later (six years after publication of the report), in Part III of the *Food and Environment Protection Act* 1985; in the passage of the legislation, the Commission was explicitly recognized as the body that had 'prompted' the debate.[28] The percolation of its ideas into this area of policy was spoken of by one senior civil servant in words that evoke Carol Weiss' (1977) concept of 'enlightenment':

> [E]ven the things ... [that] aren't worth pushing—because there is simply too much opposition—they're still there, on the table, they haven't gone away. And over time some of those also get picked up ... [The agriculture report] in particular was interesting because when it came to the legislation, it was written very much along the lines that I think the Royal Commission would have approved of ... it was MAFF who wrote the legislation. ... *it got into their consciousness* that some of the things they'd been doing, now was the time to change them.[29]

The same official reflected on the Commission's overall impact in the field of pesticides, a subject that it had touched upon in a number of different reports. His comments (in 1997) reveal a matrix of short- and longer-term, direct and more subtle effects, and at the same time offer a fascinating insight into the interplay between the Commission and the department:

> [T]he Commission has had a significant *but largely unknown*—outside the knowledgeable group if you like ... effect on pesticide policy. I actually came into [environment department] headquarters to write a report on the non-agricultural uses of pesticides directly at the request of the Royal Commission ... [The] first and [fourth] reports had said, 'there's [a] lack of information, a review is needed', and I carried out that review, which led to the non-agricultural uses of pesticides coming under the same regime as the agricultur[al] and garden pesticides ... Then [for] its seventh report I largely wrote the department's evidence on pesticides ... and I think my evidence had quite an influence on the Royal Commission, but *the ... Commission was able to create situations that I couldn't do as an insider*. ... putting [the pesticides safety system] on a proper regulatory basis as opposed to a voluntary agreement—that took some time—but *the Royal Commission pushed on that door and it gradually swung open*. Much of the transparency which is now in the system is pretty directly down to the Royal Commission. ... I think almost all the recommendations that the Royal Commission made, even the ones that were violently opposed by the Ministry of Agriculture and others at the time, have now [been] accepted as part of the everyday running of the scheme. So I think they had a big influence on that area, although as I say, unless you are an insider you probably wouldn't recognize it.[30]

Another important contribution of the agriculture report was to frame intensive livestock units as 'industrial enterprises' (RCEP 1979: para. 8.30), whose waste products could therefore be conceived of as 'pollution'. The significance of this discursive shift (discussed at greater length in Lowe et al. 1997) was described by a former member:

> I keep thinking of one conclusion, about the damage done by farm effluent, which was never taken very seriously by anybody... [A] factory... discharging coloured dye into a local stream... is wicked, wicked, wicked... but... your friendly neighbourhood farmer... not covered by the same legislation [is] turning out anything you like and doing enormous damage... oh, well that's alright, that's just farmyard smell... I thought some of the bravest remarks [in the seventh report] were about the position of modern farming in our society and the fact that to regard modern farmers as being first class stewards for the environment is perhaps a misreading of the world as it is.[31]

Concerned, in particular, with effects on the aquatic environment, the Commission urged that intensive livestock units should require planning approval irrespective of size, and that any grants should ensure appropriate provision for pollution control. It also wanted the water authorities to be consulted in all cases (RCEP 1979). The government, preferring an older narrative, insisted that the rearing of livestock was 'not an industrial activity in the general sense of that term' (DoE 1983a: para. 69). It rejected the Commission's recommendations, stressing the availability of advice on dealing with smells and other pollution problems, and promising more resources for research. But changes were afoot, propelled not only by criticism from other high-level bodies (see, for example, House of Commons Environment Committee 1987) but also by water privatization, which sharply focused attention on the regulatory arrangements for pollution. In the bigger picture, the shift was undoubtedly a product of the ecological modernization of the 1980s, combined with a growing propensity to frame agricultural (over-)production as a problem rather than a solution. Over several years from the end of the decade a more interventionist regulatory system, closely involving the National Rivers Authority (and its successors), was introduced,[32] and intensive livestock units later became subject to further controls, under the Integrated Pollution Prevention and Control (IPPC) Directive (Chapter 5).

What becomes evident, therefore, is that in the agriculture report the Royal Commission developed ideas and proposals for which the political environment became receptive over a period of a decade or two. It had influence, but the effects were gradual and (with the exception of pesticides, perhaps) difficult to disentangle from the many other forces affecting agricultural and pollution policies in the 1980s. In the case of livestock units, the Commission's influence verges upon the atmospheric, showing how delayed effects can merge along the spectrum from one kind of influence into another.

Similar points might be made about the Commission's advice on waste, which had diffuse and relatively invisible effects alongside impacts that are more readily attributable. For example, the incineration report (RCEP 1993a), as well as contributing directly to the introduction of a landfill tax, also exercised leverage in ways that were more obscure. According to one civil servant, the report gave 'a spur to [the environment department's] efforts to come up with [a more] strategic approach' to waste management[33]—efforts that came to fruition seven years later with publication of the long-awaited waste strategy for England and Wales (DETR 2000b). One official, looking back over several decades from vantage points in both Whitehall and Brussels, felt that overall the Royal Commission had done 'good work on waste', contributing to stronger institutions for pollution control in the UK: 'we are now better set up than we were . . . although it took a long time, a very long time'.[34]

Slowly Changing the Frame

The emphasis so far in this chapter has been on problem-oriented recommendations in specific policy domains—arguably the easiest form of advice to track in any assessment of influence. Over its lifetime, the Commission offered a great deal of advice of this kind, much of which had effect, sometimes quietly, sometimes visibly, over varying timescales, and almost always in conjunction with other factors. Arguably, however, it exerted at least as much influence through the processing and circulation of more loosely contextualized ideas, which were often themselves developed and applied in the investigation of particular problems. Lord Flowers (1995: 9) put it eloquently: 'One expects as the occasional distillate from more specific considerations new principles to guide our general response to environmental matters.' Most often, such ideas were already emergent in policy discourse, but the Commission's standing and constitution enabled it to refine and refract them in ways that intensified their impact. A prominent example would be its long-term (if not entirely consistent) advocacy of precaution, which, over several decades, helped significantly to change the frame for British environmental policy. Similarly, the Commission's holistic ontological perspective on the environment might be said to have had general influence, though it was manifest most visibly in the promotion of specific objectives such as those of moving to lead-free petrol or integrating pollution control functions under the auspices of a new inspectorate.

Taking the long view, we can see that another of the Commission's causes (though members might not have recognized it as such) was the opening up of policy and regulatory processes, both to proper consideration of risks to human health and the environment and to wider communities of interest and influence over time. The 'technology assessment' studies considered in Chapter 4 provide interesting examples. In the 1970s, the seminal report on nuclear power (RCEP 1976b) was important in the loosening of a tightly structured nuclear policy community, in spite of opposition from the industry (Berkhout 1991; Rough 2011; RWMAC 1995; Saward 1992): its recommendations led to the appointment of an independent Radioactive Waste Management Advisory Committee (RWMAC),[35] and to a shift in responsibility for radioactive waste strategy from a department that was promoting nuclear power to one with a remit for environmental protection. It was easier, perhaps, to embed environmental scrutiny into the regulatory structures for GMOs, a field so new in the 1980s that policy networks were still embryonic. Through its report on releases of the new organisms (RCEP 1989), the Commission helped ensure that the environment department would be involved in the necessary vetting procedures, and that the granting of consents would be overseen by an independent committee. Both measures were included in the *Environmental Protection Act* 1990. Some twenty years later, the report on nanomaterials (RCEP 2008) similarly raised awareness of the need for research, rigorous environmental scrutiny, and appropriate regulatory arrangements.

Of course, the Commission didn't win every battle on these fronts. The nuclear industry successfully deflected its proposal for an independent radioactive waste disposal body, for example,[36] and at the time of writing (mid-2014) there has been glacial progress towards the reporting and regulatory structures that the Commission was urging in the case of nanomaterials.[37] Interestingly, too, the Commission's own conceptions of inclusiveness changed over time. It didn't always envisage 'outsider' groups being represented on the relevant bodies, even if it was keen to ensure an official environmental presence—the releases committee for GMOs is a case in point (see Chapter 4). But in later reports, it came to promote a more active role for lay publics and civil society in policy and regulation. The 'standards' study (RCEP 1998b), in particular, both reflected and gave impetus to the growing pressures for inclusivity and, in doing so, had a noticeable, if 'evolutionary', effect on policy discourse (Farmer et al. 2005: 15).[38]

Overall, it seems fair to say that, while its specific interpretations changed, the Commission's instincts tended towards pluralism in environmental policy. This was reflected in another of its causes—that of access to environmental information, which, within the wider context of 'freedom of information', was the subject of energetic campaigning in the latter part of the twentieth century. The Commission, in its own way, became an important part of the coalition pressing for reform. In many of its reports it urged that information should be made more widely available, and was impatient with counter-arguments about commercial confidentiality, potential misuse, or public alarm (RCEP 1972b, 1976a, 1984, 1998a, 2005). It first tackled the subject in its second report, at a time when secrecy was the norm, making a powerful case that 'information about wastes should be available not only to the statutory bodies... but to research workers and others who can make use of it to improve the environment' (RCEP 1972b: para. 7). Maintaining that disclosure was likely to threaten commercial secrets in only a small number of cases, the Commission urged government and industry to devise measures that would 'increase the availability and flow of information' (ibid.: para. 10).[39] Coming from an authoritative source, these arguments helped to shift the terms of the debate: in the words of a prominent campaigner (whose own work later influenced the Commission), 'that little report in itself, although it didn't analyse the issues in great detail, was immediately influential... it knocked away [the] prop that [government and industry] had been relying on so heavily... Ministers were influenced by this'.[40]

The Commission kept up the pressure in its third report (RCEP 1972a). This directly influenced the *Control of Pollution Act* 1974, which provided for public registers of information on discharges into rivers, estuaries, and coastal waters (not implemented until 1985), and for more limited public information about air pollution and waste disposal to land. Unsatisfied, the Commission returned to the subject of secrecy in its report on air pollution control (RCEP 1976a) which, according to the campaigner quoted above, 'really took [the Alkali Inspectorate] to pieces... had a go at their whole philosophy of operating'.[41] In fact, in almost all of the earlier reports, some plea was made for better access to information, and such advocacy persisted as an enduring principle that was 'handed down'. The Commission noted in its tenth report (RCEP 1984), for example, that scepticism about the need for confidentiality was a view that had been 'consistently held and re-affirmed by successive Commission memberships' (ibid.: para. 2.36). It argued, further, that the 'sanctity of official data' (ibid.: para. 2.55) had become increasingly open to challenge, that secrecy had a negative effect on risk perception, and that public scrutiny could act as 'a vital discipline' for regulatory authorities (ibid.: para. 2.75). In a passage that would be widely cited, it recommended 'a presumption in favour of unrestricted access for the public to information which the pollution control authorities obtain or receive by virtue of their statutory powers, with provision for secrecy only in those circumstances where a genuine case for it can be substantiated' (ibid.: para. 2.77).[42] The Commission's Secretary at the time remembered drafting this section, aware that there were 'sympathetic ears' in the environment department for its messages:

I must confess that I wrote with a certain amount of missionary zeal... it was certainly influenced by the Maurice Frankel[43] work... Frankel was overjoyed when that particular part of the report came out... but I don't think I was pushing the Commission further than they would have gone necessarily, because I believed in it myself... it was a theme that had come up in earlier reports, it was something that Southwood [the Chair] was very keen on, and of course... in other areas it had become more and more a major issue in a European context.[44]

The government responded positively to these ideas (DoE 1984), setting up an inter-departmental working party to consider a disclosure regime, though its proposals would not be fully implemented for another decade or so (DoE 1986c; ENDS 1996). What we see here is a clear instance of the Commission being caught up in a slow but inexorable stream of change, whilst also '[putting] its might into that general trend'.[45] Different accounts—from the former Secretary, the campaigner mentioned above, and a senior scientist from the environment department respectively—are all suggestive of enlightenment or atmospheric influence:

[T]here were various attempts to promote freedom of information legislation, all of which [had] failed, resisted by the government... So I think the Royal Commission performed an important role in giving impetus to something which [was]... probably going to come anyway... it was the first time the case had been stated at such length and with such potency in the context of environmental regulation.[46]

[F]inally, after a long time, [the environment department] did get well ahead of the game... [but this]... was nowhere near the straight line of progress that you might have expected or might now assume... The Royal Commission things were helpful, but *they didn't succeed in changing things immediately, they were of assistance to people who were trying to change things and they did condition attitudes.*[47]

[The pressure for access to information was] just dripping on the stone, if you like, the whole ethos of government was changing.... That was a slow process but there were milestones along the way and certainly *the Royal Commission's interest in the subject was one of the things that kept the impetus going*; the... Freedom of Information Campaign was certainly another one; another... was the EU interest in what became the [1990] Directive on environmental information... that was the formal key that unlocked the door but a lot had been happening before that.[48]

The trends referred to by these interviewees gathered momentum in the 1990s, leading to further legislation, including the UK's *Freedom of Information Act* 2000, the Aarhus Convention of 1998, and an updated European Directive on public access to environmental information in 2003. Although its influence had become less apparent by then, the Commission had undoubtedly played a key role in promoting this policy paradigm.

In the various respects considered above—precaution, integration, pluralism, and access to information—the Commission espoused important, emergent principles of environmental governance and exemplified and reinforced them in successive reports. While it is obviously not the case that the Commission was single-handedly responsible for the changes taking place during the latter part of the twentieth century, its articulation and defence of such principles added an authoritative and *persistent* pressure to the more general impetus towards ecological modernization. Thus the Commission contributed as a cognitive and discursive agent to long-term processes of changing policy frames, even while its specific recommendations (on leaded petrol, for example, or a unified pollution inspectorate, or the publication of discharge consents) exerted visible influence with greater or lesser speed.[49]

'Doing Good by Stealth'

In the cases discussed so far, the Commission's influence is most readily apparent when its recommendations had short-term, direct effects, or when chairs and members acted like good policy entrepreneurs, promoting dormant recommendations until circumstances were right for them to spring into life. Otherwise, the Commission's role in the more gradual, diffuse, and multi-causal processes of policy evolution tended to fade from sight over time, so that only knowledgeable insiders (and policy analysts, perhaps) might recognize it as a force for change. Intriguingly, however, the Commission was also capable of exerting influence that was both direct and short term, but at the same time relatively invisible. The very fact that it was undertaking a study could raise the profile of an issue or stimulate action, not least because government departments, in preparing and giving evidence, would need to 'establish their line'. As one senior civil servant explained (when the Commission was still in existence), 'we do a lot of our best thinking on a subject not in response to a Royal Commission *report* but in response to their request for evidence, and it can easily happen that, in a sense, that's the really influential thing'.[50] Members also tended to 'talk up' the subject under investigation, and civil servants would sometimes 'get wind' of recommendations, causing them to 'change tack... so [the report would appear] to be giving retrospective advice'.[51] There are many examples of these invisible, but direct, forms of influence, aptly described by Lord Flowers as 'doing good by stealth' (see Southwood 1985: 347). In the 1970s, the announcement that the Commission intended to look at radioactive waste disposal was enough to prompt the Cabinet Office to launch its own study, to assess the adequacy of existing arrangements and prepare the government 'to answer any criticisms which might be raised'.[52] The leverage of the incineration study on waste management policy, noted earlier, was also a result of stimulation behind the scenes (in combination, in this case, with a slower, atmospheric influence).

According to one academic observer, the environment department had 'done its bit' before publication of the report (RCEP 1993a), and was able to claim afterwards that an integrated waste strategy was in preparation.[53] Similarly, in the author's own experience, the study of urban environments (RCEP 2007) encouraged both the environment department and the Department for Communities and Local Government to engage with this issue in advance of the report being published.

With hindsight, we might expand Flowers' original concept of 'doing good by stealth' to include the Commission's gentle threats and pressures behind the scenes. The seriousness of a problem, the feasibility of a potential solution, and, perhaps, the possibility that a Royal Commission report might add to political pressure, could all be communicated 'backstage', to use Hilgartner's (2000: 42) analogy. In this broader sense, 'doing good by stealth' was clearly what one member had in mind when she warned rhetorically that concentrating on the post-publication response would reveal only 'about a third of what [the Commission's] influence really is'.[54] Such work emerges, therefore, as an important way in which policy settings could be affected, not through the visible mechanism of a report, but by argumentative practices within the various networks of which the Commission, and its members and Secretariat, formed a part. It might also be seen as a form of 'anchoring', or the softening up of policy communities, whereby ministers and civil servants could be persuaded of the merits of particular approaches, and might even gain political capital by being able to say (on receipt of the formal advice) that they were 'doing this already'. The significance of networks will be considered again, in the context of important characteristics of the Commission, in Chapter 7.

The 'Dogs That Didn't Bark'

One interesting (and important) question to emerge from the discussion so far is how we might tell when the Commission had diffuse, atmospheric influence as opposed to having no discernible influence at all. There may not always be a convincing way to make this distinction, especially in the longer term and when the role of any particular agent has become obscured. Still, it is useful to separate instances in which policies evolved broadly as advocated by the Commission (even if its advice was not directly acknowledged) from those in which no legislative embodiment of its proposals was in evidence (or in prospect) many years after publication of a report.

A rationalist explanation for 'failures' of the latter kind might be that the Commission's arguments or recommendations were not only initially unconvincing but (crucially) never *came to be seen* as more persuasive or practicable as external circumstances changed. Following its influential report on GMOs, for example, the Commission proposed a system of risk assessment—GENHAZ—which drew heavily on the experience of one of its members with hazard identification in the chemicals industry (RCEP 1991b). Although the government acted quickly on the Commission's recommendation that GENHAZ be trialled with 'real proposals', it felt that the analogy with hazard identification in chemical installations had 'important limitations' (DoE 1993b: paras 2, 6). Risk estimation for GMO releases (it argued) would necessarily be more of a 'qualitative evaluation', and would thus be 'less susceptible to the prescription of methodologies' (ibid.: paras 6, 7). In practice, GENHAZ proved 'overly elaborate' in the context of GMO releases to the environment, and therefore 'had very limited impact'.[55] Similarly, a key proposal in the Commission's report on the urban environment (RCEP 2007)—that there should be an 'environmental contract' between central government and local authorities in major urban areas—was rather vague in conception and was not pursued, though the study did raise consciousness about critical questions at a time when urban and environmental issues had become the concerns of separate ministries.[56] In cases such as these, it might be said that the Commission had not fully considered the practicality of its ideas in particular contexts. The rejection of its advice on buffer zones, in connection with bystander exposure to pesticides (Defra 2006b; RCEP 2005), was rather different: here there was a view that the proposal lacked a scientific rationale (Chapter 4)—in other words, the government questioned not only the practicality but also the rationality of the advice that it was being given.[57] A reluctance to upset farmers at that particular juncture might have added a strategic element to this rejection, but nothing happened in the years that followed to reverse the department's decision.[58]

In other instances the Commission's failures accord more straightforwardly with a strategic model, in which knowledge and advice become subordinated to 'the balance of organized forces' (Kingdon 2003: 163). Whilst, as we have seen, shifting circumstances rendered some lines of thinking more palatable with the passage of time, the opposite could also be true. So, for example, in the 1990s, the Commission's contributions on transport (RCEP 1994, 1997c) added weight to an increasingly dominant coalition urging greener transport policies, and were acknowledged as 'key influences' in the 1998 Transport White Paper (DETR 1998: para. 2.52). Indeed, the first of the two transport reports generated such intense interest that it was re-issued as a paperback book (RCEP 1995c), and the conclusions of the second (RCEP 1997c) were reproduced in an annex to the White Paper itself. But as time went on, the 'new realist' discourse coalition struggled to survive in the face of fuel protests, electoral unpopularity, and a policy core belief that road transport was linked inexorably to economic growth (see Chapter 2). By 2006, a report commissioned by the Treasury and the Department for Transport was emphasizing transport networks as an 'enabler of sustained productivity and competitiveness' (Eddington 2006: 1), and acknowledging environmental considerations only insofar as they could be incorporated within a cost-benefit framework. Though it certainly helped to shape thinking in the 1990s, the Commission's material influence on transport policy seemed diminished as the 'new realist' discourse faded.

The fate of Integrated Spatial Strategies, proposed in the environmental planning report (RCEP 2002a), provides a further example of little effect, though in this case the Commission's proposals had never much impinged on policy discourse (see Chapter 5). After a decade marked by intensified efforts to reduce the 'burden' of planning and regulation, the integration of land use and environmental planning within coherent spatial strategies seems an even more distant prospect at the time of writing (early 2014) than it did in the immediate aftermath of the Commission's report.

Assessing Influence: Conclusions and Reflections

One important conclusion from the retrospective in this chapter is that policies affecting the environment within and beyond the UK would almost certainly have evolved differently—or would have developed according to a different dynamic—had the Royal Commission never been appointed. Arguably, the apogee of its visible influence can be placed at around 1990, when the *Environmental Protection Act* incorporated so many of its recommendations, though there were a number of significant impacts after that. We can say with confidence, too, that the Commission's advice became effective (if at all) in different ways and over widely varying timescales, lending support to the concept of a spectrum, or continuum, of modes of influence, as suggested at the beginning of this chapter. A third important finding is that, while the responsiveness of governments was invariably conditioned by the political climate in which advice was received, the Commission's ideas and recommendations could prove resilient, retaining (or even gaining) potency over extended periods of time.

Looking in more detail at the circumstances of influence, we see that recommendations tended to have short-term effect when they 'made sense' in reinforcing other pressures, filling a policy void, or furnishing useful legitimation for 'difficult' decisions and choices. Of course, the Commission's proposals also had to satisfy conditions of technical and economic feasibility—this is why some ideas were embraced with enthusiasm but nevertheless followed by inaction (a progress of sorts, which invariably frustrated the Commission). Advice that was unpalatable in the short term could come into its own later, in a different political climate, providing that the underlying analysis remained powerful and solutions could still be attached to problems, even if not quite as originally framed. And in a broader sense, the Commission's promotion of certain basic principles helped to change the interpretive framework of environmental policy, so that it could be seen as having a 'major impact on the climate of opinion' even when key recommendations were 'ignored or ineffectively implemented' (Frankel 1984: 43). In different contexts, then, the Commission's work led to 'enlightenment' (Weiss 1977) or exerted 'atmospheric influence' of an indirect and cumulative kind (James 2000: 163; Radaelli 1995). But it is in the important sense that *change happened*, even if on a decadal timescale, that diffuse forms of influence are distinguishable, at least in principle, from minimal impact or no effect at all. The failures—the ideas and recommendations that were never taken up—tended to be those that went deeply against the grain of social norms or policy core beliefs (Jenkins-Smith and Sabatier 1994) and, more crucially, ran counter to the general direction of change.

Looking back over forty years, it is clear that the circumstances pertaining at particular times—political, social, economic, even serendipitous—go some way towards explaining the widely varying reception of the Commission's advice. But another important factor must surely help account for the different modes and extent of its influence. There is much to suggest that the assembly of evidence by the Commission, and its practices of analysis and reasoning, were *in themselves important*—in other words, the Commission's cognitive authority had a degree of independent significance. Whether ideas emerged from its own deliberations, or were transmitted from other sources, they were more likely to have effect when they had been rigorously interrogated by the Commission, and more likely to fail when they had not. This is not to suggest, naïvely, that 'good advice' was sufficient (or even necessary) in effecting policy change. Rather, it is to recognize that the reception and ultimate influence of the Commission's ideas had something to do with the qualities of the advice itself, as well as its acceptability and perceived utility at particular moments in time.

Overall, it is clear that the Commission's relations with policy-makers can be encapsulated by neither a linear–rational model of policy advice nor its 'strategic' counterpoint, in which expertise is deployed primarily for political ends. While elements of both conceptualizations can be found among the diverse circumstances of influence examined in this chapter, neither in itself provides a remotely adequate account. Instead, what we see in the complex evolution of environmental policy, and across all modes of influence, is an entanglement of cognitive and political factors. The Commission itself emerges, for the most part, as an authoritative body, yet not as a disinterested font of wisdom, or a political pawn, or a passive recipient of circumstance. Rather, it was a proactive advisor: it 'did good by stealth'; it framed problems and policy proposals in compelling ways; and it nurtured, promoted, and maintained ownership of its ideas long after their initial conception. In many ways, it acted like a 'boundary organization' (Guston 2001; Miller 2001), and its reports might be regarded as boundary objects. It is difficult, for example, to look at the genesis, conduct, and impacts of the Commission's work on nuclear power, lead in petrol, or energy and climate change (RCEP 1976b, 1983, 2000) without invoking the idiom of co-production (Jasanoff 2006c). Such conditioning of the policy environment is as much discursive as 'rational' and the Commission acted as an effective, and often a conscious, agent of change in this respect.

Many of these findings might usefully be extended to other forms of expert advice and to comparable democratic societies. The co-existence of different modes of influence, the significance of both the qualities of advice and external circumstances, the insufficiency of explanations based on linear–rational or strategic accounts—all of these reflect the complexities and contingencies of relations among knowledge, advice, and policy, and accord well with the findings of other significant studies in this field. In particular, this in-depth, longitudinal investigation of a particular advisory body provides reinforcement for theories of policy change that afford significance to cognitive variables, as well as for those in science and technology studies (STS) pointing to the mutual constitution of science and politics. But on the issue of co-production, we can also derive important insights from the very distinctiveness of the Commission as an advisory body. Its long-standing ability to exert influence had much to do with its perceived independence, the degree of trust it commanded in environmental policy communities, and its reputation for authoritative analysis. How it acquired, developed, and retained these attributes, and how they in turn interacted with the knowledge–policy nexus, are important subjects for the next, and final, chapter.

7

Giving Good Advice?

> Most people are expert in a few things; everyone is a layman in regard to most things.
>
> Kenneth Wheare, *Government by Committee* (1955: 15)

> [T]hey [the Royal Commission] had the standing and the status that carried weight in the scene.
>
> Campaigner on freedom of information, interview (1997)[1]

> All bodies have to justify themselves, in the world we live in now.
>
> Senior civil servant, interview (1996)[2]

Introduction

The work of the Royal Commission on Environmental Pollution, and associated developments in policies, practices, and institutions, have been analysed in depth in previous chapters. There is much in this analysis to suggest that the Commission offered 'good advice', in terms of the qualities identified in Chapter 1: it was widely regarded as an authoritative, independent body; its counsel was grounded in a breadth and depth of expertise; its reasoning was closely and explicitly argued in publicly available reports; and its findings were frequently (though not invariably) influential in environmental policy domains. While the intricacies of the systems involved make the ultimate effects of the Commission's work difficult to isolate, the balance of evidence suggests that it contributed in positive ways to the protection and enhancement of the environment. Other close observers have reached broadly similar conclusions. In a thirty-year retrospective on environmental policy, the respected monthly publication *ENDS* described the Commission as 'the intellectual powerhouse responsible for a string of influential reports', and included it among the 'key organisations' that had shifted the debate and shaped the policy agenda

(ENDS 2008: 28). For Tom Burke—environmentalist, commentator, and advisor to government and industry—it was 'an extraordinary body', whose reports had been 'a seminal influence on environmental policy well beyond the borders of the UK' (Burke 2008: 6; see also Defra 2007b; DETR 2000a; Williams and Weale 1996; Wolf and Stanley 2003).

The evidence suggests, further, that certain distinctive characteristics of the Commission contributed substantially to its standing and its capacity to have effect.[3] One objective of this final chapter is to tease out these attributes and examine them in the light of wider questions about expert advisors and their role in the political sphere. It is as well to be clear at the outset, however, that the attributes most frequently mentioned—authority and autonomy—cannot be seen as independent of one another, nor are they simply given. As Hilgartner (2000: 5) reminds us, authority is something that an advisory body 'must assert, cultivate and guard', and it might do so in part by constructing its own identity as an autonomous and objective source of knowledge. But if a body is to be credible, there are two further, related, requirements: first, its critical characteristics, such as authority and autonomy, need to be conferred upon it by others as well as believed in by itself; and second, there must be an affirmation of these attributes in real material outcomes over time. Two features of the Commission helped significantly in these respects—its positioning within epistemic and policy networks, and its survival over an extended period of dynamic policy formation. All of these attributes, and the structures and practices that helped to generate them, will be considered in the sections that follow. But this chapter has two further, important tasks. One is to address the question of why it became possible, after forty-one years, to dissolve the Commission with surprisingly little fuss. The other is to summarize the findings of the study and consider any wider inferences that might legitimately be drawn. To start, however, we must go back to the beginning, when the newly established Commission was first making its influence felt.

Constructing Authority

In the early 1970s, the constitution of an environmental advisory body as a standing royal commission would in itself have signalled a capacity to deliver informed and considered advice: royal commissions were appointed to be authoritative, even when governments created them with strategic intent. Nevertheless, for this putative authority to be sustained over time it had to be emphasized and reinforced in various ways. One of the most obvious was to appoint chairs and members of high standing, as indicated by professional achievements and other established forms of recognition—though it had to be credible, too, that appointments had genuinely been made on that basis, and not with some ulterior purpose in mind. By and large (and surprisingly, perhaps, given the opacity of the procedures involved) such legitimacy seems to have been achieved: outside observers, as well as those directly involved in the process, acknowledged that members were appointed for sound reason—and this sense of an authoritative membership relates closely to the perception of the Commission as a trustworthy, independent body, considered in more detail below.

As time went on, the Commission's authority was reinforced by the direct and visible impacts of many of its reports and recommendations: in effect, and most notably in the first two decades, authority and influence became mutually constitutive in establishing the Commission's reputation. Even when governments chose not to act on its findings, they sustained an aura of authority by treating the Commission with respect, though such deference becomes less apparent in the later years.[4] Undoubtedly, governments and other actors also constructed authority through a form of 'boundary work' (Gieryn 1983, 1995): as we have seen, the Commission was typically represented as a 'scientific' body in spite of its actual, heterogeneous, composition, and even at the end of its life it was the loss of 'independent scientific advice' that influential commentators lamented.[5] The Commission itself was also an active agent of boundary work, not only in cultivating its own authority and autonomy but also (in many of its reports) in reinforcing the view that facts and values should be separated in the processes of analysis and advice.[6] Intriguingly, however, its authority derived as much in practice from the breadth of its membership as from any tendency to categorize it as a scientific committee; so much so, in fact, that its composition can be seen as a critical, positive attribute in its own right.

Interdisciplinary Deliberation

It was the Commission's structure as a 'committee of experts' that most obviously distinguished its approach from that of a more conventional 'expert committee'. With such a breadth of membership, each study in effect involved a combination of specialist expertise, diverse disciplinary lenses, and intelligent lay perspectives. Individuals saw themselves as contributing in all of these capacities depending on context, so that they served simultaneously as experts and lay members, rather than being placed ex ante into one category or the other. Kenneth Wheare (1955: 15), cited at the beginning of this chapter, observed a similar effect in his earlier analysis of commissions and committees of enquiry: 'a committee of experts may appear, from one point of view, to be a committee of laymen, for though each member is an expert in his own field, he may be a layman in regard to the field of the men sitting next to him'.

Such versatility, combined with an atmosphere in which members, for the most part, 'respect[ed] each other's judgements' and 'accept[ed] each other's expertise',[7] encouraged an open, reflective, and reflexive style of proceeding, which might best be characterized as one of 'interdisciplinary deliberation' (Owens 2012: 16). The specialists could have their say—and would carry weight in their specific fields—but other members could always ask 'the idiot question'[8] and challenge disciplinary presuppositions. As one member explained, 'you were put on the Commission because of a particular bit of expertise, but once you were on there, nearly everybody felt free to talk about anything'.[9] It is worth adding that mutual respect was aided by the frequency, duration, and intimacy of routine meetings, which enabled members (and Secretariat) to discuss matters informally, and to get to know each other individually, as colleagues, and often as friends.

The Commission's style of deliberation—engaging different perspectives and worldviews over extended periods—had a number of important effects. One, as the following comments from members testify, was to foster a special robustness in the Commission as an argumentative yet cohesive body:

> [A] mix of individuals...makes for a degree of good and common sense that is rare amongst a scientific committee (putting it, in these hallowed walls, rather crudely)...There is a strength in the Commission in that it is not an expert committee and one hopes that scientific evidence is treated in a fashion that it would have to be treated [in] by the most intelligent of lay people as well as by professional scientists.[10]

> I have never been on a commission where such a breadth of experience can be brought together, and where there is a will to produce a coherent report in the end.... There have been people fighting for various things, but I've never known a time [when] someone wouldn't sign the report.... I like it because the thing is so solid I can...be a bit of a maverick there, I don't feel I've got to keep them all together—the body does that in its own right...it's unique.[11]

Another important consequence—lending support to Sabatier's (1987: 683) assertion that 'brokers can learn even while advocacy coalitions continue to talk past each other'—was that members were 'educated' by their colleagues and became 'more knowledgeable' as their deliberations advanced.[12] This was Flowers' experience, for example, when he learnt to see the nuclear industry through the eyes of others (Chapter 4), and it was an aspect of the Commission that particularly struck a special advisor during the study of genetically modified organisms (GMOs); members, he observed, were 'shifting their ground, as they learnt more about what was going on and began to understand better the limits of knowledge'.[13] Although there were dissenting voices amongst interviewees (see Chapter 3), and one member attributed the consensus normally achieved by the Commission to an agreement to differ,[14] more spoke in terms of 'persuasion', 'conversion', or 'genuine discussion and conviction', all of which took place in the 'evolutionary process' of producing reports.[15]

A third, and crucial, effect of the Commission's particular style of deliberation was to encourage what Rein and Schön (1991: 286) have called 'frame-reflection': this was noticeable, for example, during the study of lead in the environment, when the breadth of the membership ensured that different perspectives on the burden of proof were considered, and in the energy study, when the Commission managed to frame a widely debated topic in a novel and compelling way. More generally, the rigorous testing and challenging of ideas enabled the members not only to 'bottom out' their enquiries through interrogation of the issues at hand but also, in prominent cases, to reframe the questions, open up novel policy approaches, and stimulate learning in wider policy communities. The last effect, however, was dependent on another important factor: the trust that was vested in the Commission as a function of its perceived autonomy.

The 'Independence Thing'

Towards the end of the twentieth century, it became less likely that expertise per se, however broadly based, would engender automatic trust or legitimacy. Yet the Royal Commission continued to be seen as a trustworthy body, even as these trends were gaining ground, and in this context its autonomy was at least as important as its recognized cognitive authority. Indeed, throughout its life, there was a widespread perception that the Commission was independent,[16] both in the negative sense of enjoying freedom from undue interference and in the positive sense of being at liberty to conduct business in ways of its own choosing.[17] In practice, the Commission's independence in both senses was inevitably circumscribed, so it is interesting to consider how it managed to acquire and sustain such an enduring reputation in this respect.

In interviews for this study, independence from government was widely cited as one of the Commission's most positive and empowering characteristics; and it was often associated—by policy-makers and outside observers, as well as 'insiders'—with the special status of a standing royal commission. But such status, while offering a formal independence, by no means eliminates opportunities for political meddling or bureaucratic capture; on the contrary, in the Commission's case, it could be said that multiple possibilities existed for both.[18] In spite of being 'royal', the Commission was in a very real sense created and financed by government, and ministers and civil servants were engaged with its operations in other significant ways: appointments were negotiated and approved, departments consulted about topics for investigation, and draft reports submitted for 'factual checking'. From the late 1990s onwards, the new accountability of public bodies introduced by the Blair government was inevitably in some tension with autonomy, and in the end the Commission was abolished. In what sense, then, could such a body be deemed to have been independent of government?

Provisions for consultation notwithstanding, one critical factor was certainly the Commission's extensive freedom to determine the subjects of its own investigations. Together with another feature of its constitution—its ability to outlast successive administrations and not to 'wax and wane at the convenience of Parliament and government'[19]—this freedom made it relatively difficult to control. Even so, to be seen to be independent, the Commission had to be willing to exercise its autonomy to good effect, and ministers had to be abstemious in using their powers to direct it towards particular topics. In practice both of these conditions held. Ministerial requests for studies were made on only three occasions and even then, as we have seen, the Commission tended to work to a broader remit. Otherwise, it consulted and listened but never obviously selected its subjects to suit a political agenda; indeed, on some occasions it seemed consciously to do the opposite, as in the decidedly unwelcome choices (at their respective times) of nuclear power, freshwater quality, agriculture, transport, and environmental planning.[20] Nor did the Commission's treatment of its subjects give any impression of manipulation: it was ready, when necessary, to be 'courteously but astringently critical',[21] to challenge received wisdom, and to look well beyond the short-term priorities of governments. Even when reports had been initiated by ministers, they tended to confirm Cartwright's (1975: 213) view that royal commissions were 'not an especially "safe" way' of depoliticizing contentious issues. Beyond the choice and treatment of specific topics, the Commission also resisted direction on the conduct of its studies, was vigilant about its placing within Whitehall, and valued the symbolism of having its own, non-departmental accommodation.

Without standing aloof from government, then, the Commission found ways of cultivating and asserting its independence. While its Royal Warrant conferred a formal autonomy, its credibility was generated in large part by its actual conduct and its own determination not to be 'in the pocket of the [environment] department', nor in any sense 'an arm of government...at all'.[22] What mattered was a 'scholarly' (Stone 2004: 3) or 'functional' independence,[23] which enabled the Commission to go where it wanted to go and say awkward things in spite of its resource and other dependencies. It helped, of course, that members of the Commission were not financially dependent on its existence, and that successive governments (at least until the end) were prepared to tolerate the risks of inconvenient or disruptive advice. If ministers sometimes wished the Commission away, their sporadic displeasure co-existed with a mind-set in Whitehall that found it useful to be challenged from outside. A perception of independence was crucial for this function: as one senior official in the environment department explained, 'part of the point would be lost if [the Commission] appeared to be part of "us"'.[24]

It is clearly the case, however, that political and bureaucratic interference are not the only hazards that might beset an advisory body. We must consider, too, how the Commission apparently avoided organizational or sectoral capture and how others came to be persuaded that it was not unduly influenced in such ways. Interestingly, many interviewees expressed a remarkably pure conception of the Commission's advisory role, invoking ideals of disinterest and objectivity such as one might find in rational analytical models of relations between expertise and politics. One member, himself an industrialist, was typical in seeing as 'the great benefit' of the Commission that it was 'ad hominem... rather than corporatist'—that people were not there 'to represent a particular point of view'.[25] Likewise, a former Chair saw the Commission as 'a body of disinterested experts', whose reports were the result of 'honest, objective searches for truth' (Houghton 2013: 192, 196); another insisted that members were chosen for their 'ability to listen to the evidence and to make a fair assessment and as far as possible not to have a prior axe to grind'.[26] More significantly, these dimensions of independence were noted and valued by outside observers and recipients of the Commission's advice: one special advisor found the members to be 'a very dispassionate group',[27] and a former environment minister considered the body as a whole to be 'scientific in a rather... ancient sense of... people trying to assess argumentation and not regarding themselves as representative'.[28]

Like the strong sense of independence from government, this perception of detached and dispassionate advisors raises an important and intriguing question: how could such an image be acquired and sustained, given that it must have been an idealized representation? In reality, of course, members came to the table at least with tacit commitments and 'intellectual' interests (Podger 2011: 232), if not 'hobbyhorses',[29] and whilst none was formally representative, some were there by virtue of their experience in particular sectors (Chapter 3); all, in effect, carried baggage.[30] There is evidence to suggest that sectoral interests were not always punctiliously set aside (though the line between being 'representative' and offering a legitimate sectoral perspective can be an indistinct one). A trade union leader who served during the 1980s was in no doubt that certain issues discussed by the Commission were of vital concern to his members, and that 'this [was] clearly a matter of *interest*';[31] similarly, there was a feeling among a small number of interviewees that agricultural members had been inclined to defend the interests of farmers.[32]

And there were always members with environmental sympathies, in the widest sense; indeed some within Whitehall regarded the Commission as an 'environmental' body, whose ideas might have to be countered by a broader Whitehall view. The representation of environmental concerns was seen as legitimate, since the Commission had been established for that purpose. But the sense that the environment had a strong voice among the members might explain why, in the appointments process, the green groups were not treated as a 'sector' (like industry, agriculture, and the trade unions) from which a perspective was actively sought. Alternatively, it might be that in the earlier days, environmental groups were still too marginalized to be included, while later, there was less emphasis anyway on the sectoral idea—latterly, there were no farmers or trade unionists on the Commission, for example. One member, who had himself acted in a legal capacity for environmental groups, insisted that, precisely because of the Commission's remit, it was important not to have visibly 'green' members, in order to maintain a reputation for independence and objectivity.[33] It is also interesting in this respect that the Commission had what one senior civil servant described as an 'establishment feel'; it could indeed be radical in its recommendations, but it was never 'way out'.[34]

Overall, then, it seems that perceptions of independence were shaped not so much by an absence of interests (intellectual or sectoral) but by outcomes in which no systematic dominance of particular interests could be discerned. The Commission's composition and practices, especially its robust style of deliberation, undoubtedly contributed to such outcomes by inhibiting or neutralizing any incipient special pleading.[35] But so too did a propensity among 'sectoral' members *not* to behave predictably according to their affiliations. In this way, the recommendations of the ninth report confounded the expectations of campaigners that certain members might veto a move to lead-free petrol (Chapter 4), and the sixth report rocked the nuclear industry, in spite of Flowers' close connection with one of its most prominent organizations. It is notable, too, that industrial members were among the most persistent advocates of public access to environmental information. Such non-conformity sent powerful signals about the Commission's ability to avoid capture and its capacity for independent thought.

The argument so far suggests that the Commission's reputation for delivering independent and authoritative advice was built and sustained over time through a combination of structures, practices, and visible impacts, with the latter most likely to be in evidence in propitious political circumstances. In Chapter 6, we saw that there were numerous ways in which the Commission exerted influence, beyond the most obvious mechanism of its reports: these included 'doing good by stealth', actively disseminating its messages, and adding its weight and authority to wider pressures for change. In all such activities, one of the Commission's advantages was its positioning within a variety of different networks. This has particular significance because networks can be seen both as agents of learning and as structures within which transfers of ideas and information can take place (Chapter 1).

An Intersection of Networks

The Commission was embedded in networks in two different but interconnected ways. For most of its life, and by virtue of the varied activities of its chairs, members, and Secretariat, it occupied a powerful position at the intersection of personal, professional, epistemic, and policy networks: this positioning will be considered in more detail below. But at certain times the Commission itself, as an entity, became identified with particular issue networks, or advocacy or discourse coalitions, in the world of environmental politics. So, for example, it was associated with an ecomodernist discourse coalition in the 1980s (Hajer 1995) and with a 'new realist' one (in transport) in the following decade (Chapters 4 and 6). For the most part, the existence of a wider network amplified the Commission's effectiveness; this was especially the case when its interventions acted as an 'argument stopper', or a catalyst, so that it could be credited (or closely associated) with subsequent policy change: significant examples would be its recommendations on lead-free petrol, the 'duty of care', and the 60 per cent reduction target for carbon dioxide emissions. On some occasions, however, the Commission's ideas became subsumed within a wider policy discourse, so that its specific contributions were overshadowed; this is what happened with environmental planning (Chapter 5), and it might explain, too, why a 'remarkably interesting'[36] report on the marine environment had a relatively low profile in the legislative developments that followed. In the latter case, the changes to policy were substantial, but other bodies took most of the credit and the Commission's influence, in effect, became 'atmospheric' (James 2000: 163). It is interesting, nevertheless, that when the Commission became a part of these wider movements, its role as a conduit—absorbing, interpreting, and transmitting ideas that were presented to it in evidence—could act as one of the threads helping to bind a coalition together.

One other notable way in which the Commission itself became part of a network—in this case an international policy community—merits mention here. The Commission was a founder member of the network of European Environment and Sustainable Development Advisory Councils (EEAC), initiated in 1993 to encourage official advisory bodies across the European Union to share knowledge and experience, and identify perspectives that might transcend the national (or regional) level.[37] Operating primarily through annual conferences and working groups, EEAC has been described as falling somewhere between a policy network and an epistemic community (Macrory and Niestroy 2004; see also Bulkeley 2005). For the Commission, it provided a forum for sharing ideas and information on diverse topics, arguably enhancing the quality of its own advice to government. One example of fruitful interaction was the development, in parallel, of the Commission's report on the marine environment and an EEAC position statement on European Marine Strategy (the latter prepared with significant input from members of the Commission and its Secretariat); the two documents were published almost simultaneously (EEAC 2004; RCEP 2004). Indeed, the Commission was always actively involved in EEAC's structures and activities, and at various times its members chaired the EEAC Steering Committee and the Governance Working Group. Such participation enabled the Commission not only to interact with other advisory bodies, but also to engage, indirectly, in dialogue with European policy-makers and institutions.

Being involved in various networks *as an advisory body* was, however, only part of the story. As noted above, the Commission was 'networked' through its chairs, members, and Secretariat in a variety of different ways, three of which are briefly considered here: its connectivity at a high level, through personal and professional networks, into Whitehall and government; its special relationship with the House of Lords; and its extensive reach, through the activities of a diverse membership, into a wide range of epistemic and policy communities. Although these three modes of networking are treated separately in what follows, it is important to remember that they were not mutually exclusive—and indeed they overlapped on some occasions with networks of the kind described above, in which the Commission was a participant per se.

In the earlier days of the Commission, personal and professional networks among chairs, senior civil servants, and ministers were of considerable significance (Chapter 3). Good relations at this level helped to foster the Commission's independence,[38] and provided opportunities for 'doing good by stealth'—including the 'anchoring' of ideas at the highest levels of environmental policy communities. The friendship between Brian Flowers and Tony Benn (Secretary of State for Energy) at the time of the nuclear report, for example, must have helped to promulgate the Commission's thinking,[39] as must the informal meetings in Oxford between Richard Southwood and William Waldegrave, when the latter was an environment minister in the 1980s.[40] Another early Chair recalled having 'a very good' and 'quite friendly' relationship with the Permanent Secretary from the environment department. Later, although high-level and often cordial interactions continued, they tended to be of a more formal kind, and latterly they became less frequent.[41] Throughout, however, as noted in Chapter 3, the Secretariat maintained personal and professional networks within Whitehall ('not quite fifth column, but internal channel'[42]), and was able to alert the Commission to windows of opportunity, potential obstacles, and possible policy developments.

While relations with its sponsoring department, and sometimes with other ministries, were important, one of the most potent of the Commission's network structures connected it directly into the legislature through the House of Lords. Here, for the best part of three decades, serving members of the Commission, as well as 'good old boys', were willing on many occasions robustly to represent its views.[43] So, for example, Lord Ashby and Lord Flowers (former Commission chairs) pressed in the early 1980s for the implementation of integrated pollution control,[44] while some years later, Lord Lewis (the then Chair) and Lord Nathan (a recent member) did much to align the Environmental Protection Bill with the Commission's recommendations on GMOs.[45] Members were also active within the powerful select committees of the Upper House, most notably the European Communities Committee (ECC) and the Committee on Science and Technology.[46] In the former case, the connection had a marked effect in the 1980s, when the Commission was dealing with the fundamentals of pollution control, and the ECC with a large influx of environmental legislation from Brussels (see Chapter 2). Through Lord Nathan, for example, who chaired an ECC sub-committee on dangerous discharges to water, the philosophy of the Commission's tenth report (RCEP 1984) could be brought to bear directly on the stance of the British government: the influence is reflected in the Committee's report (House of Lords Select Committee on the European Communities 1985), in the subsequent Lords debate,[47] and in a marked shift in the government's position a couple of years later (see Chapter 2). Significantly, such cross-membership meant that the Commission (in the words of one senior official) 'was not just *pushing* expert advice to government, but to a certain extent *pulling* expert advice [into] the legislature'; this brought 'a unique element to its brokering' (and, we might add, to its boundary work, discussed further below).[48]

Given the scope for cross-fertilization, it is hardly surprising that the views of the Royal Commission and those of these key select committees often converged, enhancing the prospects for progressive policy change; another instance of these bodies working in tandem, helped by overlapping membership, has been discussed in the context of hazardous waste and the 'duty of care' in Chapter 6. However, there were some occasions on which they differed: in the early 1990s, as we have seen, the House of Lords Science and Technology Committee was critical of the Commission's stance on GMOs (Chapter 4), and it fell to a former Commission member, Lord Cranbrook, to be a 'good old boy' in the debate on the Committee's report.[49] Before leaving this discussion, it should be noted, too, that the Commission enjoyed fruitful interactions with select committees of the House of Commons, though without the direct cross-membership that linked it to the House of Lords. Commission members acted as special advisors to Commons committees from time to time,[50] and there were exchanges 'backstage'—as, for example, when the Commons Environment Committee was adding its own weight to the arguments for stringent controls on waste in the 1980s.[51] And of course there were interactions with both Houses of Parliament on a formal basis, with the Commission and select committees each hearing evidence from the other, a good deal of mutual citation in reports, and occasional meetings between the Commission and committee chairs.[52] These processes, too, in a less personal way, served to strengthen networks and contributed to the framing and circulation of ideas.

In the third, and broader, sense of networking mentioned earlier, the Commission was a locus for productive exchange between the epistemic and policy networks of its members: in this it exemplified Hall's concept of advisory bodies occupying 'privileged positions at the interface between the bureaucracy and the intellectual enclaves of society' (Hall 1993: 277). Members brought knowledge from their own academic (or industrial, agricultural, or other) communities to the process of interdisciplinary deliberation, and in turn they used these communities to test the Commission's ideas and disseminate its key messages. At the very first meeting, when it was established that papers and proceedings of the Commission were to be regarded as confidential, it was made clear, nevertheless, that 'members should use their discretion in discussing [its] business with a limited number of colleagues'.[53] Much later, guidelines drawn up by the Commission for the conduct of its studies explicitly advocated such epistemic exchange: 'There is an expectation that Members will draw the Commission's attention to significant developments in their own fields.... Members may need to have either private or public discussions about matters before the Commission with experts on a particular subject' (RCEP 1998a: para. 7). Perhaps as a special case of Hall's phenomenon, whereby advisory bodies operate at an important interface, the Commission also performed the functions of a 'boundary organization' (Guston 2001), making it a site of co-production: it did, after all, exist at the frontier of science and politics, it involved actors from both sides,[54] and it helped to create 'boundary objects' (ibid.: 400), in the form of substantial reports and concepts like that of BPEO, which made sense to individuals within each of these social worlds.

Overall, then, whether the Commission acted as a node of connectivity or as a network participant in its own right (and even if it didn't take the credit for subsequent policy change), its positioning within networks helped to ensure that its arguments came to circulate, persuade, and have effect. Arguably, too, its 'privileged position' at the intersection of epistemic and policy networks provided an important means through which it became not merely a passive advisor but also, at least in part, the architect of its own impact and influence. And, as we might expect from the more nuanced of the theoretical perspectives outlined in Chapter 1, the processes involved in the interplay of knowledge and policy were at once cognitive, discursive, and social: as one well-placed official put it, reflecting on the Commission's role and the mechanisms of policy formation, 'you must understand that this is the way these things happen. It is all by evolution, it's all by talking to people'.[55]

Continuity

One feature of the Commission interacted with all of the attributes already discussed: this was the length of time for which it existed as a stable and identifiable institution. The Commission's organic evolution, with its rolling, sometimes long-serving, membership, and strong continuities of form and practice, gave it distinct advantages over bodies with more limited lives. For one thing, continuity allowed time for the mutual reinforcement of authority and visible influence that contributed so much to the Commission's reputation (and helped, in turn, to ensure its extended life). As a standing body, the Commission could sit out the vagaries of particular administrations, reassert proposals that had initially been unsuccessful, and wait for constraints to weaken as policy norms and external conditions changed. All of this was possible, in part, because the Commission's stability fostered the ownership of ideas amongst its members and helped to forge a collective identity—the latter reinforced by the 'micro-continuity' of frequent and enjoyable meetings. As one member reflected, it is the 'intellectual bonding and growth that a permanent body...acquires' that make it of 'much greater worth than the sum of its parts'.[56] One interesting effect was that the Commission developed an institutional memory and a willingness, at least on some occasions, to use it to good effect: members felt, as one with long experience described it, as if they 'were carrying forward a flame'.[57]

This is not to suggest that the Commission always followed up its reports or pressed for the take-up of its recommendations: some ideas were quietly dropped; on other occasions (as with BPEO, for example), members had to rediscover what their predecessors had been trying to achieve; and the Commission was always wary of being seen to 'lobby' the government, even if it was politically rather active.[58] But on issues that it considered to be of great importance, it could certainly be persistent in a measured way. During the 1980s, for example, both Hans Kornberg and Richard Southwood (successive chairs) pressed the government on the continuing failure to implement Part II of the *Control of Pollution Act* 1974: they were 'like terriers', according to one member from that time, 'they wouldn't let [this issue] go away'[59]—and they had reinforcements, as we have seen, in the House of Lords. The Commission was similarly tenacious in arguing (over several decades) for tighter controls on the disposal of waste, and for the introduction of integrated pollution control, an objective that it pursued for many years after the publication of its fifth report (RCEP 1976a). Such pressure could be sustained in a number of ways. Most obviously, the Commission could revisit specific issues, or refine and develop principles, in the context of later studies; or it could issue a 'chivvying' report, as it did with transport in the 1990s (RCEP 1994, 1997c). But it also employed all the other means at its disposal for pressing its particular concerns—promoting its ideas in Parliament, 'talking up' subjects within its networks, communicating through formal channels with ministers and senior civil servants,[60] and 'doing good by stealth' behind the scenes. Reflecting on situations where the government had 'dragg[ed] its feet' and the Commission had been persistent, a former Secretary observed: 'that's where this continuity of the Royal Commission is so important—because it will keep banging on at government and at least bring matters to a head'.[61] When it chose to do so, therefore, the Commission could act very much like one of Pielke's (2007: 135) 'issue advocates' or Kingdon's (2003: 179) '[p]olicy entrepreneurs... pushing their pet proposals or problems'. Its longevity and continuity were clearly very helpful in this respect.

It is important to emphasize, before leaving this section, just how much these different attributes of the Commission were interdependent. Its authority, which grew substantially over its first two decades, was reinforced by certain practices, enhanced by continuity, and closely related to its formal and perceived autonomy. Its independence, in part a function of its constitution, was fostered by the diversity of its membership, its practising of interdisciplinary deliberation, and its robustness as a standing body. Its positioning within, and at the intersection of, different networks helped to ensure that its ideas circulated, and enhanced its reputation as an authoritative and independent source of advice. And its longevity, as we have seen, enabled these other characteristics to mature and interact, contributing to its legitimacy and to its ability to have effect.

Explaining the Commission's Demise

In the end, however, none of the attributes described above, nor their mutual reinforcement, was enough to save the Commission from being abolished. The question therefore remains of how we should interpret the events following the general election of 2010. Was the Commission lost, without deep consideration, in fulfilment of manifesto commitments to cut spending and reduce the number of quangos? Had it come to be seen as a remnant of 'classical–modernist government' (Hajer and Wagenaar 2003b: 10), out of line with the more inclusive governance structures of the twenty-first century? Should we take the Secretary of State's rationalization at face value, accepting that this particular body was no longer needed, now that the environment was a mainstream issue, on which there were many other sources of 'expert, independent advice and challenge'?[62] Or had something changed in the nature of the problems that the Commission tackled, making them intrinsically less 'governable'?

It might be argued, of course, that we need look no further than the first of these possibilities—that a new government, committed to 'a bonfire of the quangos', swept the Commission into the initial tranche of those bodies that were considered expendable. The fact that the environment department (by this stage, the Department for Environment, Food and Rural Affairs) was under severe pressure to make savings adds weight to this argument: even if the Commission wasn't, constitutionally, a 'Defra body', this was the department from which its resources had to be found (to the tune, by that stage, of around £1 million per year). And, after all, there is no shortage of precedent. Many advisory committees have been terminated more or less abruptly, both in the UK and elsewhere: in Sweden and the Netherlands, for example, bodies not unlike the Commission were abolished at around the same time, and the US Office of Technology Assessment (OTA) was eliminated by Congress in 1995, 'primarily as a sacrifice to the agenda of fiscal discipline' (Guston 2001: 402–3).[63] Yet political expediency, while clearly playing a part in the Commission's ending, seems inadequate as a full explanation. For one thing, it begs an obvious question, identified in Chapter 3: why was it possible to condemn the Commission at this particular political moment, when it had emerged unscathed from existential threats on a number of previous occasions? The government elected in 2010 was not the first in the Commission's history to face financial austerity or to target the proliferation of non-departmental public bodies. It seems likely, then, that other factors were involved, and that we need to look beyond the most proximate reasons to changes in policy and political agendas and in structures of governance. Shifting expectations of expertise in general, and changes affecting the Royal Commission in particular, must also be implicated in the story.[64]

Of considerable contextual significance were developments over the Commission's lifetime in the nature of the environmental issues commanding scientific and political attention. Whereas the early Commission had focused primarily (and effectively) on the poorly regulated externalities of production, the issues occupying its twenty-first-century successors were typically less tangible, less amenable to technical solutions, and more deeply embedded in lifeworlds and economies. As Tom Burke (2008: 7–8) put it in his retrospective, the gross pollution of the 1960s and 1970s often had a technical remedy, and there were 'easily identified victims and villains'; but half a century later, the solutions to complex environmental problems seemed less obvious, and both the victims and the villains turned out to be ourselves. The 'wickedness' of such problems was explicitly recognized by the Commission in several of its later reports.[65] Of course, it would be misleading to imply that there had been a wholesale shift. The 'old' depredations didn't seem so simple at the time; they were rarely resolved without conflict, and some proved persistent or re-emerged in a different form. Nor are the complexities of the 'risk society' entirely new—as we saw in Chapter 2, more difficult, structural problems were already becoming apparent, even in those earlier decades. It is reasonable, however, to say that the issues dominating the agenda in the early twenty-first century were different in critical respects from those that were prominent when the Commission was first appointed. Almost invariably, their resolution demanded environmental policy integration and multi-level governance; and often, they were profoundly unsettling for core beliefs, such as those associated with the primacy of economic growth.

Related in part (though not exclusively) to the expansion of the environmental agenda were changes in structures of governance, which affected the Commission's working practices and challenged its capacities as an advisory body. At the beginning, its remit had been clearly circumscribed: its advice was to be directed primarily at the UK government and it focused on issues that were essentially national in scale: its 'ethos', as one member explained, was to tackle pollution problems that didn't have 'geopolitical overtones'.[66] After Britain joined the 'Common Market', the Commission directed more of its attention towards Europe—at least as a locus for environmental policy, if not directly as a target of its advice—and in the following decades, as issues and institutions changed markedly in scale and scope, it developed a wider international awareness. Looking outwards, therefore, the Commission increasingly sought to influence transnational organizations, initially through the UK government but eventually more directly and in a quasi-independent capacity. Latterly, however, it also faced a wider set of governing institutions within the UK: from around the turn of the century, it had to take account of up to four different internal policy contexts, and develop relationships with the devolved administrations in Scotland, Wales, and Northern Ireland.[67] Even at the old 'national' level, there was a more complex departmental nexus, and a shift of emphasis, rhetorically at least, towards 'localism' and smaller government.

For the Commission, the changing nature of environmental issues, the rapid growth (and contestation) of the knowledge base, the proliferation of institutions, and the existence of multiple layers of governance—all aspects of what Maarten Hajer (2003b: 176) has called 'the institutional void'—greatly complicated the tasks of information gathering, analysis, and prescription.

In combination, these trends made it difficult even to identify tractable topics—that is, subjects for which there were realistic prospects of synthesizing existing material and finding a compelling and distinctive focus for a 'Royal Commission report'.[68] It also became more challenging to draft the texts. Though members in the latter years sometimes yearned after elegant statements, it would probably have been impossible, in the twenty-first century, to produce anything with the minimalist authority of the nine-page second report (RCEP 1972b), which, in spite of its brevity, covered three separate issues and carried considerable weight.

There is a sense, too, in which the expectations of experts and advisory bodies were changing, most noticeably from the mid-1990s onwards. Although the Commission continued to enjoy an unusual degree of respect and trust, it was not immune from the pressures and adjustments associated with the quest for integrity and accountability in public life. Around the turn of the century, two changes in particular had significant implications for the Commission and its relations with government. First, as discussed in Chapter 3, the appointments process became more open, with clear intended benefits but also, in practice, some costs. Second, the Commission became more directly accountable through the instigation of Financial Management and Policy Reviews (FMPRs), a regime in which it had agreed to be included though it was not strictly a 'non-departmental public body'. As the review process tightened, the Commission was increasingly exposed to consultees who were not necessarily sympathetic to its ethos or familiar with its practices and past achievements. Those involved in the second FMPR, for example, when asked to identify the Commission's most influential work (in terms of its effects on policy or government thinking), tended to choose from reports published over the previous five to ten years; while a bias towards the present is almost inevitable in such a survey, these choices suggest a limited knowledge of the Commission's historical contributions—or perhaps a sense that they were simply no longer relevant (Defra 2007b). Specific reports attracted criticism in the review, as did the structure and ethos of the Commission itself. Some consultees even implied that it had had its day: one said that it should step off 'the "Royal Commission" pedestal', others that it was 'a bit patrician and stuffy', or even '[d]usty and anachronistic' (Defra 2007b: 34). These were in a minority—the wider view was that the Commission did valuable work and would be difficult to replace—but negative comment speaks loudly on such occasions (and one wonders how the Commission would have fared if there had been an FMPR in the aftermath of some of its earlier, more controversial offerings). It is significant, too, that some of the proposals for modernization challenged attributes and practices (such as membership structure and the slow process of interdisciplinary deliberation) which, according to the current study, were among those that had made the Commission most effective. Of course, complaints about the Commission's working practices were far from unprecedented; there was certainly scope for improvement, and scrutiny in itself can hardly be a bad thing. What is most telling, perhaps, is that the Commission seemed to have lost the resilience of its predecessors and was therefore, in its last decade, more at the mercy of critical forces.

Several factors contributed to this vulnerability. One was that the Commission had been an irritant to government—or at least to certain departments—in its studies on environmental planning, bystander exposure to pesticides, and the urban environment (RCEP 2002a, 2005, 2007). It had, of course, been inconvenient on many previous occasions; what seems to have changed was the tolerance within government towards criticism from 'outside' (in part, perhaps, for reasons unconnected to the Commission per se). Two plausible factors were, however, related to the placing of the Commission within the structures of Whitehall. First, it was no longer so securely located in the 'hard science' section of the environment department's 'mental map' (Chapter 3); a number of its later studies had a primarily institutional emphasis,[69] while the controversial report on bystander exposure to pesticides—commissioned to examine the robustness of the science—was openly seen by some critics as *un*scientific. (Certainly, in the latter case, the department preferred the advice of its expert scientific committees, and was upheld in this view by the Court of Appeal; see Chapter 4.) A second possible reason for impatience with the Commission relates to changes in the machinery of government, specifically the separation of environmental responsibilities from those relating to planning and urban areas. In its interventions on these latter issues (RCEP 2002a, 2007), the Commission might have been seen as a 'Defra body' treading on the turf of another department (Communities and Local Government and its predecessors)—an interesting possibility, given that it implies a failure on the part of Whitehall to remember (or recognize) the Commission's cross-governmental remit.[70]

Undoubtedly, one further factor in the Commission's vulnerability was that it was no longer as well networked as it once had been, with fewer personal and political connections into high levels of government, and a more formal and distant relationship with the civil service.[71] Nor did it retain its direct access to the legislature through members who were also Peers—the more significant, perhaps, if it had been reprieved in earlier incarnations for fear of causing 'uproar' in the House of Lords.[72] The wider point is that, towards the end of its life, the Commission had less capacity to promulgate its arguments and defend its views within government. Government, in turn, probably had less inclination to listen, and felt that there were plenty of other bodies to which it could turn for expert advice. Perhaps, in the twenty-first century, royal commissions in general, and this surviving, standing advisory body in particular, had simply come to be regarded with less awe.

Learning from the Commission

Despite the changes towards the end, the Royal Commission on Environmental Pollution was in many respects a remarkable body. It emerges from this detailed, longitudinal study as an authoritative and influential 'committee of experts', which enjoyed a high degree of legitimacy for most of its life. In seeking to understand the Commission, and its achievements and failures, we see that its distinctive attributes, the roles that it played, its positioning by others, and its impacts on environmental policy form a nexus of interconnecting variables in which cause and effect are not always easy to disentangle. This complexity, together with the richness of the material (covering forty-one years, thirty-three reports, and extensive policy development), makes the task of distillation in this final section particularly daunting, and individual chapters should be revisited for detailed findings. The intention here is to present key conclusions about the Commission itself, to draw wider inferences about advisory roles and knowledge–policy relations, and to consider the implications of these insights for the project of 'good advice'.

At the beginning of this chapter, it was argued that the critical attribute of authority was 'designed into' the Commission from the outset, then accentuated in the years that followed through a combination of practices and outcomes. The Commission did much to cultivate its own authority through its conduct and reports, and was helped in this endeavour by the other characteristics discussed—its formal and functional autonomy, the standing and diversity of its membership, its reach into epistemic and policy communities, and its continuities of form and practice. Governments bolstered its authority too, not only through the formal mechanisms of appointments, official responses, and Parliamentary debates, but also—and especially in the first two decades—by affording it a certain dignity and respect as a Royal Commission. For all of these reasons, the Commission enjoyed a largely positive reputation in environmental policy communities, and was widely admired for its critical, but effective, interventions.

The reality was less than perfect, of course, and in identifying 'what worked', it is important not to set the Commission on too high a pedestal. It was not at all times or by everyone regarded in a positive light, and some of its reports met with disapproval, or even sharp critique, from a variety of different directions. There were mutterings, periodically, about its utility and productivity, and by the end of its life it had lost something of the 'numinous legitimacy'[73] that surrounded it in its earlier years. We must recognize, too, that over a lengthy period of activity, successes are more likely to be remembered, while failures tend to fade from view: there is always, therefore, the risk of a rosy glow. The object of the study, however, was to contribute to an understanding of advisory practices and of the interconnections between knowledge and policy. To that end, it has been particularly important to see what worked, or didn't work, and why and when—in effect, addressing Radaelli's (1995) question of 'how and why knowledge matters' in the policy process, while being sensitive to the ways in which knowledge, advice, and authority are themselves constructed. The Commission's standing and reputation, as well as its ultimate demise, have been examined as important components of the overall picture, with interesting implications for the future of policy advice.

This is an appropriate point at which to return to the conceptual framework set out in Chapter 1. Drawing on a diverse and extensive literature, this framework offered four models—or representations—of the relationship between knowledge and policy, within each of which expert advisors have been characterized in particular ways. To simplify greatly, in a linear–rational model, advisors feature as detached and dispassionate analysts, informing the policy process; in a 'strategic' one, they are seen largely as instruments of legitimation; in 'cognitive' perspectives, they become agents of policy learning; and within 'the idiom of co-production' (Jasanoff 2006c: 1), scientific bodies, at least, play a significant part in the (constructive) hybridization of science and politics.

An important question for this study was whether the Commission, over its extended period of operation, could be aligned with some or any of these abstracted roles; the answer, at first sight, seems to be that there were elements, in its advisory practices, of all four. It was admired, as we have seen, for its rigorous and impartial analysis, which often came to be reflected in policy. It clearly had a symbolic role too, when governments (and others) used it for strategic and political ends. Throughout, it was undoubtedly a facilitator of learning, in that it constructed, organized, and brokered knowledge, and dispersed and anchored its ideas within environmental policy communities. And in its conduct and reports, it performed 'boundary work' of several kinds, and contributed, in positive ways, to the co-production of science and politics. But while there is ample evidence of this diversity of activities and functions, it would be inadequate to conclude that all four representations of expert advisors (and the wider perspectives to which they relate) are equally validated in the Commission's case—or, conversely, that none is especially useful. Instead, we need to look carefully at the interdependencies of these roles as they were played out in practice, and to this end, a useful point of entry lies in the Commission's functioning as a cognitive and discursive agent of change.

The Commission was very much the kind of body that Heclo (1974: 313), in his seminal work on public policy formation, thought to be 'consciously designed to aid in political learning'. It performed this task as a broker (sometimes a source) of knowledge and ideas, variously combining the functions of analyst, synthesist, conduit, and catalyst—roles that must have a component of 'rational analysis' if they are to be successful in any degree. Crucially, however, the rationalities deployed by the Commission were wide-ranging, and its trademark style of interdisciplinary deliberation came closer to a form of 'practical public reasoning' (Weale 2010: 266) than to any technically oriented appraisal of 'the facts'.[74] The Commission was indeed an agent of learning, of both the simpler and more complex varieties outlined in Chapter 1. In many studies (including those focusing on specific aspects of pollution), it sifted the evidence and identified better means for achieving established ends; but it also contributed in fundamental ways to the changing of interpretive frames, as we see in its interrogation of problems and potential solutions, and of the norms and principles on which environmental policies should be based. Bringing these different aspects of the Commission into focus, we see a body whose role had both cognitive and discursive dimensions, embedded within which were key components of rational analysis. But we see also, in the diversity of its deliberations and their effects, that rationalities can take different forms, and that expert advice rarely feeds into policy in a detached and de-contextualized way.

It aided legitimacy, nevertheless—and in an important sense *enabled* a strategic and symbolic role—for something of the rarefied notion of the rational analyst to be maintained. This was one of the functions of the boundary work in which the Commission was implicated, emphasizing its status as an impartial provider of sound science: we have seen, for example, how the Commission defended its scientific credentials when challenged (the reports on GMOs and crop spraying are cases in point), and how it was positioned by others, for much of its lifetime, as a 'scientific body'.[75] This is especially interesting given that a royal commission, in its very conception, differs from the scientific committees whose practices have often been analysed from a co-productionist perspective. But, in fact, this particular Commission—with its environmental remit and its distinguished scientific *and* non-scientific profile—can be seen as a hybrid, deriving authority from an emphasis on its robust science but also, simultaneously, from its breadth of perspective, which was widely recognized and valued.[76] And this hybridity gains in significance if boundary work is itself seen as multi-faceted, involved in bridging boundaries as well as defending them, and sitting within the wider repertoire of material and discursive strategies that expert advisors deploy (Bijker, Bal, and Hendriks 2009; Dammann and Gee 2011; de Wit 2004; Guston 2001; in 't Veld and de Wit 2009; Jasanoff 1990; Miller 2001; Shackley and Wynne 1996).[77] In this context, we see how the Commission reached, through its members, practices, networks, and reports, into 'the two relatively different social worlds of politics and science' (Guston 2001: 401), and how it worked to produce 'serviceable truths' (Jasanoff 1990: 250)—in its proposals on lead in petrol, for example—which could accommodate the requirements of each. In choosing its topics, 'doing good by stealth', interpreting evidence, framing its arguments, and pursuing its favourite proposals, the Commission was part of a process through which science and politics were co-produced—even as they were upheld as separate spheres—in contested areas of environmental risk and regulation.

To take the argument further, it is useful to revisit the 'circumstances of influence', as set out in Chapter 6. While the Commission can be regarded in broad terms as a successful advisory body, a fine-grained analysis reveals a more complex picture, in which the reception of its reports and recommendations varied substantially in different contexts and over time. Looking closely at the fate of its advice, we find 'direct hits' and dormant seeds, invisible levers, diffuse and long-term effects, and sometimes an absence of any discernible impact at all; this variance, it was suggested in Chapter 6, might best be conceived of as a continuum, or spectrum, of influence. What emerges most clearly, however, is the *contingency* of influence, with outcomes at any one time depending on political and ideological commitments, institutional structures, economic conditions, and external (even 'quirkish'[78]) factors and events. The Commission could have rapid effect when its recommendations 'made sense'—for whatever reason—in a receptive political climate, and tended to fail when proposals went too much against the grain; but powerful ideas could gain currency as external conditions moved, and influence could sometimes be diffused or obscured within broader trajectories of change.

There is clear evidence, then, that circumstances condition the reception of advice, much as we would expect if knowledge interacts with other variables in the process of policy formation. Yet to portray the Commission's reports simply as 'landing' in a given environment would be to present too static a picture. For one thing, the quality of the advice itself *mattered*. As noted in Chapter 6, some recommendations had little effect because they didn't, in fact, 'make sense' (in terms of practicality, for example), whilst others, as products of rigorous interdisciplinary deliberation, turned out to be more persuasive. But there is another, critical reason to reject too static an account: namely that the Commission, far from being a neutral entity hovering outside the political sphere, was in reality an active agent in the environment in which its advice was received. In assessing its influence, we find that the practices through which it negotiated, articulated, and propagated its advice also mattered a great deal—hence the power of skilfully framed reports, ideas disseminated and anchored within policy networks, and persistent advocacy in cases of initial failure. Boundary work was often interwoven with these practices, upholding the Commission's authority and helping to ensure that its advice would 'make sense'; indeed, many of the attributes and practices that made the Commission such an effective agent of learning were also those that would characterize it as a boundary organization.

In summary, it can be said that the Commission was never simply a detached conveyor of analysis and information, nor a cipher whose advice was epiphenomenal; neither of these familiar representations of advisory bodies could satisfactorily account for its actual practices or for the range of outcomes observed in this study. The evidence points, instead, to a combination of brokering and 'educative' functions, as envisaged in cognitive models of the policy process, with practices such as boundary work, associated with the idiom of co-production. These theoretical perspectives don't simply collapse into one another in this study: their focal points are different, and each offers distinctive insights on the Commission's authority, advisory work, and influence. But they clearly reinforce one another in any convincing account of the Commission's role, and are needed in tandem when it engaged—as it so often did—with the complex and contested issues of regulatory environmental science.

Turning this argument around, we can say that this analysis of the attributes, practices, and impacts of the Royal Commission supports and enriches both cognitive and co-productionist perspectives on knowledge–policy relations, and clearly demonstrates their affinity. Knowledge is affirmed in this study as an important, quasi-autonomous variable in the formation of public policy—but not in the sense that it stands untouched by politics or remains intact through the processes of articulation, transmission, and absorption. Relations between knowledge and policy are shown to be complex and contingent—affected by many other factors, played out in different venues and over varying timescales, and always shaped by social processes such as boundary work, framing, and policy learning. There is structure within this complexity—we see regularities, for example, in the 'circumstances of influence', and in the ways in which advisors can condition the receiving environment—but it would be hazardous to look for deterministic relationships of the kind that might allow for prediction. Instead, this study points to a need for tempered and realistic expectations of expert advice, and of the related project of 'evidence-based policy'. In diverse contexts, it demonstrates the significance of evidence, argument, ideas, and skilful counsel, but it shows with equal clarity that we should not be too impatient to see their effects, nor expect trump cards to settle enduring controversies, nor resort to an easy cynicism in the absence of 'direct hits'. Moreover, the study provides further, compelling evidence that effective advice depends not only on maintaining something distinctive about science but also on constructive boundary work and the production of 'serviceable truths' (Jasanoff 1990: 250); the corollary is that too blind an insistence on the separation of 'science' and 'politics' is unhelpful, and impedes an understanding of productive advisory practices.[79]

Lessons for the Future?

Looking to the future, the findings of this study lead to some further observations on the attributes of advisory bodies and the constituents—and contradictions—of 'good advice'. The positive features of the Commission have already been discussed at some length, but it is worth reiterating here that authority, autonomy, and connectivity emerge as powerful assets, as do the capacities for frame reflection and sustained, interdisciplinary deliberation. It was a strength, in the Commission's case, that it was a hybrid institution—not, in fact, a 'scientific body' but one that embodied scientific expertise; this status enhanced its deliberative capacities whilst contributing to its authority and credibility. In a broader sense, the vital qualities of legitimacy and trust are shown, in this study, to depend on the interplay of other important attributes over time; but they are a product, too, of a certain fearlessness on the part of advisors, and a propensity, at least on some occasions, to be disruptive. A body's ethos and robustness are important in this respect, and depend in turn on its having the time (and the privacy) in which to rehearse its arguments and develop a collective identity.

There is much to learn from what worked well in the Commission's case, though prescription, as always, must be approached with caution. Times and expectations have changed since the Commission was established in 1970, and in any case, too reductionist an account of 'good advice' would miss intangible elements such as the chemistry of a particular body and the charisma of the individuals who serve (in other words, assembling desirable attributes may not in itself be a guarantee of success). It must be remembered, too, that governments need many different kinds of expertise, and while certain attributes are widely expected in advisors, the value of others must depend on purpose and context. Nevertheless, it is interesting to think what made the Commission distinctive—in the UK at least—in the sense of not being readily replaceable by existing advisory institutions. Here we might look to the particular combination of three features: wide-ranging expertise, a long and continuous existence, and a remit to inform and challenge government from the 'outside' (Kennet 1970: 12). Other sources of (broadly comparable) advice typically have some but not all of these characteristics: expert scientific bodies tend to be narrower in focus and composition; ad hoc committees, such as those convened by the UK national academies, usually report and disperse; government and departmental scientific advisors have more of an 'insider' status; and select committees perform a specific, Parliamentary role.[80] It might be argued, then, that within the panoply of advisory institutions existing in the twenty-first century, there is still a place for a standing, independent, interdisciplinary committee of experts with a generous remit in a given policy domain; and this structure might be of particular value when the domain has a scientific dimension. But anyone seeking to reinvent the Commission in the twenty-first century—if not as a royal commission, then as a body combining its essential characteristics—would have to negotiate the changing expectations of expertise and advice, as well as the dangers of reductionism, noted above. In addition, the difficulties experienced by the Commission in its later years point to challenging questions about what, exactly, such a body should do and how it might be rendered both autonomous and accountable. These interesting dilemmas are worthy of some final reflections.

One issue demanding close attention would be that of remit, both in the sense of the policy domain in which an advisory body would operate, and that of the institutions to which its reports and recommendations would be directed. On the first of these, the scope of any given domain can be crucial: as we have seen, 'environmental pollution' was a well-chosen remit for a standing advisory body in the 1970s, but arguably became too complex and unwieldy in the end. It may be that, in an age of rapidly expanding information and shifting governmental structures, defining the boundaries of a suitable domain has genuinely become more difficult, though one might equally argue that there will always be fields—existing or emergent—in which advice from a 'Commission-like' body would be of value. Beyond the choice of domain, there is still the question of what *kinds* of issues a body should be given licence to explore: the dilemma here is that advice needs to be meaningful in the world as it is, but it can also serve a vital function when it challenges the policy frame. One of the strengths of the Commission was that it managed, on many occasions, to perform both of these roles: indeed, its substantial achievements derived from its ability to provide sound advice on specific problems whilst saving the 'distillate' (Flowers 1995: 9) to develop ideas of wider significance—a feature that depended, of course, on its continuity. It should be noted, too, that even its more specific studies tended not in themselves to be narrow, but rather to involve 'a narrow problem dealt with broadly'.[81] All of this suggests that a body seeking to emulate the Commission should have a field of enquiry broad enough to allow freedom of investigation, but not so broad as to be overwhelming; and that in a dynamic policy landscape, the remit might need to be kept under review. It would be crucial for such a body to be able to move seamlessly between the specific and the generic; and, in order to develop this capacity (as well as other important attributes), it would need some stability and continuity of existence—one might say, perhaps, that it should have a lifespan of at least ten years.

The second aspect of remit mentioned above—the question of to whom or what a body would offer its expert advice—would necessarily relate in part to the choice of policy domain. But if the field is a relatively broad one, transcending the standard divisions of government, there is much to be said for the kind of cross-departmental remit that the Commission's Royal Warrant allowed—and for institutional arrangements to ensure that such a remit is not 'forgotten'. Increasingly, however, the positioning of an advisory body within the machinery of government at a given level is not the only important consideration. This study suggests that in a complex field in which powers are largely devolved, it would be challenging for any British advisory body to engage fully with three devolved administrations as well as the UK government; the appropriateness of a UK-wide remit would therefore need careful thought, alongside the choice of policy domain[82] (and it is notable in this context that both the Scottish and Welsh Governments have appointed their own Chief Scientific Advisors). On the other hand, where a policy domain has cross-boundary or global significance, a twenty-first-century advisory body must necessarily be international in its outlook. The Commission adjusted to this requirement by making overtures to international organizations (particularly the European Commission and Parliament), while taking care not to go 'over the head' of the UK government. Significantly, for the second half of its life it also participated in a transnational network of European advisory bodies, which had its own connections with structures of European governance. This last point suggests that choosing the most appropriate 'level' for an advisory body, or finding ways for it to relate effectively to different legislatures and administrations, might miss an important and fundamental question: how should we think in future about the locus and practice of expert advice if, as some scholars argue, our familiar, nested hierarchies of government are likely to give way to hybrid, networked, and multi-level governance formations?

There are no simple answers to these questions about remit; they are identified here to highlight the challenges of institutional design if the characteristics of 'good advice' are to be preserved. Something similar might be said about the critical, yet ambiguous and potentially fragile, attribute of independence—one of the most widely cited correlates of legitimate and trustworthy advice. Efforts to guard independence have tended to focus on the interests and commitments of advisors, and on the avoidance of bureaucratic or other forms of capture. In fact, like the *Gezondheidsraad* in the Netherlands (Bijker, Bal, and Hendriks 2009), the Commission negotiated these issues rather well, and enjoyed a high degree of trust, deriving in large part from its actual conduct. The more challenging issue is that autonomy is inevitably in some tension with resource-dependency and modern standards of accountability. Publicly funded bodies are rightly subject to scrutiny, and governments must decide whether they remain efficient, effective, and useful; naturally, over time, needs will change and individual advisory bodies will come and go. There are potential pitfalls, nevertheless, especially for a 'free-ranging' body. One is that specific notions of accountability (and cost-effectiveness) may change the nature of advisory practices in ways that are not, in fact, productive. The Commission itself increasingly came under pressure to be more open and inclusive, more communicative, less academic, less focused on environmental aspects of sustainability, and more nimble in producing its reports. Of course, not all criticism or proposals for change were misdirected, but it is troubling that the favourite targets for modernization involved features of the Commission that had contributed substantially to its achievements. As one member wistfully observed, the Commission had many unfashionable characteristics, but 'these odd things seem to work'.[83] Perhaps the lesson for any future body is that there is a need for critical scrutiny without excessive conformity, recognizing that, as Albert Weale has argued, 'the conditions for good deliberation might be quite distinct from notions of representation and rigid accountability'.[84]

At worst, of course, troublesome advisors are vulnerable to 'Thomas Becket syndrome';[85] the worry is that they might be silenced—not for good reason, but because they become inconvenient or embarrassing to government, on which their existence must ultimately depend. This matters, because good, critical counsel can be disruptive, and indeed the best and most far-reaching advice is more than likely to be unwelcome, especially in the shorter term. In the light of this study, however, we can say that the risks posed by periodic scrutiny, resource-dependency, and vulnerability to abolition should not be over-stated. The Commission, after all, had a long and productive life, during which its ideas and advice were not always palatable to those in power but could ultimately, nevertheless, be influential. Its legacy reminds us that in the project of 'good advice', the need is to ensure that the tension between autonomy and accountability is a creative one, within a political culture that welcomes an outside stimulus and can be tolerant of occasional disruption.

APPENDIX 1
Reports of the Royal Commission and Government Responses

Note: bibliographic details for reports mentioned in the text and responses (except letters) may be found in the list of references.

Report	Date	Title and Content	Government Response
First Cmnd 4585	February 1971	*First Report* Assessed state of the environment and pollution priorities; reflected on Commission's remit.	Pollution Paper no. 4 (DoE [Department of the Environment] 1975), consolidated response to first four reports.
Second Cmnd 4894	March 1972	*Three Issues in Industrial Pollution* Identified access to information, new chemicals in products, and toxic waste as key issues.	See above: Pollution Paper no. 4
Third Cmnd 5054	September 1972	*Pollution in Some British Estuaries and Coastal Waters* Called for urgent action to protect waters from gross pollution. Provided much of the groundwork for *Control of Pollution Act* 1974.	See above: Pollution Paper no. 4
Fourth Cmnd 5780	December 1974	*Pollution control: Progress and Problems* Overview of pollution, including review of regulatory structures and manpower and training requirements.	See above: Pollution Paper no. 4
Fifth Cmnd 6371	January 1976	*Air Pollution Control: an Integrated Approach* Requested by Ministers. Introduced concepts of integrated pollution control and best practicable environmental option (BPEO); recommended unified pollution inspectorate.	Pollution paper no. 18 (DoE 1982)
Sixth Cmnd 6618	September 1976	*Nuclear Power and the Environment* Overview of issues; proposed environmental safeguards and urged more coherent waste management policy; best known for the 'Flowers Criterion'.	Cm 6820 (UK Government 1977)

(*continued*)

Knowledge, Policy, and Expertise

Continued

Report	Date	Title and Content	Government Response
Seventh Cmnd 7644	September 1979	*Agriculture and Pollution* Reviewed pollution arising from agricultural practices and effects of pollution on agriculture. Argued that intensive agriculture is an industry and should not have special treatment.	Pollution paper no. 21 (DoE 1983a)
Eighth Cmnd 8358	October 1981	*Oil Pollution of the Sea* Requested by Ministers. Proposed ways of preventing and combating pollution from routine discharges and accidental spills.	Pollution paper no. 20 (DoE 1983c)
Ninth Cmnd 8852	April 1983	*Lead in the Environment* Argued that safety margins in relation to human health were insufficient and that exposure to lead from all sources should be reduced. Best known for recommendation that lead should be phased out of petrol.	Pollution paper no. 19 (DoE 1983b)
Tenth Cmnd 9149	February 1984	*Tackling Pollution—Experience and Prospects* Wide-ranging review of priorities, also commenting on European environmental policy. Championed BPEO and public access to environmental information. Recommended ban on straw burning and more research on acid rain.	Pollution paper no. 22 (DoE 1984)
Eleventh Cmnd 9675	December 1985	*Managing Waste: The Duty of Care* Most significant recommendation was for imposition of a 'duty of care' in waste management chain.	Pollution paper no. 24 (DoE 1986b)
Twelfth Cm 310	February 1988	*Best Practicable Environmental Option* Sought to clarify and develop concept of BPEO: contribution to European Year of the Environment.	DoE (1992)
Thirteenth Cm 720	July 1989	*The Release of Genetically Engineered Organisms to the Environment* First attempt in UK to look comprehensively at the risks. Proposed framework for regulation: risk assessment on case by case basis and licensing of releases.	DoE (1993a)
Fourteenth Cm 1557	June 1991	*GENHAZ: A System for the Critical Appraisal of Proposals to Release Genetically Modified Organisms into the Environment* Developed ideas in the thirteenth report; sought to adapt a procedure for hazard identification in the chemical industry to releases of genetically engineered organisms.	DoE (1993b)
Fifteenth Cm 1631	September 1991	*Emissions from Heavy Duty Diesel Vehicles* Considered diesel emissions, especially in urban areas, and scope for further reductions. Introduced as 'a new type of report', following Commission's intention to undertake more focused studies alongside its major ones.	DoT (Department of Transport) (1992)
Sixteenth Cm 1966	June 1992	*Freshwater Quality* Reviewed quality of surface and groundwaters, and sources of pollution. Advocated precautionary approach and introduction of pollution charging scheme.	DoE (1995b)

Reports of the Royal Commission

Seventeenth Cm 2181	May 1993	*Incineration of Waste* Called for national waste management strategy; recommended landfill levy; argued that incineration with energy recovery was the BPEO for residual wastes, after reduction and recycling.	DoE (1994)
Eighteenth Cm 2674	October 1994	*Transport and the Environment* Wide-ranging, well-publicized report calling for switch to more sustainable transport policies. Recommended fuel price increases, land use–transport integration, and end to 'predict and provide'. Set a number of targets.	No direct response, but some issues addressed in Green Paper, Cm 3234 (UK Government 1996)
Nineteenth Cm 3165	February 1996	*Sustainable Use of Soil* Comprehensive review of what the Commission considered to be a neglected topic. Called for national soil strategy and for soil pollution to be treated on a par with that of water and air.	DoE (1997)
Twentieth Cm 3762	September 1997	*Transport and the Environment—Developments since 1994* Reviewed developments since the Commission's 1994 transport report and called for more urgent action.	Cm 4066 (UK Government 1998)
Twenty-first Cm 4053	October 1998	*Setting Environmental Standards* Argued for a more robust basis for setting environmental standards, with the process being informed but not pre-empted by scientific and risk assessments. Its more radical point was that 'public values' must be taken into account from the earliest stages of the process, including problem definition.	Cm 4794 (UK Government 2000)
Twenty-second Cm 4749	June 2000	*Energy—The Changing Climate* Best known for recommending policy commitment to 60 per cent reduction in UK CO_2 emissions by 2050. Explored options for achieving this through four scenarios.	Cm 5766 (UK Government 2003b)
Twenty-third Cm 5459	March 2002	*Environmental Planning* Argued for a new system of Integrated Spatial Strategies, to bring together land use planning and environmental regulation, within a context informed by environmental limits.	Overview: Cm 5888 (UK Government 2003c); also Cm 5887 (England) (UK Government 2003a), Scottish Executive (2003), Welsh Assembly Government (2003)
Special report	November 2002	*The Environmental Effects of Civil Aircraft in Flight* Expressed particular concern about climate change impacts of aviation; recommended new 'climate change charge' on air travel and limits on airport expansion.	No direct response but issues addressed in White Paper, Cm 6046 (DfT [Department for Transport] 2003)
Twenty-fourth Cm 5827	June 2003	*Chemicals in Products— Safeguarding the Environment and Human Health* Argued that assessment of existing chemicals was slow and insufficiently precautionary, even under the REACH regulation; proposed a new approach.	Cm 6300 (Defra [Department for Environment, Food and Rural Affairs] 2004b)
Special report	May 2004	*Biomass as a Renewable Energy Source* Recommended encouragement of biomass as a stable, low-carbon energy option for the UK and was critical of a lack of policy co-ordination.	Defra (2004c)

(continued)

Continued

Report	Date	Title and Content	Government Response
Twenty-fifth Cm 6392	December 2004	*Turning the Tide—Addressing the Impact of Fisheries on the Marine Environment* Recommended radical changes to fisheries policy to prevent over-exploitation and protect the environment. Proposed creation of marine reserves covering 30 per cent of seas around UK, and similar measures for all European seas.	Cm 6845 (Defra 2006c) Scottish Executive (2006)
Special report	September 2005	*Crop Spraying and the Health of Residents and Bystanders* Response to ministerial request. Concluded that link between ill-health and passive exposure to pesticides was plausible; called for precautionary measures pending further research, including 5-metre no-spray buffer zones.	Defra (2006b)
Twenty-sixth Cm 7009	March 2007	*The Urban Environment* Emphasized complexity of urban environments and sought to understand drivers and constraints. Focused on health and well-being, 'the natural urban environment', and the built environment. Recommended explicit policy for the urban environment and an environmental contract between central and local government.	Defra and Department of the Environment Northern Ireland (2008)
Twenty-seventh Cm 7468	November 2008	*Novel Materials in the Environment: the Case of Nanotechnology* Assessed implications of novel materials, focusing on nanomaterials. Considered functionalities, pathways, health and environmental impacts, and governance arrangements. Called for more research, greater transparency, and flexible and resilient forms of adaptive management.	Cm 7620 (UK Government 2009)
Special Report	2009	*Artificial Light in the Environment* Followed from twenty-sixth report, which identified light as an important issue. Called for better management of artificial light in national parks, preservation of 'dark skies' in urban areas, explicit consideration of light in planning policy.	Defra (2010b) for England. Letters to Chair from environment Ministers in Northern Ireland (15 January 2010), Scotland (25 March 2010), and Wales (9 February 2010)
Twenty-eighth Cm 7843	March 2010	*Adapting Institutions to Climate Change* Recommended that public and private decision-making processes should routinely include an 'adaptation test' to help reduce risk of damage through climate change. Also called for 'adaptation duty' to be imposed on public bodies.	Brief letter to Chair from Parliamentary Under-Secretary, Defra (23 December 2010)
Twenty-ninth Cm 8001	February 2011	*Demographic Change and the Environment* The Commission's final report. Argued that environmental implications of demographic change merited greater government attention, and that population size mattered less than patterns of consumption.	No response available at time of writing (August 2014)

APPENDIX 2

Location of Environmental Functions in UK Government, 1970–2011

1970	Department of the Environment (DoE) created; includes environmental protection, local government, planning, and transport. However, pesticides control and pollution from agriculture and fisheries remain the responsibility of the Ministry of Agriculture, Fisheries and Food (MAFF).
1976	Transport moved out of DoE to form a separate Department of Transport (DoT).
1997	Creation of Department of Environment, Transport and the Regions (DETR), reuniting transport and environment.
2001 (June)	Creation of Department for Environment, Food and Rural Affairs (Defra), combining the former MAFF with the environmental protection functions of the former DETR. Planning, local government, and transport now located within the newly created Department for Transport, Local Government and the Regions (DTLR).
2002 (May)	Office of the Deputy Prime Minister (ODPM) created to take responsibilities for planning, local government, and the regions. Transport moves once more to a separate department—the Department for Transport (DfT). Environmental protection stays with Defra.
2006	ODPM re-branded as Department for Communities and Local Government (DCLG or simply CLG). Environmental protection stays with Defra.
2008	Some climate-related functions moved from Defra to the newly created Department of Energy and Climate Change (DECC).

APPENDIX 3
Interviewees (and Other Respondents) by Category

Note
Interviewees in each category occupied the positions shown at various times—so, for example, they included both former and serving members of the Commission, the civil service, and so on. A number of individuals migrated between categories during the period covered by the study and several occupied more than one position at the same time. For the purposes of enumeration, all interviewees have been allocated to a single category (representing the main capacity in which they were interviewed). The breakdown below should therefore be seen as indicative rather than definitive.

As noted in Chapter 1, the majority of the formal interviews took place in the period 1995–7; a small number took place in the later 1990s and 2000s.

Category	Interviewees
The Royal Commission	
Chairs	5
Members	37
Secretaries	4
Other members of the Secretariat	5
Special Advisors	2
Sub-total	53
Government, Civil Service, etc.	
Ministers with environmental responsibilities	2
Civil servants, environment department	
Under Secretary	1
Deputy Secretaries	3
Chief Scientists	2
Other civil servants (environment)	10
Chief Medical Officer, Department of Health	1
Civil servants, Department of Trade and Industry	1
Sub-total	20
European Commission	
Directors General, DG XI (Environment)	1
Other officials, DG XI (Environment)	5
Sub-total	6

(*continued*)

Category	Interviewees
Continued	
Others	
Academics	3
Chairs of Radioactive Waste Management Advisory Committee	1
Confederation of British Industry officials	1
Journalists	1
Local government officials	1
NGO officials	
Environmental	4
Other	1
Regulatory agency officials	3
Sub-total	15
Total number of interviewees	94
Total number of interviews (including nine second interviews)	103

Other Sources

Three respondents—a former Commission member and former Ministers with environmental and financial responsibilities, respectively—chose to put their views in writing instead of being interviewed (these are not included in the table above). A number of interviewees sent supplementary material such as papers, lectures, or correspondence, and six wrote quite extensively about their recollections, as well as being interviewed. The study was further informed by later, semi-formal discussions and exchanges of correspondence, some with previous interviewees (including two Chairs and a Secretary of the Commission, and two Chief Scientists from the environment department), and some with individuals not previously interviewed who had connections to, or an interest in, the Commission; the latter included a Chair, Secretary, and member from the 2000s; an academic observer; and a senior civil servant from the environment department. As with the interviewees, some of these individuals fitted more than one category, but all have been counted only once. Informal meetings and conversations, over the whole period of the study, also contributed to an understanding of the Commission and its work and influence.

APPENDIX 4
United Kingdom and European Legislation Mentioned in the Text

Note
This appendix is not intended as a comprehensive list of environmental legislation in the UK or the EU. Rather, its role is to provide further details of British and European legislation mentioned in the text, including Acts of Parliament in the UK and European Acts, Directives, Treaties, and Regulations. The information below is relevant to the context in which the legislation is mentioned. So, for example, if the text refers to a European Directive of a given year, the list below does not necessarily include later updates or amendments.

UK Acts of Parliament
The text in each case below is the introductory text describing the purpose of the original Act.
Source:<http://www.legislation.gov.uk/>

Deposit of Poisonous Waste Act 1972
1972 Chapter 21, 30 March, London: HMSO
   'An Act to penalise the depositing on land of poisonous, noxious or polluting waste so as to give rise to an environmental hazard, and to make offenders liable for any resultant damage; to require the giving of notices in connection with the removal and deposit of waste; and for connected purposes.'

Control of Pollution Act (CoPA) 1974
1974 Chapter 40, 31 July, London: HMSO
   'An Act to make further provision with respect to waste disposal, water pollution, noise, atmospheric pollution and public health; and for purposes connected with the matters aforesaid.'

Endangered Species (Import and Export) Act 1976
1976 Chapter 72, 22 November, London: HMSO
   'An Act to restrict the importation and exportation of certain animals, plants and items and to restrict certain transactions in respect of them or their derivatives; to confer on the Secretary of State power to restrict by order the places at which live

animals may be imported; to restrict the movement after importation of certain live animals; and for connected purposes.'

Food and Environment Protection Act 1985
1985 Chapter 48, 16 July, London: HMSO

'An Act to authorise the making in an emergency of orders specifying activities which are to be prohibited as a precaution against the consumption of food rendered unsuitable for human consumption in consequence of an escape of substances; to replace the Dumping at Sea Act 1974 with fresh provision for controlling the deposit of substances and articles in the sea; to make provision for the control of the deposit of substances and articles under the sea-bed; to regulate pesticides and substances, preparations and organisms prepared or used for the control of pests or for protection against pests; and for connected purposes.'

Environmental Protection Act 1990
1990 Chapter 43, 1 November, London: HMSO

'An Act to make provision for the improved control of pollution arising from certain industrial and other processes; to re-enact the provisions of the Control of Pollution Act 1974 relating to waste on land with modifications as respects the functions of the regulatory and other authorities concerned in the collection and disposal of waste and to make further provision in relation to such waste; to restate the law defining statutory nuisances and improve the summary procedures for dealing with them, to provide for the termination of the existing controls over offensive trades or businesses and to provide for the extension of the Clean Air Acts to prescribed gases; to amend the law relating to litter and make further provision imposing or conferring powers to impose duties to keep public places clear of litter and clean; to make provision conferring powers in relation to trolleys abandoned on land in the open air; to amend the Radioactive Substances Act 1960; to make provision for the control of genetically modified organisms; to make provision for the abolition of the Nature Conservancy Council and for the creation of councils to replace it and discharge the functions of that Council and, as respects Wales, of the Countryside Commission; to make further provision for the control of the importation, exportation, use, supply or storage of prescribed substances and articles and the importation or exportation of prescribed descriptions of waste; to confer powers to obtain information about potentially hazardous substances; to amend the law relating to the control of hazardous substances on, over or under land; to amend section 107(6) of the Water Act 1989 and sections 31(7) (a), 31A(2)(c)(i) and 32(7)(a) of the Control of Pollution Act 1974; to amend the provisions of the Food and Environment Protection Act 1985 as regards the dumping of waste at sea; to make further provision as respects the prevention of oil pollution from ships; to make provision for and in connection with the identification and control of dogs; to confer powers to control the burning of crop residues; to make provision in relation to financial or other assistance for purposes connected with the environment; to make provision as respects superannuation of employees of the Groundwork Foundation and for remunerating the chairman of the Inland Waterways Amenity Advisory Council; and for purposes connected with those purposes.'

United Kingdom and European Legislation

Environment Act 1995
1995 Chapter 25, 19 July, London: HMSO
'An Act to provide for the establishment of a body corporate to be known as the Environment Agency and a body corporate to be known as the Scottish Environment Protection Agency; to provide for the transfer of functions, property, rights and liabilities to those bodies and for the conferring of other functions on them; to make provision with respect to contaminated land and abandoned mines; to make further provision in relation to National Parks; to make further provision for the control of pollution, the conservation of natural resources and the conservation or enhancement of the environment; to make provision for imposing obligations on certain persons in respect of certain products or materials; to make provision in relation to fisheries; to make provision for certain enactments to bind the Crown; to make provision with respect to the application of certain enactments in relation to the Isles of Scilly; and for connected purposes.'

Freedom of Information Act 2000
2000 Chapter 36, 30 November, London: The Stationery Office Ltd
'An Act to make provision for the disclosure of information held by public authorities or by persons providing services for them and to amend the Data Protection Act 1998 and the Public Records Act 1958; and for connected purposes.'

Planning and Compulsory Purchase Act 2004
2004 Chapter 5, 13 May, London: The Stationery Office Ltd
'An Act to make provision relating to spatial development and town and country planning; and the compulsory acquisition of land.'

Climate Change Act 2008
2008 Chapter 27, 26 November, London: The Stationery Office Ltd
'An Act to set a target for the year 2050 for the reduction of targeted greenhouse gas emissions; to provide for a system of carbon budgeting; to establish a Committee on Climate Change; to confer powers to establish trading schemes for the purpose of limiting greenhouse gas emissions or encouraging activities that reduce such emissions or remove greenhouse gas from the atmosphere; to make provision about adaptation to climate change; to confer powers to make schemes for providing financial incentives to produce less domestic waste and to recycle more of what is produced; to make provision about the collection of household waste; to confer powers to make provision about charging for single use carrier bags; to amend the provisions of the Energy Act 2004 about renewable transport fuel obligations; to make provision about carbon emissions reduction targets; to make other provision about climate change; and for connected purposes.'

Planning Act 2008
2008 Chapter 29, 26 November, London: The Stationery Office Ltd
'An Act to establish the Infrastructure Planning Commission and make provision about its functions; to make provision about, and about matters ancillary to, the

authorisation of projects for the development of nationally significant infrastructure; to make provision about town and country planning; to make provision about the imposition of a Community Infrastructure Levy; and for connected purposes.'

Local Democracy, Economic Development and Construction Act 2009
2009 Chapter 20, 12 November, London: The Stationery Office Ltd
'An Act to make provision for the purposes of promoting public involvement in relation to local authorities and other public authorities; to make provision about bodies representing the interests of tenants; to make provision about local freedoms and honorary titles; to make provision about the procedures of local authorities, their powers relating to insurance and the audit of entities connected with them; to establish the Local Government Boundary Commission for England and to make provision relating to local government boundary and electoral change; to make provision about local and regional development; to amend the law relating to construction contracts; and for connected purposes.'

Marine and Coastal Access Act 2009
2009 Chapter 23, 12 November, London: The Stationery Office Ltd
'An Act to make provision in relation to marine functions and activities; to make provision about migratory and freshwater fish; to make provision for and in connection with the establishment of an English coastal walking route and of rights of access to land near the English coast; to enable the making of Assembly Measures in relation to Welsh coastal routes for recreational journeys and rights of access to land near the Welsh coast; to make further provision in relation to Natural England and the Countryside Council for Wales; to make provision in relation to works which are detrimental to navigation; to amend the Harbours Act 1964; and for connected purposes.'

Localism Act 2011
2011 Chapter 20, 15 November, London: TSO
'An Act to make provision about the functions and procedures of local and certain other authorities; to make provision about the functions of the Commission for Local Administration in England; to enable the recovery of financial sanctions imposed by the Court of Justice of the European Union on the United Kingdom from local and public authorities; to make provision about local government finance; to make provision about town and country planning, the Community Infrastructure Levy and the authorisation of nationally significant infrastructure projects; to make provision about social and other housing; to make provision about regeneration in London; and for connected purposes.'

European Legislation Mentioned in the Text

Source: European Union website (<http://europa.eu/index_en.htm>), especially EUR-Lex (<http://eur-lex.europa.eu/homepage.html>).

Acts and Treaties

Treaties of Rome 1957
Established the European Economic Community and the European Atomic Energy Community, signed 25 March 1957 and entered into force 1 January 1958.

Single European Act (SEA) 1986
First major amendment of the Treaties of Rome, signed in Luxembourg on 17 February 1986 by the nine Member States and on 28 February 1986 by Denmark, Italy and Greece. The purpose of the SEA was to add momentum to European integration and completion of the internal market; it amended institutional rules and extended the Community's powers, particularly in relation to research and development, the environment, and common foreign policy. The SEA came into force on 1 July 1987.

Maastricht Treaty 1992
The purpose of this Treaty was to prepare for monetary union. It introduced elements of a political union (for example, in the areas of citizenship and foreign and internal affairs policies) and gave the European Parliament more of a voice in decision-making. Signed 7 February 1992; entered into force 1 November 1993.

Treaty of Amsterdam 1997
The purpose of this Treaty was to reform the EU institutions in preparation for the arrival of future member states; signed 2 October 1997 and entered into force 1 May 1999.

Consolidated version of EU Treaties
A consolidated version of the European Treaties, as amended by the Treaty of Lisbon, is available at <https://www.gov.uk/government/uploads/system/uploads/attachment_data/file/228848/7310.pdf> (accessed 17 June 2014).

Directives

Bathing Water Directive 1975
76/160/EEC
Council Directive of 8 December 1975 concerning the quality of bathing water.

Dangerous Discharges to Water Directive 1976
76/464/EEC
Council Directive of 4 May 1976 on pollution caused by certain dangerous substances discharged into the aquatic environment of the Community.

Drinking Water Directive 1980
80/778/EEC
Council Directive of 15 July 1980 relating to the quality of water intended for human consumption.

Environmental Impact Assessment Directive 1985
85/337/EEC
Council Directive of 27 June 1985 on the assessment of the effects of certain public and private projects on the environment. Subsequently amended.

Large Combustion Plant Directive 1988
88/609/EEC
Council Directive of 24 November 1988 on the limitation of emissions of certain pollutants into the air from large combustion plants.

Directive on the Release of GMOs to the Environment 1990
90/220/EEC
Council Directive of 23 April 1990 on the deliberate release of genetically modified organisms.

Freedom of Environmental Information Directive 1990
90/313/EEC
Council Directive of 7 June 1990 on the freedom of access to information on the environment.

IPPC Directive 1996
96/61/EC
Council Directive of 24 September 1996 concerning integrated pollution prevention and control.

Air Quality Framework Directive 1996
96/62/EC
Council Directive of 27 September 1996 on ambient air quality assessment and management. Four 'Daughter Directives' followed, defining target values for pollutants listed in the Framework Directive (except mercury, for which only monitoring requirements were specified): 1999/30/EC (sulphur dioxide, nitrogen dioxide, and oxides of nitrogen, particulate matter, and lead); 2000/69/EC (benzene and carbon monoxide); 2002/3/EC (ozone); and 2004/107/EC (arsenic, cadmium, mercury, nickel, and polycyclic aromatic hydrocarbons).

Landfill Directive 1999
1999/31/EC
Council Directive of 26 April 1999 on the landfill of waste.

End-of-Life Vehicles Directive 2000
2000/53/EC
Directive of the European Parliament and of the Council of 18 September 2000 on end-of life vehicles—Commission Statements.

Water Framework Directive 2000
2000/60/EC
Directive of the European Parliament and of the Council of 23 October 2000 establishing a framework for Community action in the field of water policy.

Strategic Environmental Assessment Directive 2001
2001/42/EC
Directive of the European Parliament and of the Council of 27 June 2001 on the assessment of the effects of certain plans and programmes on the environment.

Waste Electronic and Electrical Equipment Directive 2003
2002/96/EC
Directive of the European Parliament and of the Council of 27 January 2003 on waste electrical and electronic equipment (WEEE).

Directive on Public Access to Environmental Information 2003
2003/4/EC
Directive of the European Parliament and of the Council of 28 January 2003 on public access to environmental information.

IPPC Directive 2008
2008/1/EC
Directive of the European Parliament and of the Council of 15 January 2008 concerning integrated pollution prevention and control.

Marine Strategy Framework Directive 2008
2008/56/EC
Directive of the European Parliament and of the Council of 17 June 2008 establishing a framework for community action in the field of marine environmental policy.

Industrial Emissions Directive 2010
2010/75/EU
Directive of the European Parliament and of the Council of 24 November 2010 on industrial emissions (integrated pollution prevention and control).

Regulations

REACH
Regulation (EC) no. 1907/2006 of the European Parliament and of the Council, 18 December 2006.

The Regulation on Registration, Evaluation, Authorisation and Restriction of Chemicals (REACH). Its purpose was to streamline and improve the former legislative framework on chemicals of the European Union (EU). It also established a European Chemicals Agency. REACH entered into force on 1 June 2007.

APPENDIX 5
Chairs of the Royal Commission and Membership, 1971, 1991, and 2011

Introduction

This Appendix lists all Chairs of the Commission, with their periods of service, and the membership as it was at the beginning (1971), the mid-point (1991), and the end (2011) of the Commission's life. Information about members is drawn from reports published in the relevant years. Details of the Commission's membership for the whole of the period 1971–2011 can be found in its reports, which are available at: <http://webarchive.nationalarchives.gov.uk/20110322143804/http:/www.rcep.org.uk/reports/index.htm> (accessed 18 August 2014).

A list of members with their credentials—honours, titles, and roles—first appeared in the Commission's fourth report (1974), with the practice continuing thereafter. Prior to this, names appeared only as 'signatories' (without embellishment) at the end of the main text, preceded by the words: 'All of which we humbly submit for Your Majesty's gracious consideration' (this practice also continued). Bijker, Bal, and Hendriks (2009), in their study of the *Gezondheidsraad* (the Health Council of the Netherlands), suggest that the use of credentials is an important signifier of the authority of advisory bodies. It is interesting, therefore, that no such detail was considered necessary in the Commission's first three reports. It is clear, too, that the level of detail increased substantially over time—latterly, perhaps, because it was thought necessary to include all affiliations that might be considered to constitute 'interests'.

Chairs of the Royal Commission 1970–2011

Titles are given as in the year of taking up the Chair and the main disciplinary affiliation is shown; further details can be found in the membership lists below and in the Commission's reports.

Sir Eric (later Lord) Ashby FRS   1970–3
Botanist
Sir Brian (later Lord) Flowers FRS   1973–6
Physicist
Professor (later Sir) Hans Kornberg FRS   1976–81
Biochemist
Professor (later Sir) Richard Southwood FRS   1981–5
Zoologist
Sir Jack (later Lord) Lewis FRS   1986–92
Chemist
Sir John Houghton FRS   1992–8
Atmospheric Physicist
Sir Tom Blundell FRS   1998–2005
Biochemist
Sir John Lawton FRS   2005–11
Ecologist

The Original Commission, as in 1971

(*First Report*, RCEP 1971)

See Chapter 3, note 12, for affiliations of the early members.

Eric Ashby (Chair)
Launcelot Norvic
S. Zuckerman
John Winnifrith
Frank Fraser Darling
Neil Iliff
A. Buxton
Wilfred Beckerman
Vero Wynne-Edwards

The Commission as in 1991

(Fourteenth Report, *GENHAZ*, RCEP 1991)

THE RT HON. THE LORD LEWIS OF NEWNHAM KT MA MSC PHD DSC ScD CCHEM FRSC FRS (CHAIR)
Professor of Inorganic Chemistry, University of Cambridge
Warden of Robinson College, Cambridge

PROFESSOR HENRY CHARNOCK MSC DIC FRS
Emeritus Professor of Physical Oceanography, University of Southampton
Chair, Meteorological Research Sub-Committee, Meteorological Committee

Chairs of the Royal Commission and Membership

PROFESSOR DAME BARBARA E. CLAYTON DBE MD PhD HonDSc (EDIN) FRCP FRCPE FRCPATH
Honorary Research Professor in Metabolism, University of Southampton
Past-President, Royal College of Pathologists
Chair, MRC Committee on Toxic Hazards in the Environment and the Workplace
Deputy Chair, Department of Health Committee on Toxicity of Chemicals in Food, Consumer Products and the Environment
Chair, Standing Committee on Postgraduate Medical Education
Honorary Member, British Paediatric Association

THE RT HON. THE EARL OF CRANBROOK MA PhD DSc DL FLS FIBIOL
Partner, family farming business in Suffolk
Chair, Nature Conservancy Council for England
Chair, Institute for European Environmental Policy (London)
Non-executive Director, Anglian Water plc
Member, Broads Authority and Harwich Haven Authority
Vice-President, National Society for Clean Air and Environmental Protection

MR H. R. FELL FRAgS NDA MRAC
Managing Director, H. R. Fell and Sons Ltd.
Council Member, Royal Agricultural Society of England
Member, Minister of Agriculture's Advisory Council on Agriculture and Horticulture, 1972–81
Commissioner, Meat and Livestock Commission, 1969–78
Past-Chair, Tenant Farmers Association

MR P. R. A. JACQUES CBE BSc
Head, TUC Social Insurance and Industrial Welfare Department
Secretary, TUC Social Insurance and Industrial Welfare Committee
Secretary, TUC Health Services Committee
Secretary, TUC Pensioners Committee
TUC Representative, Health and Safety Commission
TUC Representative, Social Security Advisory Committee

PROFESSOR J. H. LAWTON BSc PhD FRS
Director, Natural Environment Research Council Interdisciplinary Research Centre for Population Biology, Imperial College, Silwood Park
Professor of Community Ecology, Imperial College of Science, Technology and Medicine
Member, British Ecological Society
Member, American Society of Naturalists
Council Member, Royal Society for the Protection of Birds

MR J. J. R. POPE OBE MA FRSA
Deputy Chairman and Managing Director, Eldridge Pope & Co plc (Brewers and Wine Merchants)

Chair, The Winterbourne Hospital plc
Deputy President, Food and Drinks Federation
Member, Top Salary Review Body

Mr D. A. D. Reeve CBE BSc FEng FICE FIWEM
Deputy Chair and Chief Executive, Severn Trent Water Authority, 1983–5
Past-President, Institute of Water Pollution Control
Past-President, Institution of Civil Engineers
Member, Advisory Council on Research and Development, Department of Energy

Emma Rothschild MA
Senior Research Fellow, King's College, Cambridge
Research Fellow, Sloane School of Management, Massachusetts Institute of Technology
Associate Professor of Science, Technology and Society, MIT, 1978–88
Member, OECD Group of Experts on Science and Technology in the New Socio-Economic Context, 1976–80
Board Member, Stockholm Environment Institute

Mr W. N. Scott OBE BSc FICheme FInstPet FInstD
Director, Shell International, 1977–85
Non-Executive Director, Anglo and Overseas Investment Trust
Non-Executive Director, Shell Pension Trust
Consultant, UK and Japanese companies
Past-Chair, CONCAWE

Professor Z. A. Silberston CBE MA
Senior Research Fellow, Management School, Imperial College of Science, Technology and Medicine
Professor Emeritus of Economics, University of London
Secretary-General, Royal Economic Society
Member, Restrictive Practices Court
President, Confederation of European Economic Associations

Dr C. W. Suckling CBE PhD DSc DUniv CChem FRSC Senior Fellow RCA FRS
General Manager for Research and Technologies, Imperial Chemical Industries, 1977–82
Consultant in science, technology and innovation
Member, Electricity Supply Research Council
Treasurer, Council of the Royal College of Art
Honorary Visiting Professor, University of Stirling

Chairs of the Royal Commission and Membership

The Final Commission, as in 2011

(Twenty-ninth Report, *Demographic Change and the Environment*, RCEP 2011)[1]

PROFESSOR SIR JOHN LAWTON CBE FRS (CHAIR)
President, Council of the British Ecological Society, 2005–7
Chief Executive, Natural Environment Research Council, 1999–2005
Director [founder], Natural Environment Research Council Centre for Population Biology at Imperial College, Silwood Park, 1989–99
Vice-President, Royal Society for the Protection of Birds, 1999–
Fellow, WWF-UK, 2008–
Chairman, Yorkshire Wildlife Trust, 2009–
Foreign Associate, US National Academy of Sciences, 2008–
Foreign Honorary Member, American Academy of Arts and Sciences, 2008–

PROFESSOR JON AYRES
Professor of Environmental and Respiratory Medicine, University of Birmingham, 2008–
Honorary Professor of Environmental Medicine, University of Warwick, 2004–
Honorary Professor of Environmental Medicine, University of Aberdeen, 2008–
Honorary Consultant Physician, UHB NHS Trust, 2008–
Chair, Department of Health's Committee on Medical Effects of Air Pollutants [COMEAP], 2001–
Chair, Defra's Advisory Committee on Pesticides, 2006–
Member, Defra's Expert Panel on Air Quality Standards [EPAQS], 1996–2009
Chair, European Respiratory Society's Environment and Health Committee, 2010–
Chair, UK Indoor Environments Group [UKIEG], 2008–
Member, UK Asthma Task Force, 1991–2002
Member, UK Respiratory Research Collaborative, 2006–

PROFESSOR MICHAEL H. DEPLEDGE
Chair of Environment and Human Health, Peninsula Medical School, Universities of Exeter and Plymouth, 2006–
Keeley Visiting Fellow, Wadham College, Oxford, 2006–7
Honorary Visiting Professor, Department of Zoology, University of Oxford, 2006–
Senior Science Advisor, Plymouth Marine Laboratory, 2005–7
Chief Scientific Advisor, Environment Agency of England and Wales, 2002–6
Vice-Chair, Science Advisory Committee, European Commission, DG-Research, 2006–

PROFESSOR MARIA LEE
Professor of Law, University College London, 2007–
Member, London Sustainable Development Commission, 2007–10
Member, academic panel, barristers' chambers Francis Taylor Buildings, 2005–

PROFESSOR PETER LISS CBE FRS
Professor of Environmental Sciences, University of East Anglia, 1985–
Chair, Scientific Committee, International Geosphere–Biosphere Programme [IGBP], 1993–7
Chair, International Scientific Steering Committee for Surface Ocean–Lower Atmosphere Study, 2002–7
Council Member, Natural Environment Research Council, 1990–5
Chair, Royal Society Global Environmental Research Committee, 2007–
Council Member, Marine Biological Association UK
Guest Professor, Ocean University of Qingdao, China
Chair, European Research Council Advanced Grants Panel in Earth System Science, 2008–

PROFESSOR GORDON MACKERRON
Director, Science and Technology Policy Research (SPRU), University of Sussex, 2008–
Director, Sussex Energy Group, SPRU, University of Sussex, 2005–8
Chair, Committee on Radioactive Waste Management, 2003–7
Deputy leader, UK Government's Energy Review team, PIU, Cabinet Office, June–December 2001
Visiting Professor, Imperial College, London

PROFESSOR PETER MATTHEWS OBE
Board Member, Port of London Authority, 2006–
Chair, Northern Ireland Authority for Utility Regulations, 2007–
Chair, Northern Ireland Authority for Energy Regulations, 2006–7
Board Member, Environment Agency; Chair, Audit Committee, 2000–6
Deputy Managing Director, Anglian Water International, 1997–8
Chair, Society for the Environment, 2005
Visiting Professor, Anglia Ruskin University, 2007–
Fleet Warden, Worshipful Company of Water Conservators

PROFESSOR JUDITH PETTS
Dean, Social and Human Sciences, University of Southampton, 2010–
Chair, Environmental Risk Management, University of Southampton, 2010–
Pro-Vice-Chancellor (Research and Knowledge Transfer), University of Birmingham, 2008–10
Head, School of Geography, Earth and Environmental Sciences, University of Birmingham, 2002–7
Chair, Environmental Risk Management, University of Birmingham, 1999–2010
Member, Engineering and Physical Sciences Research Council Societal Issues Panel, 2007–
Member, Environmental Advisory Board, Veolia Environmental, 1999–
Member, Environmental Advisory Group, Onyx Environmental plc, 1999–

DR MICHAEL ROBERTS CBE
Ministerial appointee, Veterinary Residues Committee, 2008–
Non-Executive Director, National Non-Food Crops Centre [NNFCC], 2008–
Chair, Partnership Executive Committee, Scottish Government Rural and Environment Research and Analysis Directorate, 2008–
Chief Executive, Department for Environment, Food and Rural Affairs Central Science Laboratory, 2001–8
Director, Centre for Ecology and Hydrology, 1999–2001
Director, Natural Environment Research Council's Institute of Terrestrial Ecology, 1989–99

PROFESSOR JOANNE SCOTT
Professor of European Law, and Vice-Dean for International Links, Faculty of Laws, University College London
Member, UCL/Lancet Commission on Managing the Health Effects of Climate Change, 2008–9
Member, editorial boards, *Journal of Environmental Law*; *Journal of International Economic Law*
Visiting Professor, Columbia Law School, 2002–3; spring 2004, 2005, 2007
Visiting Professor, Harvard Law School, 2005–6

PROFESSOR MARIAN SCOTT OBE FRSE
Professor of Environmental Statistics, University of Glasgow
Chair, Royal Statistical Society, Environmental Statistics Section, 2005–7
Member, Natural Environment Statistics Advisory Committee, SG, 2007–
European Chair, International Environmetrics Society, 2009–
Member, Scottish Science Advisory Council, 2010–
Member, Particles Retrievals Advisory Group [Dounreay], 2009–

PROFESSOR LYNDA WARREN
Emeritus Professor of Environmental Law, Aberystwyth University
Deputy Chair, Joint Nature Conservation Committee
Chair, Wildlife Trust of South and West Wales
Chair, Wales Coastal and Maritime Partnership
Member, Committee on Radioactive Waste Management
Board Member, British Geological Survey
Board Member, Environment Agency, 2000–6
Chair, Salmon and Freshwater Fisheries Review, 1998–2000
Member, Radioactive Waste Management Advisory Committee, 1994–2003
Visiting Professor, Birmingham Central University

Endnotes

Chapter 1

1. Macmillan was speaking in a debate (16 August 1950) about the European Coal and Steel Community (Council of Europe Consultative Assembly 1950: 231; see also Beloff 1963: 59; Hennessy 1989: 158–9).
2. HC Deb 26 October 1989, vol. 158, c1044. Citations of this kind refer to *Hansard*, the official report of proceedings in the UK Parliament; HC indicates the House of Commons, HL the House of Lords.
3. Precise numbers are difficult to ascertain because of changes in the advisory landscape and nomenclature. To take the British example, there were around 400 'non-departmental public bodies' (NDPBs) with advisory roles in 2009 (Cabinet Office 2009), including some seventy-five 'science advisory councils' (House of Commons Innovation, Universities, Science and Skills Committee 2009; for an earlier picture, see House of Commons Public Administration Select Committee 2001). In 2010, a 'bonfire of the quangos (quasi-autonomous non-governmental organizations)' reduced the number of advisory NDPBs by around a fifth. However, many of those culled were reconstituted as departmental expert committees, which perhaps explains why a similar number of 'scientific advisory committees' were in existence several years later (House of Lords Select Committee on Science and Technology 2012; see also Cabinet Office 2012; Government Office for Science 2013). These included scientific expert committees, some NDPBs, science advisory councils for certain government departments, and the cross-cutting Council for Science and Technology. Overall, this represents a substantial input of advice from bodies operating 'to a greater or lesser extent at arm's length from Ministers' (Cabinet Office 2009: 5)—and the official statistics take no account of external organizations such as independent think tanks, or of the many academics and other experts who act as advisors in an individual capacity. There are internal advisors too, including a Government Chief Scientific Advisor (an office instituted in 1964) as well as Chief Scientific Advisors (mostly academic scientists, appointed for a fixed period) in all government departments (for a discussion of this institution, see Doubleday and Wilsdon 2013).
4. Cases include particular bodies, selected reports, and national arrangements for policy advice. Useful examples can be found in edited collections such as Barker and Peters (1993); Bulmer (1980b); Lentsch and Weingart (2011); Maasen and Weingart (2005); Peters and Barker (1993a). See also special issues: *American Behavioural Scientist* 1983, vol. 26, no. 5; *Science and Public Policy* 1995, vol. 22, no. 3;

Endnotes

Science, Technology, and Human Values 2001, vol. 26, no. 4. Several authors review earlier work on expertise and advisory bodies: for example, Jasanoff (1990) and Nelkin (1975) discuss North American studies and Bulmer (1983a) considers contributions in an Anglo-American context.

5. See, for example, Barker (1993); Boehmer-Christiansen (1995); Clark and Majone (1985); Collins and Evans (2007); Everest (1990); Jasanoff (1990); Macrory and Niestroy (2004); Nichols (1972); Pielke (2007); Renn (1995); Wheare (1955).
6. Examples of the former include Cartwright (1975); Chapman (1973b); Everest (1990); Herbert (1961); Rhodes (1975); Wheare (1955). Martin Bulmer (1980a, b, c, 1983a, b, 1993) has written extensively about the functioning of commissions and committees of enquiry, and in particular has analysed their use of social research. Others have approached the question of scientific advice primarily from the perspective of science and technology studies (STS—see note 29, this chapter). Examples include: Bal, Bijker, and Hendriks (2004); Bijker, Bal, and Hendriks (2009); Hendriks, Bal, and Bijker (2004); Hilgartner (2000); Jasanoff (1990). See also note 14 (this chapter).
7. For discussion of think tanks in a range of national contexts, see, for example, Fischer (1993); Gaffney (1991); Stone (1996); and the essays in Stone and Denham (2004).
8. 'Science' is interpreted in some writings on knowledge, advice, and policy in a broad and inclusive sense, as a systematized form of knowledge (see, for example, Renn 1995; Schmant 1984). More often it refers to the natural and sometimes the social sciences, though in common usage it is usually synonymous with the former; in either case, it can be seen as 'a body of knowledge', including both what is known and what remains in dispute (Fisk 1998: 3). Bijker, Bal, and Hendriks (2009: 27) note that in the 'standard view' of science the quantitative social sciences have been regarded by some as meeting demarcation criteria that qualify them as 'scientific' but the interpretive social sciences and humanities have not. The meaning of other terms, as used in this book, is considered in the following section.
9. Known as the 'Nolan principles', after the Committee's first chair, Lord Nolan, these are: selflessness, integrity, objectivity, accountability, openness, honesty, and leadership. The Committee was set up in 1994 by the then Conservative Prime Minister, John Major, in response to concerns about the conduct of certain politicians and advisors. Nolan established new rules governing (inter alia) the conduct of advisory committees, including appointments procedures and the obligations placed upon members. Later, guidance was issued for government departments on obtaining and using scientific advice (OST 1997, 2000) and on non-departmental public bodies (for an overview, see Cabinet Office 2006b). A code of practice for scientific advisory committees was published in 2001 and subsequently updated (Office of Science and Technology 2001; Government Office for Science 2007, 2011). There was also more general guidance, and a code of practice, on public appointments (see, for example, Cabinet Office 2006a; Commissioner for Public Appointments 2012). Similar codes and guidance can be found elsewhere. For a review of rules and standards for holders of public office in European countries, see Demmke et al. (2007), and for a set of principles and guidelines governing the use of expertise by the European

Commission, see Commission of the European Communities (CEC) (2002). Hilgartner (2000: ch. 2) reviews the extensive procedures for maintaining the quality and effectiveness of advice from the National Academy of Sciences in the United States. For further discussion of 'quality assurance' in scientific advice, see the essays in Lentsch and Weingart (2011).

10. Important exceptions include book-length studies of the US Office of Technology Assessment (OTA, an independent analytical support agency of Congress which operated during the period 1973–95) (Bimber 1996), the *Gezondheidsraad* (the Health Council of the Netherlands) (Bijker, Bal, and Hendriks 2009), and the UK Monopolies and Mergers Commission (Wilks 1999), and a shorter work charting the history and policy evolution of the International Commission on Radiological Protection (Clarke and Valentin 2005, 2009). Sheila Jasanoff (1990) and Stephen Hilgartner (2000) each draw the empirical evidence for their studies of expert advice from a small number of scientific advisory committees (of Federal agencies and the US National Academy of Sciences respectively). UK Parliamentary select committees, while not advisory bodies in the sense considered here, have also been scrutinized: see, for example, Derek Hawes (1993) on committees dealing with environmental issues in the House of Commons, and Roger Williams (1993) on the work of the House of Lords Select Committee on Science and Technology. David Hart (2014) has assessed the performance of the US Office of Science and Technology Policy, a long-standing scientific advisory institution within the Executive Office of the President. See also note 14 (this chapter).

11. Other standing royal commissions that survived into the twenty-first century (and still exist at the time of writing [early 2014]) are different in character. The Royal Commission on Historical Manuscripts (first appointed in 1869) was, in effect, subsumed into The National Archives (TNA) in 2003, when its warrant was amended to allow the Keeper of Public Records to become the sole Commissioner. The Royal Commission on the Exhibition of 1851 remained constituted after the Great Exhibition and set up an educational trust in 1891; it makes charitable disbursements but is not an advisory body. The Royal Commissions on the Ancient and Historical Monuments of Scotland and Wales are bodies with various responsibilities for historic environments and collections, sponsored by the Scottish and Welsh Governments respectively; the equivalent body in England was merged with English Heritage, an executive non-departmental public body, in 1999.

12. In this book the Royal Commission on Environmental Pollution will also be referred to as 'the Royal Commission' or simply 'the Commission'.

13. The original Royal Warrant is reproduced in full in the Commission's First Report (RCEP 1971: iii–iv).

14. For shorter accounts relating to the study on which this book is based, see Owens (1995, 2003, 2006, 2010, 2011a, 2012); Owens and Rayner (1999).

15. Alvin Weinberg (1972) used the term 'trans-scientific' to characterize questions that are frequently asked of, but cannot be answered by, science, either because there is no prospect of uncertainties being reduced on a policy-relevant time scale (if at all), or because the issues are fundamentally ethical or political rather than scientific in nature. Other scholars have questioned whether it is meaningful, in

Endnotes

deeply contested areas such as that of environmental policy, to draw a firm line separating what is scientific from what is not (see, for example, Jasanoff 2006a). This issue will be discussed in the next section of this chapter.

16. Overviews of developments in environmental policy in a range of national and international contexts are provided by Doherty and de Geus (1996); Dryzek et al. (2003); Gray (1995); Haigh (1992); Jordan (2002); Lafferty and Meadowcroft (2000); Lowe and Flynn (1989); Lowe and Goyder (1983); Lowe and Ward (1998b); McCormick (1991); Paehlke (1989); Rees (1990); Sandbach (1980); Vig and Faure (2004); Yearley (1991). Among the authors using environmental issues to test and develop policy and political theories are Blowers (1986); Crenson (1971); Hajer (1995); Litfin (1994); Nilsson and Eckerberg (2007); Sandbach (1980); Weale (1992); Weale et al. (2003).

17. See note 8 (this chapter). 'Science' and 'scientific' are normally used in this book to mean the natural or the 'hard' sciences, unless otherwise specified. However, their meaning as applied to the Royal Commission on Environmental Pollution is in itself an interesting phenomenon, by no means always fixed and yet related to the construction of the Commission's authority. More will be said about this in Chapters 3 and 7.

18. The separation typically derives from the Weberian ideal in which the design and implementation of policy take place in an apolitical, administrative sphere.

19. A useful overview of different approaches to the interaction of knowledge and policy is provided by Radaelli (1995).

20. A substantial body of research has been concerned with the structure and rationality of the policy process, and with the factors involved in decision-making and organizational choice. Ham and Hill (1993: ch. 5) provide a useful overview.

21. As opposed to analysis *of* policy and policy-making—the wider sense in which this term is often used. It was Harold Lasswell who made the important distinction between knowledge *in* and knowledge *of* the policy process, seeing both as important components of the policy sciences (Lasswell 1951, 1971, 1976; for further discussion, see Hajer 2003b; Torgerson 1985). This book engages with policy analysis in both senses: it offers an analysis *of* the environmental policy processes with which the Royal Commission was involved, but at the same time is centrally concerned with the Commission's analysis *for* policy.

22. Beginning with the classic works of Simon (1947) and Lindblom (1959), which offered theories of policy-making as a less linear, less 'rational' process, and developing in a number of directions, as discussed later in this chapter.

23. The critique from these perspectives is often bound up with calls for more deliberative and inclusive approaches to policy-making, and for the democratization of expertise. See, for example, Fischer (1990, 2000, 2009); Fischer and Forester (1993); Hajer (1995); Hawkesworth (1988); Jasanoff (1990, 2003b); Rein and White (1977); Schön and Rein (1994); Torgerson (1986); Tribe (1972); Weingart (1999); Wynne (1975). See also the vigorous exchanges arising from the work of Harry Collins and Robert Evans on the so-called 'Third Wave' of science studies, for example: Collins and Evans (2002, 2007); Collins, Weinel, and Evans (2010); Epstein (2011); Fischer

Endnotes

(2009, 2011); Forsyth (2011); Jasanoff (2003a); Owens (2011b); Rip (2003); Wynne (2003).

24. A critical contribution of Michel Foucault's work, drawn upon by many scholars of interpretive policy analysis (using qualitative methods to explore the construction and communication of meaning in the policy process), has been to identify the great variety of sites, institutions, and mechanisms—some of them mundane—through which power is exercised, including discursive practices that come to define which knowledges count. In such perspectives, government is seen as a 'regime of practices'—a set of 'mechanisms and techniques, which, in the name of truth and the public good, aspire to inform and adjust social and economic activities' (Gottweis 2003: 255). The appeal to science as truth would be an important example of such a mechanism. See Foucault (1972, 1979, 1991a, b); also Dreyfus and Rabinow (1982: 196).

25. Theories differ in their primary units of analysis, and in the degree to which power is seen as concentrated or dispersed. In an analysis of politics and policy-making in the domain of pollution, Albert Weale (1992), for example, identifies four (non-exclusive) theoretical frameworks—or, as he prefers, 'idioms'—the first three of which might be considered conventional in their treatment of knowledge. In the 'rational choice' idiom, the pursuit of interests (by individuals, groups, and bureaucracies) is the focal point; in the 'systems idiom' (exemplified by neo-Marxist analysis) it is systemic structures that matter; and in the 'idiom of institutions', explanatory power lies with established practices, rules, and norms. Weale's fourth idiom, that of 'policy discourse', aligns more with cognitive and discursive perspectives on the policy process, which this chapter will go on to discuss.

26. Similarly cited in Wheare (1955: 89). Gladstone served four periods as British Prime Minister, between 1868 and 1894.

27. Faced with agitation about urban smog, Macmillan (at that time Minister of Housing and Local Government) wrote: 'I would suggest that we form a Committee.... We cannot do very much, but we can seem to be very busy—that is half the battle nowadays' (Cabinet Office Minutes [Memoranda]: CAB 129/64, C(53) 322, 18 November 1953 [The National Archives, Kew], cited in Hyam [2007: 241]).

28. Wheare (1955: 91) noted the phenomenon of setting up a committee purely 'for form's sake' to sell a policy to interested parties and the public, and Peter Hennessy (1989) provides examples of the 'rigging' of royal commissions in his account of 'Whitehall auxiliaries' ('Whitehall' is the term often used in the UK as a metonym for the civil service and the administrative machinery of government in London). Manipulative intent has most frequently been associated with 'one-off' commissions and committees, appointed to report on specific issues, but the concept of a 'safe pair of hands' may also be applicable to standing bodies.

29. 'Science and technology studies' is an umbrella term for investigations concerned with the place of science and technology in society. STS draws on a wide range of disciplines (including cultural studies, history, human geography, law, philosophy, politics, and sociology), and adopts a variety of theoretical perspectives and methodological approaches.

Endnotes

30. Important contributions have been made by organizational theorists as well as political scientists. For useful discussions of the literature on learning, see Bennett and Howlett (1992); Jachtenfuchs (1996); May (1992); Parson and Clark (1991, 1997).
31. The proliferation of terminologies in a now extensive literature can be confusing. Here, for example, May uses 'social learning' in a more specific sense than many other authors and makes a further distinction between *policy* and *political* learning, using the latter (in a different way to Heclo) to mean a process through which actors become more adept at promoting their own point of view and challenging the views of others.
32. Advocacy coalitions are relatively open networks, involving actors from diverse backgrounds, who co-operate to a non-trivial degree and operate within a 'policy sub-system'. The latter is defined to include the actors, interest groups, and institutions participating in the policy process in a specific policy area, and is similar to Heclo's concept of an 'issue network'. (Networks, and their role in the policy process, will be discussed in more detail later in this chapter.) In the advocacy coalition framework, rival coalitions within a policy sub-system pursue divergent objectives related to their basic values and causal beliefs. The belief systems of the coalitions are seen as having a tri-partite structure. Deep core beliefs (about human nature, for example) are likely to be shared across most policy domains and are highly resistant to change. Policy core beliefs (concerning issues such as the relative importance of economic and environmental goals) are the principal 'glue' of the coalitions and are also relatively stable, so that the policy core aspects of specific programmes are unlikely to change unless the dominant coalition is replaced, or change is imposed by a 'hierarchically superior jurisdiction' (Jenkins-Smith and Sabatier 1994: 191). Secondary aspects of a belief system—concerning suitable regulatory instruments, for example—and, over time, certain empirically based policy core beliefs are, however, susceptible to policy-oriented learning, 'in the light of new data, experience, or changing strategic considerations' (ibid.: 182). See also Sabatier (1987, 1988, 1998); Sabatier and Jenkins-Smith (1993).
33. These changes involve the 'settings' of policy instruments or the instruments themselves respectively, and are associated with the 'broad continuities' usually found in policy domains (Hall 1993: 279).
34. Frank Baumgartner and Bryan Jones have sought to explain why the evolution of public policy tends to be characterized by long periods of stability punctuated by intensive periods of innovation and change. They have argued that new developments (for example, new research findings) often have little impact until it is acknowledged that something needs to be done. When conditions are right, however, positive feedback mechanisms can result in dramatic change, and shifting 'policy images' (or frames) can contribute substantially to this process.
35. While the dominant frame at any one time may be so familiar as to be tacit, *framing* can also be a discursive strategy, as Hall found in his analysis of radical policy change. Rein and Schön (1991: 275) argue that 'the sponsors of a frame seek to develop [it], explicate its implications for action, and establish the grounds for argument about it. They may also devise metaphors for communication of the

frame'. See also Fischer and Forester (1993); Hajer (1995, 2003a); Hajer and Versteeg (2005a); Hajer and Wagenaar (2003a); Hall (1993); Hisschemöller et al. (2001); Jachtenfuchs (1996); Laws and Rein (2003); Schön and Rein (1994).
36. Kingdon's widely cited analysis, in which he draws on Cohen, March, and Olsen's (1972) 'garbage can model' of organizational choice, focuses on agenda-setting and the generation of alternatives in relation to key issues of public policy.
37. Litfin (1994: 4) defines 'knowledge brokers' as 'intermediaries between the original researchers, or the producers of knowledge, and the policy-makers who consume that knowledge but lack the time and training necessary to absorb the original research'. Advisory bodies, in this role, might offer '[k]ey locations from which members of epistemic communities [can] gain significant leverage over policy' (Haas 1992: 31). We can imagine them in the advocacy coalition framework, for example, furnishing the technical information, empirical evidence, and policy analysis on which policy-oriented learning is deemed to depend. (Note that individuals on the inside of policy teams—in the UK, for example, the Chief Scientific Advisors and specialist staff in government departments—can also perform such roles; see Parker [2013].) In the advocacy coalition framework *policy* brokers are presented in a rather different light, as facilitating learning across advocacy coalitions, with 'blue-ribbon commissions' being seen as particularly effective in this respect (Jenkins-Smith and Sabatier 1994: 184). Pielke (2007) identifies another form of broker, suggesting that one of four possible roles for scientists in relation to policy is that of 'honest broker of policy alternatives' (the others being 'pure scientist', 'science arbiter', and 'issue advocate').
38. In a discussion of ways in which think tanks influence governments, James (2000: 163) distinguishes 'atmospheric influence' from the effects of targeting short- to medium-term agendas and from 'micro-policy research'. The concept of 'enlightenment', developed by Carol Weiss in her work on the influence of social scientific research, refers to the time lags often observed before new knowledge comes to have effect. Enlightenment might still involve an essentially rationalistic form of learning—as in the advocacy coalition framework, for example, where the gradual alteration of certain policy core beliefs by empirical evidence might be characterized in this way. However, it can also be used to mean a conceptual or discursive form of learning, whereby the policy frame, but not necessarily 'the science', changes over time.
39. The idiom of co-production, as set out by Jasanoff (2006a), gives primacy to neither the natural nor the social; rather, it provides a helpful conceptual framework for discussion of 'relationships between the ordering of *nature* through science and technology and the ordering of *society* through power and culture' (p. 14). Such an approach is not to be confused with relativism: Jasanoff's (2006a: 43) volume on co-production 'freely acknowledges the cultural uniqueness of science and technology, insisting only that their specialness arises from repeated, situated encounters between scientific, technical and other forms of life'.
40. Jasanoff focuses particular attention on committees attached to the Environmental Protection Agency (EPA) and the Food and Drug Administration (FDA).

Endnotes

41. Jasanoff (1990: 236) notes that 'the most politically successful examples of boundary work are those that leave some room for agencies and their advisers to negotiate the location and meaning of the boundaries'.
42. The *Gezondheidsraad*, which has been in existence for over 100 years, is an independent scientific body advising the Dutch government on matters of public health. Like the Royal Commission (as was), it is charged with providing independent advice to government in a complex area of public policy. It is, however, a larger and differently structured body, with a secretariat of around eighty and a membership network of more than 200 experts. It works primarily in responsive mode through ad hoc committees (with standing committees providing internal peer review), and produces reports with greater frequency than the Commission did during its lifetime. As in the study of the Commission, Bijker, Bal, and Hendriks (2009) were concerned with the practices of this advisory body, its role in policy-making, and the attributes and reputation that it enjoyed. There is overlap, though not complete correspondence, in the methodologies of the two studies. The Dutch investigators used archival work, interviews, focus groups, and ethnographic observations (especially attendance at meetings) to enable them to study the *Gezondheidsraad* 'from the inside out' (ibid.: 10); they focused on ten of its studies in the period 1985–2000, while offering some broader historical context. The study of the Royal Commission covered its whole lifespan (1970–2011), with a particular focus on selected aspects of its work; this study, too, drew upon documentary evidence, interviews, and participant observation (in this case, through the author's own membership for ten years). Detail on methods will follow in the penultimate section of this chapter, and on the Commission itself in Chapter 3.
43. See also Shackley and Wynne's (1996) discussion of 'boundary-organizing devices' and the special issue of *Science, Technology, and Human Values* 2001, vol. 26, no. 4, on boundary organizations.
44. Wheare (1955), for example, said this of functions like delay, pacification, and camouflage, and maintained that abuses of advisory bodies had been 'neither flagrant nor frequent' (Wheare 1955: 92; see also Bulmer 1983a; Rhodes 1975).
45. It comes at one end of what Litfin sees as a spectrum of power relations, with force at the other.
46. Simplifying greatly, one might say that in developing such critiques, political scientists have tended to focus on the assumed linearity and rationality of the policy process, while STS scholars have questioned claims about the objectivity of knowledge and the neutrality of expertise. Some scholars have sought to combine these critical traditions (see, for example, Fischer 1990, 2000, 2009). Torgerson (1986) offers an interesting discussion of perspectives on policy analysis.
47. The concept of autonomy as it relates to the Royal Commission is developed in greater depth in Chapters 3 and 7.
48. The riskiness of symbolism has long been recognized. Committees set up to placate 'may not produce the desired result. They may add fuel to the fire and force governmental action' (Wheare 1955: 89; see also Cartwright 1975; Richardson and Jordan 1979). Torgerson's (2003) account of a 'scientific panel' established during an apparently intractable controversy over logging in Clayoquot Sound on

Endnotes

Vancouver Island, Canada is an interesting modern example. Though initially greeted with much scepticism, the panel developed its own agenda and proved to be an arena for mutual learning and conceptual innovation.

49. Networks are not a new concept in policy analysis, nor is their significance restricted to cognitive or discursive approaches. There is an established literature in which policy is seen as emerging from interactions within or between networks; Marsh and Rhodes (1992a, b) provide a useful overview. Many network structures have been identified, ranging from tightly drawn 'policy communities' held together by resource dependencies (see, for example, Jordan and Richardson 1983; Richardson and Jordan 1979) to looser formations such as 'issue networks' (Heclo 1978) and groupings that coalesce around interests, values, or discourses (including advocacy and discourse coalitions). The concept of epistemic communities is most familiar from the literature on international environmental negotiations (for example, Haas 1990, 1992; Litfin 1994), but such networks are not necessarily international in scope. As Rhodes and Marsh (1992: 2) point out, the network concept is a useful 'meso-level' one, compatible with different models of (macro-level) power distribution in liberal democracies.

50. In this useful overview of policy analysis and governance, Hajer and Wagenaar draw on Castells' influential and wide-ranging concept of the network society (Castells 1996). Note that while the older, more exclusive forms of network were often associated with policy stability, as structures have shifted networks have increasingly been seen as agents of change.

51. Bovine spongiform encephalopathy (BSE), or 'mad cow disease', was discovered in cattle in the UK in the 1980s, and was attributed to the recycling of animal protein in ruminant feed. On the basis of expert advice, the government took a number of measures to prevent infected beef entering the food chain (though later it became clear that these had been inadequately implemented and enforced). The government was also persuaded that the likelihood of the disease being transmitted to humans was remote, and it was anxious to avoid what might be seen as an overreaction. Assurances were given to the public that British beef was safe (and to emphasize the point, the then Minister for Agriculture, Fisheries and Food, John Gummer, famously encouraged his young daughter to eat a beefburger in front of the cameras). In the mid-1990s, however, people began to die of a new variant of Creutzfeldt-Jakob Disease (vCJD), subsequently shown to be linked to BSE; 177 people have died at the time of writing (late 2013). The British beef industry was also severely damaged by a worldwide export ban, which has only gradually been lifted. The 'BSE fiasco' was an important spur to the publication of the guidelines and codes of conduct for expert advice, discussed above, not least because, as the report of the Phillips enquiry into this episode observed, '[c]onfidence in government pronouncements about risk was a further casualty of BSE' (Phillips 2000: xviii; see also Jasanoff 1997).

52. The Royal Commission granted full access to its archives, including minutes, other papers, and correspondence, for the purposes of the study. Certain environment department files relating to the Commission's work were also made available. Note that, to avoid confusion, this book generally refers to 'the environment department'

Endnotes

when discussing the Commission's interactions with the UK ministry holding the main responsibility for environmental affairs. This remit has always been combined with other portfolios including, at various times, local government, planning, transport, and agriculture. In the early years of the Commission, environmental responsibilities lay primarily with the Department of the Environment (DoE), created in 1970; at the end, they were primarily with the Department for Environment, Food and Rural Affairs (Defra), created in 2001. In the intervening period there were several other combinations and changes of name, details of which are given in Appendix 2.

53. In total, 103 semi-structured interviews (including nine second interviews) were conducted, mainly in the period 1995–7, but with some taking place later than this. A breakdown of interviewees by category is given in Appendix 3. Interviews lasted between half an hour and two and a half hours. The majority were conducted face-to-face, recorded, and later transcribed, though seven took place (and several were supplemented) by telephone. For reasons of preference or practicality, a small number of respondents were not (tape-) recorded, and in these cases notes were taken instead. Three additional respondents chose to put their views in writing in lieu of being interviewed. Formal interviews were supplemented by semi-formal discussions and/or correspondence with interviewees and others, with these exchanges extending up to the time of writing (July 2014) (see Appendix 3). To protect anonymity, extracts from interviews and related material are not attributed to named individuals in this book, except where the identity of the individual would be readily apparent (for example, as Chair of the Commission during a particular study) and/or the person had indicated that they were content to be quoted. The interviewee's position (usually the role connecting them to the Commission in the context under discussion) is indicated, though (again for reasons of anonymity) precise job titles and dates are not always given. Informal meetings and conversations, over the whole period of the study, also contributed significantly to an understanding of the Commission and its work.

54. Interview (November 1996), Deputy Secretary, environment department (1990s). In interview citations, the month and year of the interview are given, and the period during which the individual held the relevant role is indicated (see note 53, this chapter). For administrative civil servants, the hierarchy in UK Government departments is as follows (traditional titles first, with modern equivalents in brackets): Permanent Secretary—the Civil Service Head in each Department; Deputy Secretary (Director General); Under Secretary (Director/Grade 3); Assistant Secretary (Deputy Director or Director/Grade 5); Senior Principal and Principal (Head of Unit/Grade 6, Head of Branch/Grade 7). For scientific or professional and technical civil servants (often working alongside policy teams), the (rare) equivalents to Deputy Secretary (Director General) are Chief Scientific Advisors. At Under Secretary (Director) level are Chief Scientists or Chief Scientific Officers, and so on through Deputy Chief Scientist or Deputy Chief Scientific Officer (Grade 5); Senior Principal Scientific Officer or Senior Professional and Technical Officer (Grade 6); and Principal Scientific Officer or Professional and Technical Officer (Grade 7). Sometimes a more neutral title (e.g. 'senior civil servant') is used to protect anonymity (see note 53).

Endnotes

55. Triangulation is a widely recognized strategy in qualitative research, though it may not always provide a safeguard. Raab and McPherson (1988) maintain, for example, that it may not work when a community is caught in the grip of an idea.
56. Not everyone agrees. There is lively dispute among scholars of the policy process between those who advocate a search for general laws and those who favour more interpretive analysis in context. In the former camp, Paul Sabatier, for example, maintains that theory-building must involve falsifiable hypotheses, and he and his colleagues present the advocacy coalition framework as 'a causal theory of the policy process' (Jenkins-Smith and Sabatier 1994: 175; Sabatier 1999, 2000). (For an interesting commentary on this perspective, see Parsons [2000]). Others argue that the complexity of the social world means that 'universally applicable generalizations are not to be expected' (Litfin 1994: 7), and that an interpretive approach will ultimately be more productive; see also Hajer and Laws (2006); Hajer and Wagenaar (2003a); Hisschemöller et al. (2001); Kingdon (2003); Parsons (2000); Radaelli (2000); Yanow (2000).
57. John Kingdon (2003: 222), for example, characterizes his model of agenda-setting and the generation of alternatives as probabilistic, with elements of randomness but also 'quite a bit of structure' in the policy process.

Chapter 2

1. No single source provides a comprehensive picture, but useful accounts include Barry and Patterson (2003, 2004); Haigh (1984, 1987, 1992, and loose-leaf updates); Hajer (1995); Johnson (1973); Jordan (2002); Lowe and Goyder (1983); McCormick (1991); Weale (1992); Weale et al. (2003); and the edited volumes by Gray (1995); Jänicke and Weidner (1997); Lafferty and Meadowcroft (2000); Lowe and Ward (1998b).
2. The Commission interpreted its remit broadly, ranging well beyond issues of 'pollution' in the narrow sense (see Chapter 3). However, for the most part it did not concern itself with landscape or nature conservation, the main exceptions being two of its later reports dealing with environmental planning (RCEP 2002a) and marine ecosystems (RCEP 2004). Therefore, although the protection of landscapes and biodiversity have been important aspects of UK and European environmental policy, they are not among the key areas covered in this chapter.
3. The much-quoted figure of 80 per cent was widely attributed to John Gummer, UK Secretary of State for the Environment, 1993–7.
4. It did, however, seek to influence such bodies via national institutions, for example by urging the UK government to pursue a variety of measures within Europe; latterly it also launched several of its reports in Brussels as well as in the UK. In addition, from the early 1990s, it was an active participant in a transnational network of European environmental advisory bodies, discussed in more detail in Chapter 7.
5. At the time of writing (2013), the devolved administrations are known as the Scottish and Welsh governments and the Northern Ireland Executive.
6. The pace had accelerated even by the mid-1990s, when Nigel Haigh (1995: 14), Founding Director of the influential Institute for European Environmental Policy,

Endnotes

observed that it was 'increasingly difficult to see the field [of environmental policy] as a whole and still know the detail'. In an analysis of the function and achievements of the UK Department of the Environment for the occasion of its twenty-fifth anniversary, Haigh demarcated four periods: 1969–73, 'Enthusiasm'; 1973–9, 'Financial restraint'; 1979–83, ' "Freedom of choice" and deregulation'; 1983–8, 'A turning tide'; and 1988–95, 'The global challenge recognised'.

7. Mol (1996: 312) notes that there has been wider acceptance of ecological modernization as a means of 'characterising and analysing the way in which contemporary industrial societies are trying to cope with the environmental crisis' than as a normative programme which could lead to the crisis being resolved. For more on the origins, development, and application of the concept, see Dryzek (1997); Hajer (1995); Huber (1982, 1985, 1991); Jänicke (1985, 1988); Weale (1992); and the special issue of *Journal of Environmental Policy and Planning* 2000, vol. 2, no. 4.

8. These shifts have been associated with particular storylines in policy discourse: environmental considerations must be integrated into other policy sectors; negative impacts can be decoupled from economic growth; markets have a key role in achieving environmental objectives; and, rather than being in tension with the needs of the economy, robust environmental policy can deliver substantial economic benefits.

9. Lowe and Flynn (1989) suggest that Britain had perhaps the most highly organized environmental lobby of any European country at this time. The emergence of new groups like the Conservation Society (in 1967) and Friends of the Earth (in 1970) built upon a longer tradition of public and private interest in issues such as nature conservation and clean air, but the changes also invigorated existing organizations, which often adopted a more campaigning stance (ibid.).

10. Expert bodies on the environment established at about the same time as the Royal Commission included, for example, the Council on Environmental Quality (CEQ) in the United States (1969), the Canadian Environment Advisory Council (1972), the German Advisory Council on the Environment (*Sachverständigenrat für Umweltfragen*, SRU) (1972), and the Swedish Environmental Advisory Council (*Miljövårdsberedningen*, MVB) (1968). See also Weale (1992). Some followed a little later; for example, the Dutch Advisory Council for Research on Spatial Planning, Nature and the Environment (RMNO) was established in 1981.

11. Sweden was something of an exception, being one of the first countries to adopt a more integrated approach to pollution control (see Hinrichsen 1990).

12. Bugler's description of the Mersey estuary is typical of his colourful prose: 'on the tide excreta, balls of fat, sewer scum, rotting vegetable waste, contraceptives. In summer children have been seen bathing amid flotillas of these objects. On winter days people strolling on the promenade itself have had to watch their feet to keep their shoes clear of crude sewage' (Bugler 1972: 32). Like Bugler, the Royal Commission was appalled by state of the Mersey estuary, even into the 1980s. A member of the Secretariat from that time recalled a visit to the area when the Commission was working on its tenth report (RCEP 1984): 'there was no doubt, no controversy whatever, that here was something grossly wrong, and something had to be done' (interview, July 1996).

Endnotes

13. According to Solesbury (1976: 384), '[i]ssues have particularity if they can be clearly exemplified by particular occurrences or events'. Kingdon (2003: 94–5) similarly argues that problems may 'need a little push' to get attention and that the push 'is sometimes provided by a focusing event or disaster that comes along to call attention to the problem, a powerful symbol that catches on, or the personal experience of a policy maker'.
14. Storms in the Irish Sea had driven thousands of dying seabirds ashore, without any obvious cause for their mortality. Analysis found that many birds had high concentrations of PCBs in their livers. These chemicals were then identified in the fat of healthy birds collected elsewhere. An enquiry concluded that PCBs had moved to the liver and kidneys of the affected birds when they had used their fat reserves under stress, thus weakening them even further (Holdgate 2003: 172–3).
15. With its roots in the nineteenth century, the system consisted of 'an accretion of common law, statutes, agencies, procedures and policies', most of which had been 'pragmatic and incremental responses to specific problems and the evolution of relevant scientific knowledge' (Lowe and Flynn 1989: 256; see also Lowe and Ward 1998a).
16. See also Jordan (1998); Lowe and Flynn (1989); McCormick (1991). The legislation covering the most serious air pollutants, dating back to the Victorian era, stipulated the use of the 'best practicable means' (BPM) to prevent emissions or to render them harmless. Determination of what was 'practicable' was a matter for the pollution inspectorate, operating at local level and taking account of economic considerations as well as the state of technology. A nineteenth-century Chief Alkali Inspector, Alfred Evans Fletcher, famously likened the system to an elastic band, 'ever tightening as chemical science advanced and placed greater facilities in the hands of the manufacturer' (Alkali Inspectorate 1888: 40, cited in Ashby and Anderson 1981: 66). The twentieth-century critics who became vociferous in the 1960s and 1970s considered the system far too cosy and consensual (see Chapter 5). BPM was, however, a technology-based framework, focusing on the source of the pollutant. In contrast, discharges to water were regulated with reference to the quality of the receiving environment—not necessarily requiring deployment of the best technical methods of control; this, too was to become a point of controversy. A summary of the arrangements that prevailed can be found in Owens (1990).
17. The Alkali Inspectorate was re-named the Alkali and Clean Air Inspectorate (ACAI) in 1971, though it continued to be referred to by its shorter name. In 1983 it became the Industrial Air Pollution Inspectorate (IAPI).
18. Acts of Parliament mentioned in the text are listed, with further details, in Appendix 4.
19. CoPA contained new provisions for public registers of discharges. Otherwise, it was mainly a codification of existing practice (Haigh 1987; Weale 1997). In 1981, the *Deposit of Poisonous Waste Act* was repealed and replaced by the 1980 *Control of Pollution (Special Waste) Regulations,* within the framework of CoPA (for more detail, see Porter 1998: 196–7; Haigh 1987: 127–9).
20. This individual later became a Secretary of the Royal Commission.

Endnotes

21. See, for example, Bugler (1972); Frankel (1974); Tinker (1975). The Inspectorate's approach to implementing the legislation also had stout defenders, including Eric Ashby, the first Chair of the Commission (see Ashby and Anderson 1981). In the mid-1970s, at the height of the controversy, the Commission was asked by the Secretary of State for the Environment to investigate arrangements for air pollution control—one of the few occasions on which an issue has been referred to the Commission in an (ultimately unsuccessful) attempt to placate and depoliticize. See Chapter 5 for further discussion.
22. Interview (June 1997), Chief Scientific Advisor, environment department (1990s). This interviewee recalled that when he had begun working in Whitehall in the 1960s—on motorway noise barriers—'you were given a form to sign the *Official Secrets Act*...this might seem like a rather excessive thing to have to do'.
23. There was no Directorate General with specific responsibility for environmental matters until 1981, and it was not until the *Single European Act* of 1986 that Community action in this area first acquired a firm legal basis. European legislation mentioned in the text is set out in more detail in Appendix 4.
24. The Dangerous Discharges to Water Directive, proposed in 1974, the Bathing Water Directive, proposed in 1975, and the Drinking Water Directive, also proposed in 1975, were all controversial.
25. Under a photograph of Blackpool (a resort in the North West of England), showing both beach and sea crowded with holiday-makers on a brilliant summer's day, the caption read: 'The bathing water at Blackpool beach—not identified as a stretch of water where bathing was "traditionally practised by a large number of bathers"' (RCEP 1984: Plate 5).
26. 'Secret documents reveal Tory strategy', *The Sunday Times* (Business News) 18 November 1979: 53.
27. Interview (May 1997), Deputy Secretary, environment department (1980s–90s). The issues were not confined to the environment department, as will become clear.
28. In Article 130r: 'Environmental protection requirements shall be a component of the Community's other policies.' The *Act* came into force in 1987. It is noteworthy, too, that an OECD conference on *Environment and Economics* in June 1984 explicitly focused on ecological modernization (Hajer 1995).
29. Interview (May 1997), Director of CUEP (1970s). With key industries such as electricity and water still in public ownership, government and industry were effectively one and the same in such cases. Costly measures to reduce pollution had direct implications for public expenditure, which the government was determined to hold down. In the case of acid rain, a former Assistant Secretary in the environment department (1980s) explained that 'the idea of doing anything about curbing sulphur dioxide emissions was pretty low on the Treasury agenda, to say the least' (interview, June 1996). The water industry came under the remit of the Water Directorate in the environment department. According to the former Director of CUEP, 'it was a very cosy relationship...All the industrial dischargers would have the ear of the ministers who could argue that business would be damaged'.

Endnotes

30. As recalled in interview (January 1997) by one environment department official (1980s).
31. Interview (September 1997), Chief Inspector, Air, Water, and Wastes, Her Majesty's Inspectorate of Pollution (HMIP) (late 1980s); previously with Industrial Air Pollution Inspectorate (IAPI). The formation of HMIP is discussed later.
32. Interview (May 1997), Deputy Secretary, environment department (1980s–90s).
33. The Alliance took 25 per cent of the popular vote in the 1983 general election, though won only 5 per cent of seats. In 1988, the majority of the Social Democratic and Liberal Parties merged to form the Social and Liberal Democrats (soon shortened to Liberal Democrats).
34. There was a shift from voluntary to statutory regulation (Lowe and Flynn 1989), exemplified by replacement of the voluntary Pesticide Safety Precautions Scheme with statutory controls under the *Food and Environment Protection Act* 1985 (a change that was influenced by the Royal Commission—see Chapter 6). There was also a shift towards standards being laid down in legislation.
35. An interesting perspective on the intransigence that preceded this shift was provided by a civil servant who had been in the environment department in the 1980s. His comments nicely illustrate the non-linear nature of the policy process (Kingdon 2003), in which sometimes a phenomenon can only be framed as a problem once an acceptable solution has become available: 'We were going to... get our sulphur emissions down, which we have, not by the dirty business of fitting FGD [flue gas desulphurization] but by stuffing the coal industry and burning gas, and that was not a thing which could be said publicly by any government in 1983 to '84... if we had said we were going to join the 30 per cent club [a group of European countries committed to emissions reductions], people would ask how we were going to do it and then it would be quite clear, if you are honest about it... it was because we were switching fuels and that would have not gone well with the coal lobby which was still extremely powerful.... all in all it was environmentally preferable to switch to natural gas because there was plenty about but you couldn't say so at the time. Since then we have accepted a number of facts on the damage... but again I have been a bit cynical—*we can now accept these after we got to a position where our SO_2 emissions were going down fast*' (interview, November 1996, emphasis added).
36. The dispute went back to 1974, when the European Commission published a draft directive on dangerous discharges to water, proposing that 129 substances (the 'black list') should be subject to strict, uniform emission limits at source. The UK objected vigorously, arguing for a system based on environmental quality standards set for the receiving waters; this would allow for emission limits in different locations to vary, according to assimilative capacity. Industry and government were as one in insisting that Britain should not be denied the comparative advantage of a robust aquatic environment. The European Commission, and other member states, maintained that variable limits would themselves represent a distortion of competition, and expressed concern that for some substances harm thresholds (and thus assimilative capacities) could not readily be determined. An uneasy compromise prevailed until (and to some extent after) the UK shifted its position in 1987, though the red list to which it then agreed was considerably

211

Endnotes

shorter (with twenty-six substances) than the original black list. This issue, and the Royal Commission's involvement, is discussed further in the context of the latter's treatment of risk and precaution in Chapter 4.

37. The NRA was conceived in the face of resistance from both environmentalists and industry to the government's plan for the newly privatized water utilities to inherit the environmental protection functions of the regional water authorities (see Owens 1990, Appendix 8.2, for a brief account).

38. The Royal Society of London is the UK's prestigious national academy of sciences. It is interesting to note that, on ozone, the UK had initially been obstructive within Europe (see McCormick 1991: 104).

39. Interview (January 1997), environment department official, latterly Head of Central Unit on the Environment (part of the Central Directorate of Environmental Protection) (1980s).

40. Interview (November 1996), chief scientific officer, environment department (1970s–80s).

41. Interview, May 1997.

42. Interview (August 1997), Chief Inspector, IAPI (1980s). The individual had also previously been in CUEP, then in Air and Noise Division, in the environment department.

43. Interview (January 1997), Head of Central Unit on the Environment (late 1980s).

44. The principle requires that all those who produce wastes should have a 'duty of care' to ensure that their wastes are subsequently managed and disposed of without harm to the environment.

45. HL Deb 18 May 1990, vol. 519, cc504–5.

46. Interview (May 1997), Deputy Secretary, environment department (1980s–90s).

47. This marked a break with tradition because in previous years the Treasury had powerfully reasserted the principle of the neutrality of the tax system, 'thus rendering it difficult to justify a policy of taxes on public "bads" and subsidies on public "goods"' (Weale 1997: 106).

48. As Holdgate (2003: 327) observes, 'Agenda 21 and the new Conventions on Biological Diversity and Climate Change were important documents. Never before had well over a hundred Heads of State or Government sat down together to discuss the links between environment and development. Never before had two environmental conventions attracted over 150 signatures each within a few days of their being opened for signature.' There was also a third convention, on combating desertification.

49. The Maastricht Treaty amended the 1957 Treaty of Rome and advanced the integration agenda set out in the *Single European Act* of 1986 (see Appendix 4). The term 'sustainable development' was not in fact used in the Maastricht Treaty, but Article 2 set out the objective of 'sustainable and non-inflationary growth respecting the environment'. According to one observer, those drafting the text had 'manifestly wished to avoid an outright commitment to "sustainable development"'; he considered the amendment to be significant, nevertheless (von Moltke 1995: 9). The wording was later strengthened in the Amsterdam Treaty (1997), which included sustainable development as an overarching objective of the EU. Further information

Endnotes

about the evolution of EU policy in this area can be found at <http://ec.europa.eu/environment/eussd/>(accessed 25 February 2014).

50. See, for example, Hulme (2009); Jäger and O'Riordan (1996); Paterson (1996); Schneider, Rosencranz, and Niles (2002); Stern (2006).

51. The Kyoto Protocol was adopted in Kyoto, Japan, in 1997 and came into force in 2005, with the first commitment period being 2008–12. However, it was never ratified by one of the world's biggest polluters, the United States. (For further detail on parties to the Protocol, see <http://unfccc.int/kyoto_protocol/status_of_ratification/items/2613.php> [accessed 25 February 2014]). Negotiating a successor to Kyoto was to prove intractable in the final period covered by this chapter.

52. This had interesting effects internally within Whitehall too. By the mid-1990s, as Nigel Haigh (1995: 15) pointed out, the environment department 'had levers in the form of international commitments which gave it some purchase on other Government Departments, even if they remained difficult to shift'.

53. Including a climate change levy package, establishment of the Carbon Trust to recycle levy receipts, a domestic carbon trading scheme, renewables targets (with an obligation on electricity suppliers to increase the proportion of electricity coming from renewable sources), and measures to improve energy efficiency.

54. Developments in climate science, analysis of potential impacts, and trends in emissions and atmospheric concentrations were also important (Committee on Climate Change 2008; IPCC 2001, 2007). The Royal Commission's influence on energy and climate policies is discussed in Chapter 6.

55. For further discussion of the development of climate change policy in the UK, see Anderson, Bows, and Mander (2008); EAC (2010b); Jordan et al. (2003); Lovell, Bulkeley, and Owens (2009); Matthews (2011); RCEP (2000); SDC (2003, 2006); Smith (2004).

56. Directives on air quality established frameworks and set limit values for particular pollutants (for further detail, see <http://ec.europa.eu/environment/air/quality/legislation/existing_leg.htm> [accessed 30 April 2014]). Important legislation on waste included directives on landfill, waste electronic and electrical equipment (WEEE), and end-of-life vehicles. See Appendix 4.

57. For the initial air quality strategy and later revisions, see DoE and The Scottish Office (1997); DETR et al. (2000b); Defra et al. (2003, 2007). On waste, the Conservatives had produced a White Paper in 1995 (UK Government 1995), but it was not a formal strategy for the purposes of the *Environment Act*. The first such strategy was published in 2000, followed by a number of revisions (DETR 2000b; Defra 2007c, 2011). An account of the evolution of waste policy in the devolved administrations can be found in Department of the Environment Northern Ireland (2006); Scottish Government (2010); Welsh Assembly Government (2002).

58. The REACH Regulation was approved by the European Parliament and adopted by Council in December 2006; it entered into force on 1 June 2007, with a phase-in period up to 2018. For more information about REACH, see Appendix 4 (also European Commission 2007), and for an interesting account of the conflict and lobbying in the policy process that led to its adoption, see Hey, Jacob, and Volkery (2006).

Endnotes

59. Transport, as a major source of CO_2 emissions, was inevitably linked to climate change, but many non-climate related impacts of traffic growth also caused consternation. Some commentators argued that startling road traffic projections at the end of the 1980s (DoT 1989) had acted as a focusing event, helping to ensure a firm place for 'transport and the environment' on the political agenda during the following decade (Goodwin et al. 1991; Owens and Cowell 2011).
60. For perspectives on the enduring GM controversy, see, for example, AEBC (2001); Gaskell (2004); GM Nation Steering Board (2003); Grove-White et al. (1997); Levidow, Carr, and Wield (2000); S. Rayner (2003).
61. UK governments made strenuous efforts to streamline and 'modernize' the land use planning system, which became the subject of numerous reviews and several Acts of Parliament after the turn of the century. Key objectives were to facilitate economic development, increase the rate of house-building, and reduce 'delays' to major infrastructure projects. Legislation to streamline procedures for the latter (given effect in the *Planning Act* 2008) was driven in large part by the desire, towards the end of the third era, to instigate a new nuclear power programme in the UK. For a more detailed account of these developments and associated conflicts, see Cowell and Owens (2006); Owens and Cowell (2011).
62. The principle had been given a legal basis in the *Single European Act*, as noted above, and was reinforced in the Maastricht and Amsterdam Treaties. The Cardiff European Council of 1998 emphasized the need for 'integration of environmental protection into Community policies, in order to achieve sustainable development' and sought to begin this process in the energy, transport, and agricultural sectors (<http://www.consilium.europa.eu/uedocs/cms_data/docs/pressdata/en/ec/54315.pdf> [accessed 29 January 2014]). However, many considered the Cardiff process to have been eclipsed by an agreement reached in Lisbon in 2000, when a Special European Council sought urgently to respond to the 'quantum shift' of globalization and rapid, knowledge-based, technological development. A strategic goal was adopted for the coming decade: 'to become the most competitive and dynamic knowledge based economy in the world, capable of sustainable economic growth, with more and better jobs and greater social cohesion'. See <http://www.europarl.europa.eu/summits/lis1_en.htm> (accessed 29 January 2014). For overviews of European progress on integration, see EEA (2005); Nilsson and Eckerberg (2007).
63. 'Environmental policy appraisal', never implemented with much enthusiasm, was later absorbed into a new system of regulatory impact assessment, where the environmental component was all too easily marginalized (NAO 2006; Russel 2005; Russel and Jordan 2007). Similar points have been made about integrated appraisal of European policies (EEAC 2006; Owens 2007).
64. The Sustainable Development Unit was established in 1997. Later, the environment department continued to lead, working with the Cabinet Office and the Department for Energy and Climate Change, but the emphasis shifted even more towards embedding sustainability across government. The Environmental Audit Committee—a committee of the House of Commons—began work in November 1997 and continues as this book is nearing completion (2014). Its full remit is 'to consider to what extent the policies and programmes of government departments

and non-departmental public bodies contribute to environmental protection and sustainable development; to audit their performance against such targets as may be set for them by Her Majesty's Ministers; and to report thereon to the House'.

65. Later codified as the IPPC Directive of 2008, then replaced by the Industrial Emissions Directive of 2010 (see Appendix 4 and CEC 2007b).
66. This broad shift was emphasized by a senior civil servant in the UK environment department, reflecting on changes in environmental policy over ten years or so from the mid-1990s (interview, March 2006).
67. The SEA Directive came into force in 2004, significantly extending the remit of environmental assessment from that of the earlier Directive on environmental impact assessment, which had been in force since 1988.
68. A brief overview of water policy developments, and information about the key provisions of the Directive, can be found at: <http://ec.europa.eu/environment/water/water-framework/info/intro_en.htm> (accessed 1 May 2014). The Directive, which came into force in 2003, requires that all inland and coastal waters should be of 'good status' by 2015.
69. The European Environment Agency and the OECD have compiled a data base of 'market-based instruments': see <http://www2.oecd.org/ecoinst/queries/> (accessed 1 May 2014).
70. The enduring significance of interests and power in environmental decision-making is well illustrated by the troubled history of road fuel duty in the UK. In September 2000, road hauliers and farmers, angry about rising fuel prices, blockaded oil refineries and threatened to bring the country and the economy to a halt. The government largely capitulated, first by freezing fuel duty and then by abandoning the policy of above-inflation annual increases (the escalator) in 2001. Emphasis subsequently shifted to differentials in vehicle excise duty (related to CO_2 emissions) and, during the remainder of the third era, fuel duty failed even to keep pace with inflation. Later—in the 2009 budget—the escalator would be re-introduced (at a lower rate than previously), only to be abolished again two years later.
71. The 1990 Freedom of Environmental Information Directive was implemented in the UK in 1992. The 1998 Aarhus Convention (Convention on Access to Information, Public Participation in Decision Making and Access to Justice in Environmental Matters), drawn up under the auspices of the United Nations Economic Commission for Europe, came into force in 2001 and was ratified by both the UK and the EU in 2005. The 1990 Directive was brought into line with the Convention by the 2003 Directive on Public Access to Environmental Information. Legislation specifically concerned with environmental information was reinforced by the more general *Freedom of Information Act* 2000 and *Freedom of Information (Scotland) Act* 2002, giving members of the public the right to request disclosure of information held by public bodies.
72. The Round Table was a 'stakeholder' body, co-chaired by the Secretary of State for the Environment, John Gummer, and a former Chair of the Royal Commission, Sir Richard Southwood. In 2000 it was merged with the smaller, high-level Prime Minister's Panel (also set up in 1994) to create the SDC. This too was a stakeholder body, with around sixteen members, which later became an executive

Endnotes

non-departmental public body. Its remit was to monitor progress towards sustainability across all government departments—in a sense acting as an external watchdog to complement the activities of the Environmental Audit Committee. Like the Royal Commission, the SDC was abolished after the general election of 2010.

73. The government was forced to step in and, when no other buyer could be found, Northern Rock was nationalized in 2008.
74. The Climate Change Committee was initially set up in shadow form so that its first report could make an input to the Climate Change Bill, which gained Royal Assent (becoming the *Climate Change Act*) at the end of 2008.
75. The package included three key targets for 2020: a reduction in greenhouse gas emissions (from 1990 levels) by at least 20 per cent; a renewables objective (20 per cent of energy consumption); and a 20 per cent reduction in primary energy use compared with projected levels (Council of the European Union 2009; European Commission 2009).
76. Within Europe, the Marine Strategy Framework Directive established a framework for community action in the field of marine environmental policy. At a wider international level, the setting up of marine protected networks, including representative networks by 2012, had been one of the few firm commitments to emerge from the World Summit on Sustainable Development held in Johannesburg in 2002 ('Rio+10').
77. This Act affected mainly England and Wales. The *Marine (Scotland) Act* 2010 was passed in the following year.
78. The pledge was made by Prime Minister David Cameron, speaking to staff at the Department of Energy and Climate Change (DECC) on 14 May 2010. See DECC Press Release: <http://www.decc.gov.uk/en/content/cms/news/pn10_059/pn10_059.aspx> (accessed 29 January 2014).
79. The security of a server at the UK's Climatic Research Unit (University of East Anglia) was breached and climate scientists were accused of manipulating data, though they were later cleared of any scientific misconduct (House of Commons Science and Technology Committee 2010; Oxburgh 2010; Russell 2010). In 2010, considerable publicity was given to errors in an IPCC (2007) report, especially to a claim that the Himalayan glaciers might melt by 2035. Though the errors were found to be relatively minor, the IPCC's structures and procedures were investigated by an Inter-Academy Council set up by the UN (IAC 2010). For further discussion, see Beck (2012); Hulme (2010); *Nature* (2010); Pearce (2010); Sarewitz (2010).
80. The 'Durban Platform' was notable for including countries such as India, China, and the United States. It also allowed for the Kyoto Protocol to continue in the interim, though not all countries would commit to this. The conference agreed that the new Treaty should be implemented by 2020.
81. Similarly, the European Council (at which heads of state or government meet periodically) had agreed to an increase in the European Community's 2020 emissions reduction target from 20 to 30 per cent 'on condition that the other developed countries undertake to achieve comparable emission reductions and that the economically more advanced developing countries make a contribution commensurate with

Endnotes

their respective responsibilities and capabilities' (see Brussels European Council, 11–12 December 2008, Presidency Conclusions, conclusion 22 <http://consilium.eur opa.eu/uedocs/cms_data/docs/pressdata/en/ec/104692.pdf> (accessed 14 July 2014).

82. George Osborne, UK Chancellor of the Exchequer: speech at Conservative Party Conference 2011, 3 October 2011, <http://www.telegraph.co.uk/news/politics/georgeosborne/8804027/Conservative-Party-Conference-2011-George-Osborne-speech-in-full.html> (accessed 29 January 2014).

83. A new-found enthusiasm for nuclear power as a component of UK energy policy was already evident in the third era (see, for example, DTI 2006, 2007), became the subject of a dedicated White Paper in the fourth (BERR 2008), and was not significantly diminished by the disaster at the Fukushima nuclear power complex following the earthquake and tsunami in Japan in 2011. It is interesting to note that by the end of the 2000s, the environmental movement itself was split both on nuclear power (which some now saw as a lesser evil than climate change) and on renewables, such as wind and tidal power, which had major implications for landscape and habitats.

84. In the UK, for example, improvements in air quality were leveling off and becoming increasingly costly to achieve (EAC 2010a).

85. This point was made by a senior civil servant in the environment department (2000s), who emphasized (in an interview in March 2006) that individual behaviour was seen as increasingly significant in the context of environmental policy.

86. In the environmental domain, different orders of change have been intertwined and interdependent. For example, at various times certain forms of pollution have been recognized as problematic (this recognition in itself involving a cognitive shift), and policy instruments adjusted to deal with the 'new' problems (first or second order change). But the environmental implications of whole sectors have sometimes come to be seen as structural, leading to further re-conceptualization of both problems and policy solutions.

Chapter 3

1. This lack of coherence was emphasized at the inaugural meeting of the Royal Commission (minute 13a, RCEP first meeting 1970, 25 February); it was confirmed by two individuals who had engaged with environmental policy at that time, one as a member of the government and the other as a civil servant (interviews, June 1996). See also Kennet (1970).

2. Memorandum from Lord Kennet to the Rt Hon. Anthony Greenwood MP, entitled 'The Environment', 18 February 1970 (copy kindly made available by Lord Kennet).

3. Kennet had discussed environmental issues with his half-brother, the naturalist Peter Scott, and with Max Nicholson of the Nature Conservancy Council (Holdgate 2003).

4. The Central Council for Science and Technology existed for three and a half years from 1967, with Zuckerman in the chair. It reported directly to the prime minister and had the authority to scrutinize the scientific activities of departments as well as

Endnotes

the work of the research councils. Zuckerman (1988: 425) writes in his autobiography that 'it was on [the Council's] recommendation that the Royal Commission on Environmental Pollution was set up'. But the impetus came from more than one source. Holdgate (2003: 173) records that Greenwood, prompted by Kennet, had written to Harold Wilson in November 1968 to suggest the establishment of a committee or commission to review arrangements for controlling pollution. Kennet himself claims that the setting up of the Royal Commission was the biggest internal fight of his years in government (Kennet 1972: 79). He later reinforced the point in a letter to the editor of the journal *Science in Parliament* (Kennet 2007).

5. Changes in machinery of government and departmental responsibilities were announced in the House of Commons on 13 October (HC Deb 13 October 1969, vol. 788, cc31–44). Some commentators at the time suggested that Wilson had ulterior motives for giving Crosland a new role: Edward Heath, for example, then Leader of the Opposition, suggested that Wilson had 'created a vague overlordship of the environment in order to keep Mr Crosland inside the government but outside the inner conclave' (Heath 1970: 473). Zuckerman (1988: 414) recalls that a few days after taking up his new post, Crosland had asked him, 'What is Harold up to? Am I being sold down the river?' Zuckerman had responded with the view that Crosland 'had a major and responsible job on his hands'.

6. HC Deb 11 December 1969, vol. 793, c639. According to Zuckerman (1988), it was Wilson's idea that the body should be established as a royal commission, which could not be dissolved without the agreement of the monarch. One member of the government from that time recalled that Crosland, Kennet, and Holdgate met to decide on the form of the Commission in October 1969 (interview, June 1996).

7. Johnson refers here to the extensive salt desert and seasonal salt marsh that had been the subject of border disputes between India and Pakistan in the 1960s.

8. As described by Sir Martin Holdgate himself in an interview (June 1996). He records in his autobiography that Zuckerman saw the arrangement 'as giving him a considerable measure of control over what we [the Central Unit] did' (Holdgate 2003: 177). The Unit was initially serviced by the Cabinet Office.

9. The new department merged the Ministries of Housing and Local Government, Transport, and Public Building and Works. Paul McQuail (1994), in his fascinating history of the department, argues that the 'environment', in its modern sense, was not the determining force in setting it up. But, given the temper of the times, it was both attractive and presentationally important to have a department with responsibilities covering the built and natural environments, including a central coordinating responsibility for pollution control.

10. Interview (June 1996).

11. For the first year or two the Commission's meetings were held in offices in Great George Street (the location of the inaugural meeting) or Marsham Street in Westminster, central London, and thereafter mainly at nearby Church House, which was to remain the Secretariat's base until the late 1990s. The location of this base—never merely a neutral consideration—will be discussed later (see note 67 and the penultimate section of this chapter).

Endnotes

12. The appointment of a serving Government Chief Scientist to the Commission was never repeated. Alongside Ashby and Zuckerman (whose background was in zoology), were Wilfred Beckerman, an economist; Aubrey Buxton, Director of Anglia Television and co-founder of the World Wildlife Fund; Frank Fraser Darling, ecologist and conservationist; Neil Iliff, Deputy Chairman and Managing Director of Shell Chemicals UK; The Right Reverend Launcelot Scott, Lord Bishop of Norwich; Sir John Winnifrith, retired Permanent Secretary of the Ministry of Agriculture; and Vero Wynne-Edwards, a zoologist.
13. Minute 1, RCEP first meeting 1970, 25 February.
14. The first female member was in fact Shirley Paget, Marchioness of Anglesey, at that time Vice-Chair of the Civic Trust for Wales; she served from 1973 to 1979. The Commission remained predominantly male throughout its lifetime, though with an increasing proportion of women, rising to about one-third of the membership for the last five years or so of its existence.
15. As noted in Chapter 1, the UK government department with primary responsibility for environmental affairs changed in name and composition several times over the period 1970–2011 (Appendix 2). For consistency it is therefore referred to as 'the environment department', or simply 'the department', throughout this book, unless specifically named in extracts from interviews or other sources.
16. Interview (June 1996). These arrangements essentially remained in place, with minor variations in the Commission's last few years. The Chair was responsible for writing annual reports on the Secretary, but there was provision for reports to be countersigned and confirmed by senior officials in the environment department, so it could be said that the Secretary also had a line manager there. Interestingly, there was some concern about the effects of joining the Secretariat on its members' career progression. One official (who had held various high-level posts in the department) thought it important for the Secretariat not to be seen as a destination for 'amiable eccentrics' (personal communication, December 1997), and a member of the Commission (1980s–90s) recalled a general feeling 'that the Secretariat was an end posting and that they [had], as it were... burned their boats' (interview, March 1996). Some secretaries—Heads of the Secretariat—did indeed serve in the final years of their civil service careers, but for more junior members the posting could in fact be an important career step—and indeed, for most members a 'revolving door' with the department was typical. This sometimes had curious effects; on returning to the environment department, former members of the Secretariat might find themselves working on the government response to a Commission report that they had previously helped to draft—the experience of one interviewee (July 1996) who had worked on the incineration report (RCEP 1993a). In later years, the Secretariat was larger and movement in and out more fluid, with not all members coming from (or returning to) the department, though this remained the dominant pattern.
17. Interview (March 1997), division head in environment department (1970s–80s). One member of the Commission (1990s) described the links to the department as an 'umbilical cord' (interview, August 1996).
18. Minute 4, RCEP sixth meeting 1974, 16 September.

Endnotes

19. Interview (June 1996), senior civil servant (1970s–80s) and former director of CUEP, environment department.
20. The frequency of 'routine' attendance (in the sense of briefing and participating in discussion, as opposed to giving oral evidence) varied: it was relatively high, for example, during 1989 and 1990, when both the Environmental Protection Bill (ultimately the *Environmental Protection Act* of 1990) and the Environment White Paper were being drafted. The Director of what was then the Central Unit on the Environment (and the Commission's 'sponsor' within the department at that time) recalled that he would regularly 'go and ... explain to [the Commission] what was what' on policy issues, and give 'a situation report on how things were in the department' (interview, January 1997); this recollection is confirmed by the Commission's minutes. The Secretary of State himself (at that time Chris Patten) was present for part of the April meeting in 1990. During the period of the author's own service (1998–2008) attendance by environment department officials had ceased to be routine, though officials, and occasionally ministers (or their Opposition shadows), would join the Commission for lunch or dinner from time to time.
21. Interview (June 1996), Secretary (1980s).
22. Interview (July 1996), Assistant Secretary (1980s).
23. Interview (August 1996), member of Secretariat (1990s).
24. Minute 22, RCEP fourth meeting 1974, 6 June.
25. Letter from Sir Hans Kornberg to Prime Minister Margaret Thatcher, 11 November 1980, and response dated 28 November 1980. The proposal was that the costs of running the Commission should become part of the environment department's expenditure under the Administration Vote.
26. Interview, August 1996.
27. Minute 10, RCEP tenth meeting 1991, 31 October–1 November.
28. Minute 11, RCEP tenth meeting 1991, 31 October–1 November.
29. Minute 13, RCEP tenth meeting 1991, 31 October–1 November; also minute 2, RCEP eleventh meeting 1991, 5–6 December. After the tenth meeting, the Secretary wrote to the responsible official to set out members' concerns, reminding the department that the Commission was 'not responsible to the Secretary of State or any other Minister', whilst confirming that it would nevertheless consider some of the suggestions made in the audit report (letter, 8 November 1991). The meeting with the Director General took place on 26 November, after which he wrote to the Chair to assure him that the department was 'keenly aware both of the value of the Commission's published work and of the substantial voluntary contributions of time and expertise which members make' (letter, 4 December 1991). He confirmed the department's view that 'the content and direction of the Commission's study are for the Commission itself', though in practice there was 'full and constructive informal consultation' with the department, which was greatly valued. Further, the recommendations of the audit report were to be regarded as 'recommendations for the Commission itself to consider, and not in any sense as purported instructions from the Department to the Commission'. One agreed outcome was that the Commission would consider producing public-facing annual reports—a practice that did in fact commence later in the 1990s. Despite the emollient words, this was

Endnotes

a period of some dissatisfaction with the Commission, as will be discussed in the penultimate section of this chapter.

30. The guidelines (RCEP 1998a) stemmed from a review of working methods carried out by the Commission in preparation for the FMPR. These essentially set out its established working practices but also laid the ground for new procedures—for example, regarding which of its papers should be placed on its incipient website.
31. Interview (May 1997), Deputy Secretary, environment department (1980s).
32. The origins and evolution of the 'Central List' (its official name) are vividly described in Peter Hennessy's *Whitehall* (Hennessy 1989: ch. 13). The list was managed by the Treasury in the post-war years but taken over by the Civil Service Department in 1968. The Public Appointments Unit (PAU) was established within this Department in 1975 (later transferred to the Cabinet Office) to take a proactive role in broadening the list, which was further reformed in 1980 when self-nomination (or nomination of others) became possible. However, according to a former Secretary of the Royal Commission, the PAU was not enormously useful as a source of potential members, because it tended to provide 'vast lists of people, most of whom [were] totally unsuitable' (interview, August 1997).
33. Interview (January 1997), environment department official (1980s); emphasis in original. The interviewee was referring to the former Ministry of Agriculture, Fisheries and Foods (MAFF), absorbed into Defra in 2001 (see Appendix 2).
34. Interview (August 1996), Secretary (1980s–90s).
35. Interview (June 1996), Secretary (1980s).
36. Interview, July 1997.
37. Interview (November 1996), Deputy Secretary, environment department (1990s).
38. The same group would generally sift, shortlist, and interview. Panels usually included a senior official from the environment department (chairing the panel on behalf of the minister); the Chair of the Commission; a representative of at least one of the devolved administrations; someone representing the Civil Service Commission, which oversees public appointments; and an independent member. According to a civil servant who had chaired such panels, the Chair of the Commission had a 'major choice' in the appointments made (personal communication, March 2014).
39. Interview (August 1997), Secretary (1990s–2000s).
40. Interview (November 1996), Deputy Secretary, environment department (1990s). According to one former Chair, terms of service were initially interpreted quite strictly but became more open-ended from around the mid-1980s (Sir Richard Southwood, personal communication, March 1998).
41. With an overall limit of ten years' membership.
42. Interview, March 1997.
43. Interview (June 1996), Secretary (1980s).
44. This was in the Thatcher era. The member recalled that an appointment had eventually been made after 'a long period of arm wrestling' (interview, December 1996).
45. Interview (May 1997), Deputy Secretary, environment department (1980s–90s).
46. Interview (August 1997), Secretary (1990s–2000s).

Endnotes

47. Interview (August 1997), Secretary (1990s–2000s), referring to the practice of consulting retiring members about their replacements. Another interviewee (May 1997)—a Deputy Secretary in the environment department in the 1980s—thought that chairs were well-practised in identifying 'good' people, but this carried some risks: 'you have to make sure that the Chairman doesn't just cast everyone in his own image'.
48. Personal communication (March 2011 and June 2014).
49. Interview, August 1997. The possibility was also raised by Martin Holdgate (2003: 343), who had followed the Commission's development from its inception, and seen it from the perspective of senior positions in Whitehall as well as that of a member: 'Could the process of appointment change without reducing the ability to recruit Commissioners with both broad and deep knowledge, and willingness to give the considerable amount of time that was needed?'
50. Natural scientists over the years included biologists, chemists, ecologists, geologists, oceanographers, and physicists. Engineering was almost always represented. Among the members with backgrounds in the social sciences and humanities, economists and lawyers were almost invariably present; at different times there were also philosophers, human geographers, planners, and anthropologists.
51. Interview (November 1996), Director General for Environmental Protection, environment department (1990s).
52. In 2010, for example, Lord Rees, then the outgoing President of the Royal Society, mentioned the Commission when expressing concern that the 'independence of scientific advice' would be compromised by the abolition of advisory bodies (see Chapter 1): <http://www.bionews.org.uk/page_83514.asp> (accessed 10 January 2014).
53. Interview (July 1996), Assistant Secretary (1980s).
54. The term 'collegiate body' was used to describe the Commission by one of its longest serving members (1990s–2000s) (interview, February 1996).
55. Interviews (March, April, and June 1996), four members, serving at various times during the 1970s, 1980s, and 1990s. Many others made similar points.
56. Interview (March 1996), member (1980s–90s).
57. Interviews (March, April, and June 1996), three members, serving at various times in the 1970s, 1980s, and 1990s.
58. Interview (April 1996), member (1970s–80s).
59. Interview (May 1996), member (1970s–80s).
60. Interview (August 1996), member of Secretariat (1990s).
61. Written submission (25 May 1996) for this study from an industrialist, a member in the 1980s and 1990s, who had been frustrated by what he saw as the overly laborious process of producing reports. Interestingly, this member, together with another with a background in industry, had sought to modify the working practices of the Commission, urging, for example, more rapid production of reports and greater use of working groups, rather than plenary sessions, in their preparation. He acknowledged, however, that 'the ideas that I advanced were not attractive to many of the Commissioners'.
62. Interview (March 1996), member (1970s).

Endnotes

63. Interview (June 1996). This particular Secretary had experienced communication problems between economists and natural scientists; on other occasions, there were tensions between economists and philosophers (see Chapter 4).
64. Command papers were in practice laid before Parliament by the Ministry of Justice (and its predecessor departments). The post-2000 special reports were distinguished from the main reports of the same period by their shorter length, more specific focus, and relatively rapid production. In fact they were no shorter and no more specific than a number of the Commission's earlier outputs. The change in nomenclature was due in part to perceived pressures to be more productive—to be seen to run major and special reports in parallel—and in part because these special reports were not published as Command papers. The author recalls some consternation that the move to publishing certain studies in a non-traditional way might diminish the Commission's authority; an argument in favour, however, was that it could reduce the publication timetable by several months.
65. Interview (August 1997), Secretary (1990s–2000s).
66. In 1997, for example, the Commission felt that a statement of its views on the sulphur content of diesel and petrol (RCEP 1997b) should be issued in advance of the European Council of Ministers' meeting in June of that year, rather than awaiting publication of the twentieth report, its second look at transport and the environment (RCEP 1997c). The Commission had previously published a report specifically concerned with emissions from diesel engines (RCEP 1991a).
67. Additional meetings were sometimes organized during periods of intensive drafting when reports were nearing completion. For the first decade or so, the routine meetings were less frequent and, until the early 1980s, one day seemed to have sufficed to cover the Commission's business. For nearly three decades, as noted earlier, the Commission met, and the Secretariat was based, at Church House in Westminster. Because of rising rents and budgetary constraints, it became less settled in the late 1990s, moving its base to Steel House in nearby Tothill Street in November 1998; to The Sanctuary, adjacent to Westminster Abbey, in December 2001; and to 55 Whitehall (its final home) in May 2008. For part of the time at The Sanctuary, the lack of a meeting room forced the monthly meetings to become peripatetic (though they most frequently took place at the Institute of Mining, Minerals and Materials in Carlton House Terrace). As noted above, after a short initial period, the Commission managed to maintain its own offices until 2008, resisting co-location with the environment department and, later, with the Sustainable Development Commission (SDC). This was seen as a material expression of its independence: in a discussion about the first FMPR, for example, members 'were strongly of the view that the Commission's independence implied its offices should be separate from [the environment department's]' (minute 9, RCEP eleventh meeting 1999, 2–3 September). However, in the delicate negotiations about its budget that followed the second FMPR (discussed in the penultimate section of this chapter and in Chapter 7), the Commission agreed, with some reluctance, to move into a departmental building (55 Whitehall)—though it was one that the department regarded as 'special territory', housing only advisory bodies like the

Endnotes

Commission and the SDC and no departmental functions (personal communication, senior official, February 2014).
68. Interview, July 1997.
69. Interview (March 1996), member (1980s).
70. Professor (later Sir) Malcolm Grant, former Chair of the Agriculture and Environment Biotechnology Commission (a UK advisory body, 2000–5), used this analogy in reflecting on his experiences of the functioning of that body (Symposium on *The Cultural Politics of the Global Knowledge Economy*, University of Lancaster, July 11–12, 2005). The significance of the 'invisible wiring' cannot be overstated, and it will be discussed again in the context of the Royal Commission's positive attributes in Chapter 7. Congenial working conditions, and opportunities for social interaction ('nice dinner in the evening'), paid dividends in terms of building trust and facilitating rigorous deliberation. Indeed, these conditions were probably more important than the daily fee, about which there was ambivalence, at least initially, as we have seen.
71. 'Anchoring' refers to the 'softening up' of particular policy communities for the reception of new knowledge and ideas: 'Any new idea, especially any idea that results from reframing, is tested on the potentially involved actors well in advance so as to secure their cooperation or at least minimize their resistance' (Czarniawska 2000: 7).
72. Interview (February 1996), member (1990s–2000s).
73. The scoping stage was criticized by some interviewees, who considered it unduly restrictive, though it was not unusual for the scope to be broadened as a study progressed (in the lead study, for example, human health impacts of lead from petrol were initially excluded—see Chapter 4).
74. Interviews (July 1997 and July 1996), member (1980s–90s) and member of the Secretariat (1990s).
75. Interview (May 1996), member (1980s–90s).
76. Interview, July 1997.
77. There were some reservations about visits, in part because not everyone had time to participate. One member from the 1980s perceived a danger of creating a distorted picture—'it's very easy to be affected by the last visit you have', while another (1970s–80s) felt that visits had predominantly served a public relations purpose (interviews, June and December 1996).
78. One member (1970s–1980s) reflected in an interview (December 1996) that the ability simply to refer issues to a standing commission was a 'cracking idea for a government who might be in trouble'. It is not clear why ministers made so little use of this provision, though possibly there was a reluctance to impose on the Commission in ways that might be thought to interfere with its independence. See Chapter 7.
79. Interview (June 1996), member (1980s–90s).
80. Interview (June 1997), member (1990s).
81. Interview, March 1996.
82. Interview (April 1996), member (1970s–80s).
83. When asked whether the environment department had ever opposed a study, one Deputy Secretary (1970s–80s) replied, 'I'm sure we would never have been so crude' (interview, June 1996).

84. Interview (August 1996), member (1990s).
85. Interviews (July 1997 and June 1996), former Chair and former Secretary respectively.
86. Interview (January 1997), civil servant, environment department (1980s).
87. As described by one member (1970s–80s) in an interview (April 1996). Note, however, that a senior scientific officer in the environment department (1980s–90s) said that the oil pollution report had been 'timely' and had led to significant changes in emergency procedures (interview, March 1997).
88. Interview (March 1996), Sir Richard Southwood. According to a former Secretary of the Commission, the head of what was then the Central Directorate on Environmental Pollution in the department (Martin Holdgate) was also anxious to ensure the Commission's survival (interview, June 1996).
89. Minute 59b, RCEP sixth meeting 1981, 2 October.
90. Minute 8, RCEP second meeting 1999, 3–5 February.
91. Calls for evidence were published in the London, Edinburgh, and Belfast *Gazettes* and latterly also on the Commission's website.
92. These sessions should be distinguished from the less formal, invited presentations that often informed the Commission's thinking, especially in the early stages of an investigation.
93. Interview (August 1996), Secretary (1980s–90s). On one occasion, the Commission was embarrassed by its apparent neglect of a key player. Failure to engage with the Peat Producers' Association during the soils study (RCEP 1996) led to a threat of judicial review, though this was later withdrawn. The Association complained that the Commission had not made it clear at the outset of the study that peat extraction would be considered, nor had it taken evidence from the Association, but it had nevertheless made important recommendations that were relevant to the Association's interests (see Barron 2001).
94. The Royal Warrant did in fact grant powers to the Commission to summon witnesses. These were not invoked in any formal sense, though the Secretariat sometimes had to pursue potential witnesses assiduously. One interviewee, a former Confederation of British Industry official (1990s), suggested that it had been easy to persuade industry representatives to appear before the Commission but that they had sometimes been less enthusiastic about preparing detailed written evidence (interview, November 1997).
95. Interview (October 1995), former Chair.
96. Interview, August 1996.
97. Interview (June 1996), member (1980s–90s).
98. Written submission for this study (25 May 1996).
99. The point about 'guests' was made in interview (July 1997) by a former Chair. It is telling that the few sessions that were not congenial stood out in the recollections of interviewees. For example, several interviewees mentioned an incident during the incineration study (RCEP 1993a), when a witness from Greenpeace 'just blew his top', as one member put it; this was considered counterproductive—'he didn't do his cause any good' (interview, August 1996). Even so, according to the same member, the evidence from Greenpeace impressed upon the Commission 'the

Endnotes

need to design out waste'. On other occasions, including one in the author's own experience, civil servants exasperated the Commission beyond politeness by adhering doggedly to a government or departmental line (this was one reason for the tetchy sessions during the urban study [RCEP 2007], mentioned later). The Secretariat were more likely to feel a degree of sympathy for civil servants placed in a difficult position.

100. Interview (August 1996), member (1990s).
101. In the author's own experience, when witnesses adopted very different perspectives on an issue, they tended to engage in a form of boundary work, characterizing the views and strategies of their adversaries as 'political' or 'emotional', but seeing their own as factual, objective, and rational.
102. Interview (March 1996), member (1980s).
103. Interview (July 1997), former Chair.
104. A member who served in the 1980s and 1990s described working groups in this way, noting that the rest of the Commission then acted essentially 'as a referee' (interview, May 1996). On some occasions the Commission split into groups to consider specific issues, or different chapters of a report. Unusually, the report on incineration (RCEP 1993a) was substantially the product of a working group, though the draft was considered and approved by the whole Commission (see Chapter 4).
105. Interviews (June 1997, August 1996, June 1996) with members who served at various times in the 1980s and 1990s. An extreme case of re-drafting 'right up to the end' took place in the context of the environmental planning study (RCEP 2002a). At a meeting early in 2002, when the text of the report was due to be cleared, members demanded 'some re-ordering and redrafting' of one chapter. As the minutes record with subtle irony, '[t]he Secretary produced a revised version overnight, which was then approved, with some further amendments' (Minute 6, RCEP third meeting 2002, 31 January–1 February). As we shall see in Chapter 5, significant changes were also made at the following meeting, even after the report had gone to press.
106. Interview (September 1995), member (1980s).
107. Interview (February 1997), member (1980s–1990s).
108. Interview (March 1996), member (1970s).
109. Paper RCP(99)137-4 (para. 13), response to draft FMPR report (RCP[99]137-1), sent to review team following discussion at the Commission's eleventh meeting 1999 (2–3 September, minute 9).
110. Minute 6, RCEP fifth meeting 2001, 5–6 April.
111. Agendas and minutes became available from 1998. It is difficult to judge whether or how potential public scrutiny affected the minutes, because the level of detail varied in any case over time and under different secretaries. The availability of evidence (including written submissions and professional transcriptions of oral evidence) on the website became the default assumption unless particular considerations of confidentiality were involved.
112. The framework document discussed earlier (DETR 2001: para. 2.20) provided that: 'In future, RCEP reports will be presented to the Scottish Parliament, National

Endnotes

Assembly for Wales and the Northern Ireland Assembly at the same time as they are presented to the UK Parliament.' It also set out details of a rather complex relationship with the devolved administrations.

113. Interview (June 1997), member (1980s).
114. Interviews (March and May 1996), members serving in the 1980s–90s and 1970s respectively.
115. Written submission (25 May 1996) from a member who served in the 1980s and 1990s.
116. The largest report, at 377 pages of A4, was the twenty-fifth, on fisheries and the marine environment (RCEP 2004). It should be acknowledged, however, that the increase in the physical size of reports can be accounted for in part by differences in format, print, and layout. The last report in A5 format was the eleventh (RCEP 1985). There was also a change in style: the last of the main reports to have the traditional, plain blue cover was the nineteenth (RCEP 1996), after which covers became more decorative—another reflection, perhaps, of the perceived need to be accessible.
117. Full and summary reports from the twenty-second onwards were published online (free to download) as well as in hard copy. Scanned versions of all of the earlier reports were also made available; see <http://webarchive.nationalarchives.gov.uk/20110322143804/http:/www.rcep.org.uk/reports/index.htm> (accessed 10 August 2014). File download figures for a two month period shortly before abolition (December 2010 and January 2011) are interesting, though such figures can be influenced by many factors. Copies of twelve reports were downloaded during that period. The majority were relatively recent (2000 onwards), the most 'popular' being the (full) reports on the urban environment (RCEP 2007, c. 5,800 downloads), energy and climate change (RCEP 2000, c. 3,600 downloads), and novel materials (RCEP 2008, c. 3,500 downloads). But several earlier reports—on environmental standards (RCEP 1998b, c. 3,300 downloads), BPEO (RCEP 1988, c. 1,000 downloads), and oil pollution of the sea (RCEP 1981, 265 downloads)—also feature in the list (data provided by the Commission, March 2011).
118. Interview, July 1997.
119. Interview (August 1996), member (1990s).
120. Interview (November 1996), Deputy Secretary, environment department (1990s).
121. Interview (January 1997), senior civil servant, environment department (Central Directorate of Environmental Protection) (1980s).
122. On various occasions in 1989 the Chair (sometimes accompanied by other members) had met the Secretary of State, the Minister for the Environment, and the Permanent Secretary to discuss these issues (minute 3, RCEP fifth meeting 1989, 4–5 and 10 May; minute 5, RCEP sixth meeting 1989, 1 June; minute 20, RCEP seventh meeting 1989, 6–7 July). Deliberations within the Commission itself (revealing some dissent among members about desirable ways forward) are recorded in the minutes of the seventh (6–7 July), eighth (7–8 September), and eleventh (30 November–1 December) meetings of that year.
123. Both the fourteenth and fifteenth reports focused on tightly defined topics (RCEP 1991b, a), but the practice of producing such reports in parallel with the major

Endnotes

studies went into abeyance when criteria for choosing topics (Table 3.1) were formally adopted in 1993. The practice was revived in the form of 'special studies' some ten years later.

124. The Commission also re-affirmed its reluctance to conduct 'state of the environment' reviews (minute 34, RCEP eighth meeting 1989, 7–8 September). A similar idea had been resisted in the early 1980s, as the Chair from that time recalled: 'I said...that is the function of the [department]...we don't have the resources...we don't have the information and we don't have the expertise...it seemed to me a way of neutralizing the Royal Commission' (interview, July 1997). This is interesting given that in its early days, commenting on the state of the environment was seen as one of the functions of the Commission. The government response to its first four reports noted that 'the Commission have seen their role as requiring them not only to investigate in depth particular topics of special environmental significance, but also to make more general information on the state of the environment readily available' (DoE 1975: 1). Clearly this had been thought appropriate at that time, when such information had not previously been presented, but the later Commission did not want the onerous responsibility of regular state-of-the-environment reporting.

125. Letter from Brian Glicksman to Lord Lewis, 19 May 1989. Later the reports did develop a more international flavour and the Commission became an active member of the network of European Environment and Sustainable Development Advisory Councils (EEAC); see Chapter 7.

126. The report of the first FMPR (DETR 2000a) corroborates these points, looking back to the department's concerns of the early 1990s, and noting also those of the Treasury, which had questioned the purpose of the Commission at that time.

127. Interview (January 1997), civil servant in the environment department (1980s).

128. Interview (August 1996), member (1990s).

129. PricewaterhouseCoopers sought views from a range of governmental and non-governmental actors who had been involved in some capacity with the Commission's work since the previous review. In the report (Defra 2007b), they drew upon twenty-seven individual consultations and forty-seven responses to an on-line survey. This was more extensive than the consultation for the first FMPR, for which the in-house team had interviewed twenty-three individuals (including members of the Commission) and issued a more general invitation for views to which twenty-one organizations had responded.

130. Personal communication, August 2011.

131. Although the review recommended that there should be fewer academics, the department recognized in its response 'that the RCEP is a committee of experts...which will lead it to consist mainly of academic members' (Defra 2008: 4). There was also a softening of the consultants' emphasis on speed: 'The RCEP is already exploring new ways of working in its current study to increase the speed at which drafts are prepared without compromising the quality of outputs' (ibid.: 8).

132. In 2006–7, the Commission's budget was running at £940K per annum. In the interviews conducted for this study, many members refuted any suggestion that it was expensive to run, arguing instead that it represented outstanding value for

money compared with many other forms of expert advice. A former Chair saw it as a cheap way of providing advice 'which would have been difficult to obtain by other means', and one member (1970s–80s) commented that 'you couldn't get that intellectual firepower together for nothing very easily, you couldn't get it together for large sums of money' (interviews, July 1997 and December 1996). Both FMPRs tended to concur with the view that the Commission represented good value for money.

133. Parliamentary Written Statement on Departmental Arm's Length Bodies by the Rt Hon. Caroline Spelman MP (HC Deb 22 July 2010, vol. 514, c32WS).

134. It was lamented, for example, by Geoffrey Lean, Britain's longest-serving environmental correspondent, in his *Telegraph* blog (Lean 2011); by the outgoing President of the Royal Society (see note 52); and by *ENDS* magazine, which had so often covered the Commission's work (ENDS 2010). Prospect (2011), a union for professionals, complained of a 'cavalier disregard' for expert advice, undermining claims that the coalition would be the 'greenest government ever', while Scientists for Global Responsibility regretted the passing of the level of scrutiny that independent expert bodies like the Commission (and the Sustainable Development Commission) had provided: <http://www.sgr.org.uk/resources/scientists-condemn-government-over-cuts-environmental-watchdogs> (accessed 10 January 2014).

Chapter 4

1. This widely adopted exposition was explicitly endorsed in the first two versions of the UK Sustainable Development Strategy (UK Government 1994c, 1999). A principle along similar lines had in fact been accepted by the government in 1988 (HL Deb 13 January 1988, vol. 491, c1311), and had appeared in the 1990 White Paper on the environment (UK Government 1990: para. 1.18). Interestingly, the 2005 Sustainable Development Strategy did not repeat the Rio definition, but presented the precautionary principle as a means of 'taking into account scientific uncertainty' under the general principle of 'using sound science responsibly' (UK Government 2005: 16). At the European level, the *Maastricht Treaty* of 1992 (Article 130r; later, Article 191[2] in the consolidated version of the Treaty on European Union [Council of the European Union 2012]) established that Community policy on the environment should be based on the precautionary principle, though the principle itself was not explicitly defined. The European Commission later detailed procedures for the application of the principle, noting that it had been accepted as a risk management strategy 'when there are reasonable grounds for concern that potential hazards may affect the environment or human, animal or plant health, and when at the same time the available data preclude a detailed risk evaluation' (CEC 2000: 9). The European Council, broadly endorsing this Communication, considered that 'use should be made of the precautionary principle where the possibility of harmful effects on health or the environment has been identified and preliminary scientific evaluation, based on the available data, proves inconclusive for assessing the level of risk' (Nice European Council, 7–9 December 2000,

Endnotes

Presidency Conclusions: Annex III <http://www.european-council.europa.eu/council-meetings/conclusions/archives-2002-1993> (accessed 16 July 2014).

2. The term 'preventive' is used in this chapter in the sense proposed by Wynne (1992: 111). Note, however, that Levidow and Tait (1992) and Tait and Levidow (1992) use it differently, essentially to mean reactive attempts to prevent or ameliorate known harms.
3. The field is vast, and only a brief outline can be given here. Useful overviews of developments at different times can be found in Fischhoff, Slovic, and Lichtenstein (1982); Glickman and Gough (1990); Kasperson et al. (1988); Lash, Szerszynski, and Wynne (1996); Lewens (2007).
4. The emergence is often said to have been marked by Chauncey Starr's paper in *Science*, in which he asked what society was willing to pay for safety (Starr 1969).
5. For further discussion see Adams (1995); Boholm (1998); Dake (1992); Oltedal et al. (2004); Rayner and Cantor (1987); Thompson and Rayner (1998); Wildavsky and Dake (1990).
6. Factors considered included the degree to which risks were concentrated or diffuse, voluntary or involuntary, familiar or unfamiliar, natural or unnatural, and so on.
7. There are different typologies of 'incertitude' and the various terms are not always used consistently. One widely accepted version distinguishes between risk (known or knowable probabilities and impacts), uncertainty (impacts identifiable, probabilities not known), ambiguity (impacts or 'harms' poorly characterized, or contested), and ignorance (unknown unknowns; even the questions asked may be problematic). The term 'indeterminacy' has also been employed, to reflect the ways in which scientific knowledge and the social world are constructed, as well as 'the open-endedness in the process of environmental damage due to human interventions' (Wynne 1992: 119). For further discussion, see Expert Group on Science and Governance (2007); Funtowicz and Ravetz (1993); Hunt (1994); Ravetz (1990); Spiegelhalter and Riesch (2011); Stirling (2003, 2007). I am grateful to James Palmer for discussions on the different qualities of incertitude.
8. Wynne (1992: 123) argues that conventional risk science fails in respect of 'second order risks', which incorporate such aspects as 'institutional demeanour and forms of social control'. Similarly, Jasanoff (1995: 313) identifies concerns about social and political risks—including commodification of nature, economic dislocation in developing countries, and the loss of meaningful control over potentially transformative technologies—as important in shaping responses to biotechnologies. See also Grove-White (2001).
9. Such failures were obvious in the case of the BSE crisis, which seriously dented trust in institutions (Chapter 1), and in controversies over the Brent Spar oil platform (see Chapter 5), nuclear waste disposal, and GM crops. Reports on risk produced at different times by the Royal Society (1983, 1992, 1997) are indicative of the shift in thinking. The first acknowledged emergent research on 'risk perception' but was essentially technocratic. The reports of 1992 and 1997 included extensive discussion of social scientific perspectives, though the convenor of the former, Sir Frederick Warner (who had been a member of the Royal Commission), observed that the 'bridge' between natural and social scientists had yet to be put in place. By 1997,

John Ashworth (convening the third report) could argue that 'there had been a considerable change in the attitudes of all concerned in this debate' (Ashworth 1997: 4). Fischhoff (2000) traces similar changes in the work of the US National Research Council.

10. Notable examples include the *GM Nation* debate coordinated by the Agriculture and Environment Biotechnology Commission (GM Nation Steering Board 2003), the nationwide public deliberation initiated by the Committee on Radioactive Waste Management (CoRWM 2006), and a range of dialogues and engagements relating to nanotechnologies (reviewed in RCEP 2008). For an interesting discussion of outcomes and responses in the context of GM, see Doubleday and Wynne (2011).

11. This formulation had been influenced by the second annual report of the US Council on Environmental Quality, established in 1969—a case of policy ideas being in the air on both sides of the Atlantic (see RCEP 1972b: para. 12; US CEQ 1971: 260). Two other issues were covered in the second report: confidentiality (discussed in Chapter 6 of this book) and the disposal of toxic wastes on land. The Commission was in no doubt that the latter was a source of risk to the public, and called on the government to take urgent action—a point that it would press in a number of subsequent reports.

12. Interview (August 1996), Secretary (1980s–90s).

13. In these earlier reports the Commission tended to use 'ethical' in a vernacular way, counterpoising it with 'scientific' or 'economic', rather than couching its discussion more formally in terms of different ethical theories.

14. Other members acknowledged that economic instruments had potential but were less convinced about their likely efficacy in practice; even the two dissenters (Wilfred Beckerman and Solly Zuckerman) accepted that there would be practical difficulties (at the very least) in determining and achieving the 'socially optimum level of pollution', and agreed that in certain circumstances a prohibition of discharge would be more appropriate. In both the third and the fourth reports, the Commission called for more research on the economics of pollution (RCEP 1972a, 1974).

15. Sir John Hill, Chairman of the UK Atomic Energy Authority, quoted in Cook (1976: 11).

16. As noted in Chapter 3, members had been greatly affected by a visit to the Windscale [later Sellafield] reprocessing plant in the North West of England, where much of the existing waste was stored.

17. Interview (May 1996), member (a chemical engineer) who served during the nuclear power study. Another member, also a chemical engineer, recalled how he 'nearly blew up' when he read a draft of the report, which, in his view, overemphasized the problems of ultimate disposal (interview, April 1996). In contrast, a biologist serving at the time had become *more* concerned about waste after seeing the liquid waste storage tanks at Windscale and grasping the implications of the long-term need for cooling: as he said with feeling when interviewed, 'it wasn't *right*' (interview, August 1996).

18. Later Lord Marshall of Goring, at that time Chief Scientist at the Department of Energy and Deputy Chair of the UKAEA.

Endnotes

19. Interview, August 1997. Flowers had also previously been Head of the Theoretical Physics Department at Harwell, the Atomic Energy Research Establishment in Oxfordshire (1952–8). The point that he makes here is corroborated by Rough (2011), and indeed in the Parliamentary record. In a House of Lords debate on nuclear power and the environment, a few months after publication of the report, Baroness White (herself a member of the Royal Commission 1974–81), observed that Flowers' knowledge of the subject and his continuing membership of the UKAEA had given him 'an unrivalled advantage' in directing the Commission's enquiries. She went on to say: 'The widespread, indeed world-wide interest, which the report has commanded has, as he himself publicly admitted, set him at odds with certain of his colleagues' (HL Deb 22 December 1976, vol. 378, c1335).
20. It is interesting that lead in the public water supply (dissolved from lead pipes in areas of soft water) was also recognized as a threat to health, but it was a problem with a much lower profile. The Commission, too, afforded it less attention, though recommended that financial considerations should not constrain a pipe replacement programme.
21. Because of traffic growth, this policy already required progressive reductions in the lead content of petrol. On the basis of advice from its Chief Medical Officer in 1971, the government had introduced a 'phased three stage reduction from 0.84 to 0.4g/l to be achieved in 1986' (Haigh 1998: 137).
22. While the official government line was to reduce rather than remove, there were Whitehall departments with sympathies in both camps. Transport, Energy, and Industry had resisted moves to eliminate lead, and had prevailed on that point over Health and Environment in the compromise decision of 1981.
23. Interview (December 1996), member who served at the time of the lead study.
24. Paper RCP(82)301, 23 August 1982, para. 8. CLEAR had submitted written evidence (RCP[82]174) and gave oral evidence at the July meeting in 1982. The Commission seemed keen, however, to disclaim any 'conduit' role. The minutes record that 'quite independently of CLEAR's pressure, [it] had grounds for reconsidering its earlier decision not to review medical evidence' (minute 96a, RCEP seventh meeting 1982, 1–2 July).
25. Letter from the Chair, Professor Richard Southwood to the Secretary of State for the Environment, the Rt Hon. Michael Heseltine MP, 16 December 1982.
26. At the beginning of 1982, for example, when the focus of the study was still to be decided, Barbara Clayton said that she and Acheson (who was absent) might agree to look at lead pollution if the investigation were to concentrate on 'certain aspects' of the problem, but they 'would not be happy if the Commission were to attempt a full scale review' (minute 49, RCEP first meeting 1982, 15 January). Clayton argued, further, that 'whatever the effect of lead [on children], it must be small when account was taken of all the other relevant factors' (minute 50).
27. Interview (March 1996), former Chair.
28. RCP(82)345, 29 November 1982, a paper on the outline and conclusions of the ninth report, discussed at the eleventh meeting 1982, 2–3 December (minute 19).
29. Note by the Secretariat (RCP[82]336, 19 November 1982: para. 9), discussed at the eleventh meeting 1982, 2–3 December (minute 33).

30. Interview (February 1996), member (1980s–90s).
31. Minute 48, RCEP first meeting 1983, 6–7 January. The argument was reflected in paragraphs 5.23 and 5.25 of the report (RCEP 1983).
32. Critics argued that the non-linearity of the dose-response curve made the margin of safety greater than the Commission was suggesting (Harrison 1993; Williams 1983).
33. Interview (March 1996), member (1980s–90s).
34. Interview (June 1996), member (1970s–80s). The Commission was well aware that the science was unlikely to resolve the controversy. At a meeting in 1982, for example, when one member asked what sort of evidence could 'settle' the issue of low levels of lead and health, and how long it would take, Acheson responded that this was unlikely within five years, 'possibly never', and Clayton's view was that 'a judgment would have to be based on what was prudent' (minute 67, RCEP seventh meeting 1982, 1–2 July).
35. Interview (March 1996), member (1980s–90s).
36. He felt that even small additional costs should not be incurred unnecessarily (interview, June 1997).
37. Interview, June 1996. There is an interesting comparison with lead in drinking water (see note 20), on which the ninth report's recommendations were less than compelling (Millstone 1997). Here the Commission seems to have accepted that replacing lead pipes would be difficult and expensive, and perhaps preferred not to dilute its radical recommendation on lead in petrol with other controversial proposals.
38. Interview (June 1996), official who held various senior posts in the environment department in the 1980s.
39. The Surface Waters Acidification Programme (SWAP) was conducted jointly by the Royal Society, the Royal Swedish Academy of Sciences, and the Norwegian Academy of Science and Letters, and was funded by the British Central Electricity Generating Board and National Coal Board. All participants had agreed to abide by its findings, which, in the end, did lend support to the view that British emissions were contributing to the acidification of Scandinavian freshwaters (Mason 1990).
40. Interview, March 1996; also personal communication, September 1998. The official version was that the Commission felt unable to make detailed recommendations on the basis of a relatively brief consideration of acidification in the context of its wider overview (RCEP 1984: para. 5.96); a member of the Secretariat from the time also offered this reason in interview (July 1996). But this logic stands in some contrast to that of the ninth report, where the recommendation on lead in petrol was not dependent on a detailed interrogation of the medical evidence. It is notable, however, that the costs of abatement had been shown to be modest in the case of leaded petrol, but were likely to be high in the case of acid emissions from power generation (see Chapter 2).
41. Interview (June 1997), Chief Scientific Advisor in the environment department (1990s), alluding here to Wynne's critique of the construction of lay publics in conventional risk discourse. Wynne had argued that the contextual and experiential

Endnotes

knowledge of lay communities had much to contribute to the framing and assessment of risks, and that the trustworthiness of institutions was an important, and rational, consideration (see, for example, Wynne 1996; Lash and Wynne 1992).

42. The Commission seems likely to have had indirect impact, via the UK Government, during negotiations leading up to the 1990 Directive on releases of GMOs. There were also opportunities to make its views known more directly: for example, members visited Brussels during the study of GMOs for discussions with the Environment and Research Directorates (RCEP minutes of discussions with representatives of the Commission of the European Communities, 23–4 June 1988). According to one member serving at that time, the European bodies 'were thinking very much along the same sort of lines and for the same sort of reasons' as the Royal Commission (interview, April 1996).

43. Unusually, there were five special advisors for the GMO study. These included Professor John Beringer, chair of the Advisory Committee on Genetic Manipulation's Intentional Release Sub-committee, and subsequently the first chair of the statutory Advisory Committee on Releases to the Environment (ACRE). Certainly, in the use of advisors, learning was being facilitated by intersecting epistemic networks.

44. For example, HL Deb 18 May 1990, vol. 519, cc480–586.

45. Interview, October 1995. Another member who had served at that time expressed the view that 'it wasn't for the Commission to take an Olympian view about the morals or ethics of the situation' (interview, June 1996).

46. Minute 16G.1, RCEP tenth meeting 1988, 3–4 November. This was when a legislative slot was in prospect and the environment department had expressed a desire to have the Commission's report as soon as possible. According to a civil servant involved in the drafting of the Bill, Part VI (on GMOs) was at one time 'at risk of being thrown out'; in his recollection it was during the various manoeuvrings that the Commission came into play—'the whole thing was an iterative sort of situation' (interview, March 1997). The fact that Brussels was about to move on GMOs intensified the pressures on the department.

47. Interview (February 1996), member (1980s–90s).

48. Open letter from the Chair, Sir John Houghton, to the Secretary of State for the Environment, the Rt Hon. John Gummer MP, 29 November 1993 (RCEP 1993b). The Commission was also defended by one of its members in the House of Lords—see Chapter 7.

49. This framing was influenced by an impression of badly managed landfill sites carried forward by several members from the earlier investigation of waste; one said in interview (March 1996): '*nothing* would make me live next to a landfill, I think they're *disgusting* things, causing us untold problems...we're sitting on time bombs'. Interestingly, the Commission's distaste for landfill caused consternation within the environment department, at a time when delicate negotiations over the European Landfill Directive were in prospect (letter from Head of Waste Management, Department of the Environment, to Secretary of the Royal Commission, 25 January 1993). In any case, it was clear that landfill would have to be lived with for some time. Although the department welcomed the conclusions on

Endnotes

incineration, it felt that the Commission had overlooked improvements in landfill practices and taken insufficient evidence before 'pronouncing', as one official involved at the time put it (interview, July 1997; this individual later became Secretary of the Commission). This was one instance where a draft report was moderated after 'factual checking', though the same official insisted that the department would have drawn the line at 'leaning' on the Commission. Indeed, the department would have preferred no mention of landfill at all, as is clear in the letter referred to above.

50. Written evidence from Greenpeace, Annex C, paper RCEP(92)401, 27 May 1992. The incineration report acknowledged that public confidence might be increased by a more strategic approach to waste management, but the wider argument about framing was not developed.
51. It is significant that environmentalists had failed to persuade the Commission to adopt a wider framing (and the evidence session with one group had been unusually confrontational). More than ten years later, one scientist from a prominent environmental NGO, in welcoming a recent Commission report, told the author that he had 'finally forgiven' the Commission for its stance on incineration. Such hostility was not universal amongst environmentalists; indeed, one of the most prominent, Jonathon Porritt, rounded on the anti-incineration coalition in an article in *The Daily Telegraph* a few months after publication of the report (Porritt 1993).
52. Interview (July 1997), official who had dealt with the incineration report in the environment department.
53. *Vorsorgeprinzip* was associated, in 1980s West Germany, with emissions control based on state-of-the-art technology, though in application it was qualified by the principle of proportionality. The ecomodernist lessons were being transmitted within transnational epistemic communities.
54. Personal communication (April 2004), Secretary (1980s).
55. Minute 8, RCEP eighth meeting 1997, 3–5 September.
56. The tensions were still fresh in the minds of members when the author joined the Commission in 1998. 'Conflicting views' (minute 3, RCEP sixth meeting 1998, 13 May) are apparent from the minutes of meetings held earlier that year, when proponents and critics of cost-benefit analysis were drafting sections for the report. Minute 3 of the sixth meeting recorded members' appreciation for 'the way in which [the Secretariat] have tried to ensure that the views expressed in the Commission's discussions are reflected fairly and impartially in drafts prepared for Members to consider'.
57. The author remembers a conscious effort to revisit the *Standards* study and apply the ideas within it. In one case, a decision was made to reproduce a figure showing how public values should inform all stages of the standard-setting process (Figure 8.1 in the Standards report [RCEP 1998b] and Figure 1–II in the Chemicals report [RCEP 2003]). In meetings, however, the arguments about different approaches to risk had to be rehearsed all over again.
58. Commenting on the incorporation of safety factors in toxicological risk assessment, for example, the Commission suggested that '[c]laims that this process is scientific must be doubted—the basis for using these factors is often a political or

235

Endnotes

regulatory decision, and subjective interpretation can dominate the risk assessment' (RCEP 2003: para. 2.144).
59. In similar vein, the Commission felt that the use of tonnage thresholds of production for determining testing requirements was 'unsatisfactory' (RCEP 2003: para. 2.80).
60. The ACP, in existence at the time of writing (2013), is a statutory committee charged with advising ministers on matters relating to the control of pests. It is a classical 'expert committee', being composed of individuals with expertise in areas such as health, animal and plant sciences, and agriculture, plus two 'lay' members.
61. The Committee on Toxicity of Chemicals in Food, Consumer Products and the Environment (COT) and the Committee on Carcinogenicity of Chemicals in Food, Consumer Products and the Environment (COC), asked to comment by Defra and the ACP, were also critical of a number of aspects of the Commission's report (COT/COC 2006) and argued that there was no scientific basis for the adoption of a more precautionary approach (for further discussion, see van Hemmen 2006; Warren 2009).
62. The brief overview presented here draws on Linda Warren's (2009) interesting and more detailed analysis of the legal process and its implications.
63. Downs v. Secretary of State for Environment, Food and Rural Affairs [2009] Env L R 19.
64. Secretary of State for Environment, Food and Rural Affairs v. Georgina Downs [2009] EWCA Civ 664: 43.
65. Secretary of State for Environment, Food and Rural Affairs v. Georgina Downs [2009] EWCA Civ 664: 51–2.
66. The Commission defined a nanomaterial as 'one that is between 1 and 100 nm [nanometres, where one nm = 10^{-9}m] in at least one dimension and which exhibits novel properties' (RCEP 2008: para. 2.5). Gold, for example, is an inert material in its natural bulk form, but at a particle size of 2–5nm, it becomes highly reactive. The particular properties are often the reason for the utility and production of nanomaterials.
67. Either because there was some evidence of harm, or because harm seemed plausible on the basis of analogy with more familiar materials, such as asbestos fibres, which were known to be harmful.
68. That is, an experiment in which (because of the large numbers and multiple variants of nanomaterials), 'the possible combination of conditions rapidly becomes daunting' (RCEP 2008: para. 4.53).
69. This point about breadth was made in interview by former members from different disciplines, including, for example, a philosopher (1970s–80s) and a biologist (1990s) (interviews, June 1996 and June 1997).

Chapter 5

1. Paper RCP(75)111, 25 April 1975, written by one of the Associate Members for the air pollution study (Jon Tinker), and discussed at the Commission's ninth meeting 1975, 1 May.

Endnotes

2. Extract from *The Rubáiyát of a Royal Commissioner*, presented to Sir Richard Southwood by fellow members of the Commission on his retirement as Chair in 1985. *The Rubáiyát* was written by Professor Gordon Elliott (Tony) Fogg (member, 1979–85) and presented on a scroll, with calligraphy by John Schyvers.
3. Minute 14, RCEP fourth meeting 1974, 6 June.
4. Interview (June 1996), Secretary (1970s).
5. Minute 29, RCEP fifth meeting 1974, 4 July.
6. Interview (May 1996), member (1970s), a chemical engineer with an industrial background.
7. Tinker was a journalist who had run a prominent, critical campaign about the Alkali Inspectorate in *New Scientist*—see Chapter 2.
8. RCEP(75)111, 25 April 1975, p. 2.
9. Interview (March 1996), member, later Chair (1970s–80s).
10. Interview (June 1996), Secretary (1970s).
11. Interview (May 1996), member (1970s). The change to which he referred was the move of the Alkali Inspectorate from the environment to the employment department in 1975, discussed later.
12. Interview (June 1996), Secretary (1970s).
13. The phrase 'elephantine gestation period', referring to the time taken to assemble a government response to the fifth report, was used in an internal environment department circular from an Assistant Secretary in the Central Directorate on Environmental Pollution (CDEP) to various others, 16 November 1982 (CDEP/10/29 Part 5). The department's archives were helpful in tracking the extended development of the response to the fifth report. References of the form CDEP/10/29/Part X denote the file consulted. Individuals are not identified by name.
14. Minute 18, RCEP twelfth meeting 1974, 5 December. Tensions came to a head at the end of 1979 when it was proposed to move the HSE wholesale to Bootle in the north west of England. The environment department opposed any move that included the headquarters of the Alkali and Clean Air Inspectorate (ACAI), complained about lack of consultation, and wondered whether the HSE's behaviour warranted 'declaring war and seeking the return of ACAI to DoE' (memorandum from official, Air and Noise Division, to Assistant Secretary, Air, Noise and Wastes Directorate (and others), 4 March 1980 [CDEP/10/29A Part 2]).
15. Memorandum from Under-Secretary, CDEP to Deputy Secretary, Environmental Protection, 2 August 1979 (CDEP10/29A/Part 1). The quote is from an attached draft minute for the Deputy Secretary to send to the Private Secretary of the Minister for Local Government and Environmental Services (henceforth, 'the environment minister', at that time Tom King).
16. Memorandum from Under-Secretary, CDEP to the department's Chief Scientist and Deputy Secretary, Environmental Protection, 12 August 1982 (CDEP 10/29/Part 5).
17. Memorandum from official, Air and Noise Division, to Under Secretary, Air, Noise and Wastes Directorate, 22 December 1980 (CDEP 10/29/Part 3).
18. Memorandum from Head of Air and Noise Division, to Under Secretary, Air, Noise and Wastes Directorate, 4 December 1980 (CDEP 10/29/Part 3). This view is

Endnotes

surprising, given that evidence from a study conducted by the department and the Inspectorate in 1976, based on visits to nearly 50 plants, suggested that the BPEO approach would lead to environmental improvements in about 1,200 registered plants (Annex to note from Deputy Secretary, Environmental Protection to Private Secretary, environment minister, 3 August 1979 [CDEP 10/29/Part 1]). Later, the department used this evidence as an argument in favour of IPC, by which time it was claiming that 'over half' of IAPI's scheduled processes might be suitable for cross-media control (DoE 1988: para. A2.2).

19. Memorandum from Head of Air and Noise Division to Under Secretary, Air, Noise and Wastes Directorate, 8 December 1980 (CDEP 10/29/Part 3).
20. Memorandum from Deputy Secretary, Environmental Protection to Private Secretary, environment minister, 29 February 1980 (CDEP 10/29A/Part 2, 5).
21. Interview (March 1997), Deputy Chief Scientific Officer, CDEP (1970s–80s).
22. Interview (January 1997), environment department official (1980s).
23. It had been suggested, for example, that a letter from the Secretary of State to the Chair of the Commission might be sufficient, perhaps supplemented by a Written Answer to a Parliamentary Question. But by 1982 the view was that 'this would look decidedly as if we were trying to brush our embarrassment under the carpet' (memorandum from Assistant Secretary to Director, CDEP, then to Deputy Secretary, Environmental Protection, 11 March 1982 [CDEP 10/29M/Part 4]). There was concern to respond to the Commission with proper dignity, and the department was keen not to draw attention to the long delay. Handwritten alterations to a near-final draft are revealing: 'The ~~Government greatly regrets the~~ time that has elapsed in responding to this report. ~~This~~ in no way reflects on ~~the~~ [its] value ~~of the report~~ as seen by Government'. A note scribbled in the margin reads: 'The Government should not emphasise the belatedness of the response by "greatly regretting"' (CDEP 10/29/Part 5). In the end, the response acknowledged the 'considerable time' that had elapsed and attributed the delay to 'the far-reaching nature of certain recommendations' (DoE 1982: para. 1).
24. In a circular to various colleagues on follow-up consultations with the pollution control authorities, an Assistant Secretary in CDEP sounded exasperated: 'I have simply tried to make the best of a bad job: we do seem to be at sixes and sevens over how to progress the BPEO philosophy or whatever', 16 November 1982 (CDEP/10/29 Part 5).
25. Interview (May 1997), Deputy Secretary, environment department (1980s), emphasis added.
26. Interview (May 1997), Deputy Secretary, environment department (1980s); see also Weale (1996).
27. Interview (May 1997), Deputy Secretary, environment department (1980s).
28. In 1980, for example, the Deputy Secretary, Environmental Protection, had informed the environment minister that '[t]he setting up of a thorough-going HMPI with wide-ranging statutory powers would be likely to call for substantial extra resources' (memorandum, 29 February 1980; CDEP 10/29A/Part 2). Interestingly, in the end, the Department of Trade and Industry (DTI) also supported the creation of the unified inspectorate. A senior official in the environment department, Head

of the Central Unit on the Environment (CUE) in the late 1980s, joined forces with his opposite number in DTI, who was keen to change that department's traditionally negative reaction to environmental regulation (interview, former Head of CUE, January 1997). The two officials had a 'mutuality of interest', and the Central Unit 'brought the DTI along' into a cooperative relationship (interview, former DTI official, March 1997). Indeed, according to the first interviewee, the DTI was more easily persuaded of the benefits of integration than some parts of his own department, notably the Water Directorate.

29. Letter to environment minister from Deputy Director General of the CBI, 13 October 1980; also 24 March 1981 (CDEP/10/29/Part 3). There is ample evidence from archives and interviews that industry wanted intelligible, predictable controls and favoured a strong unified inspectorate located in the environment department (whereas the trade unions preferred it to stay with HSE). Some within the department, however, felt that the CBI's support stemmed more from opposition to the HSE than from any conviction about the intrinsic merits of integrated pollution control and BPEO (memorandum from official in CDEP to Assistant Secretary, 27 November 1979, assessing potential support for HMPI proposals outside the department [CDEP/10/29A/Part 1]).

30. HMIP brought together the existing Industrial Air Pollution, Radio-Chemical, and Hazardous Wastes Inspectorates with a new inspection branch for water pollution. Initially, there was no statutory basis for integrated pollution control, and for several years the new inspectorate was beset by difficulties and controversy (Owens 1989, 1990; Weale 1996). The government's acceptance, in 1987, of technology-based controls for certain dangerous discharges to water meant that differences between traditional approaches to 'difficult' air and water pollution could (begin to) be reconciled. As noted in Chapter 2, this move helped to clear the way for the functioning of the new inspectorate.

31. Interview (May 1997), Deputy Secretary, environment department (1980s), emphasis added. BATNEEC—'best available technology not entailing excessive costs'—was a concept that was 'coming over the hill from Europe' (according to the same official), with some similarities to BPM, but incorporating the European emphasis on best *available* technology. The interviewee implied that BATNEEC (across media) was essentially what HMIP would seek to achieve, thus contributing to European objectives. This would indeed be the case when HMIP was placed on a statutory footing in 1990, though the relationship between BATNEEC and BPEO was conceptually rather complicated (see Owens 1989, 1990).

32. Minutes 12 and 27, RCEP eighth meeting 1986, 4–5 September.

33. Minute 8, RCEP second meeting 1986, 6–7 February. European Year of the Environment (EYE) was due to commence in March 1987.

34. Minute 23, RCEP fourth meeting 1986, 4 April.

35. The department's report had claimed to be an assessment of BPEOs but the Commission considered its methodology to be seriously deficient. The issue was discussed extensively at the fourth and fifth meetings in 1986 (4 April and 1–2 May). It led to correspondence between the Chair and the environment minister

Endnotes

(Waldegrave), and a meeting in October of that year between certain members of the Commission and environment department officials.

36. Interview (January 1997), Head of CUE, environment department (late 1980s).
37. Interview (June 1997), senior scientist (latterly Chief Scientist), environment department (1980s–90s).
38. Smith (2000: 107) claims that it was Chris Patten, appointed as Secretary of State for the Environment in July 1989, who 'rode the political wave' and secured a 'big slot' for integrated pollution control in the Environment Bill (introduced at the end of that year). One environment department official involved at the time suggested that it was Patten's predecessor, Nicholas Ridley, who had initiated thinking about the Bill in the aftermath of Thatcher's speech (interview, January 1997).
39. Minute 13, RCEP ninth meeting 1987, 1–2 October.
40. Minute 66, RCEP seventh meeting 1987, 2–3 July.
41. Interview (April 1997), former Director, HMIP (1980s–90s). A similar point was made by the Head of Pollution Policy in HMIP (later in the Environment Agency) (1990s), who had been in the environment department in the 1980s (interview, March 1997).
42. A particularly prominent controversy was that over disposal of the Brent Spar North Sea Oil Platform in 1995: the protagonists (led by Shell UK and the UK government on the one hand, and Greenpeace on the other) adopted deeply divergent interpretations of what was 'best' and 'practicable', bound up with different conceptions of risk and 'harm' (Collins, Weinel, and Evans 2010).
43. At the time of writing (2014) the Environment Agency cites the definition *verbatim* in advice on Strategic Environmental Assessment <http://webarchive.nationalarchives.gov.uk/20140328084622/http://www.environment-agency.gov.uk/research/policy/32949.aspx> (accessed 23 January 2014), and it continues to be cited in the context of waste management (see, for example, Environment and Heritage Service Northern Ireland 2005).
44. Interview (January 1997), Head of CUE, environment department (late 1980s); the individual had been closely involved with the development of the Environmental Protection Bill.
45. Interview (March 1997), Head of Pollution Policy, HMIP (later the Environment Agency) (1990s), formerly in the environment department; emphasis added.
46. Nigel Haigh, Director of the Institute for European Environmental Policy, made this comment about the Environment Agency (of which he was at that time a Board member) at the Royal Commission's twenty-fifth anniversary seminar (RCEP 1995b: 45).
47. The Directive applied to 'installations' rather than processes, and was broader in scope in a number of other ways (Mills 1998: 10). For example, it included considerations of energy efficiency, raw materials consumption, and decommissioning within the remit of the regulatory authorities, and in defining the 'best available technology' did not limit 'costs' to those of the operators but could include wider economic and social considerations (arguably coming closer than the *Environmental Protection Act* 1990 to the Royal Commission's concept of BPEO).
48. As seen by one environment department official in the 1990s (interview, March 1997). See also Skea and Smith (1998); Weale et al. (2003).

Endnotes

49. Interview (March 1997), environmental department official (1990s). A civil servant from the department had been dispatched to Brussels to help draft the IPPC directive.
50. Interview (March 1996), member (1980s–90s).
51. Interview (May 1997), senior advisor in DG environment, European Commission (1980s). The point about multiple origins was also made by Michael Rogers (former advisor to the Bureau of European Policy Advisors, European Commission) at a Symposium on *Quality Control in Scientific Policy Advice*, the Berlin–Brandenburg Academy of Sciences and Humanities, January 2006 (Lentsch and Weingart 2011).
52. Interview (March 1997), civil servant, environment department (1990s). This official remembered being puzzled by his colleagues' eagerness to take this 'new thing' and 'inflict it on the rest of Europe'. For some years, however, there had been a view that BPEO could be a useful concept in defending British pragmatism. This was evident, for example, during the Commission's discussion of the concept with Martin Holdgate, then Chief Scientist (and Deputy Secretary) in the department, at a meeting in 1986. Holdgate observed that UK governments had 'consistently rejected the extremism implicit in the *Vorsorgeprinzip*' (minute 46, RCEP ninth meeting 1986, 2–3 October), but BPEO held out the prospect of 'a compatible and rational approach to environmental protection throughout the European Community' (minute 48). One member (1980s–90s) felt that the elaboration of BPEO was 'all part of this argumentation against the continental practice' (interview, March 1996), though the minutes reveal a concern within the Commission that the concept should not be seen as a 'soft option' (minute 23, RCEP second meeting 1987, 5–6 February; minute 45, RCEP fifth meeting 1987, 30 April–1 May). Similarly, an official in the department when the IPPC Directive was being drafted claimed that the British version of integrated pollution control was seen as a counterpoint to the 'Germanic' approach to regulation; a former Director of HMIP made much the same point (interviews, March and April 1997). This tactic may have worked, in that 'the creature which emerged from the negotiating process was very different from the European IPC of original conception' (Mills 1998: 10; see also Pallemaerts 1996).
53. Interview (March 1997), Head of Pollution Policy, HMIP (later the Environment Agency) (1990s), formerly in the environment department.
54. This ability was in part dependent on 'torchbearers': a Secretary from the 1980s, for example, recalled that Richard Southwood (Chair 1981–5) had seen himself as a 'custodian of BPEO, allied to HMPI, and he really did see it as part of his mission in life to keep up the pressure on the government to go for integrated pollution control' (interview, June 1996). See Chapter 7 for further discussion of such tenacity.
55. Interview (June 1996), emphasis added.
56. Minute 7, RCEP thirteenth meeting 1998, 3–4 December. There was a counter-feeling, however, that the study could take things forward by showing how principles set out in the standards report might be applied in practice and how an integrated approach to environmental objectives could be achieved within a given area (former Secretary, personal communication, March 2010).

Endnotes

57. Agricultural and forestry operations were not deemed to be 'development' in this context so were largely excluded from control by the planning system. It should be noted that Scotland, even before devolution, had a distinctive land use planning system of its own.
58. Land use had been considered extensively in a report on transport (RCEP 1994), and touched upon in the Commission's work on waste (RCEP 1985, 1993a), water (RCEP 1992), and soils (RCEP 1996).
59. As noted in Chapter 3, there were significant changes to the text at the meeting early in 2002 at which it was supposed to be approved. At the next meeting, members decided that the final chapter, a crisp resumé of key messages, should be brought to the front to replace a summary which they considered 'inferior to the concluding chapter' (minute 4, RCEP fourth meeting 2002, 28 February–1 March). Although the Secretariat protested that this change might endanger the publication date (the first part of the report having already been paginated by the printer), it was pointed out that the page length of the two sections was the same, and members prevailed. The drafting of reports, especially in the final stages, was always challenging, but the author recalls this meeting as a particularly intense and fraught one.
60. The term 'Integrated Spatial Strategies' was settled upon at an extra meeting in January 2002 (minute 4, RCEP second meeting 2002, 24 January), not long before publication of the report. In including agriculture and forestry within the remit of such plans, the Commission was not recommending that these activities be brought within the definition of 'development'; rather, its aim was that they should be considered 'at a strategic level and on a spatial basis alongside other relevant factors' (RCEP 2002a: para. 10.16).
61. Strategic planning, providing the statutory framework for more detailed land use planning at the local scale, had previously been the responsibility of elected County Councils (outside metropolitan areas) or Unitary Authorities (usually covering major cities and their hinterlands). The proposed changes meant that the County Structure Plans and Unitary Development Plans (Part I) prepared by these authorities would be abolished. At the next level up, non-statutory Regional Planning Guidance would be replaced by statutory Regional Spatial Strategies, which would become the strategic plans for the areas that they covered and would be prepared by Regional Assemblies. These bodies, created in 1998, were seen by their critics as less accountable than directly elected local authorities, being composed entirely of appointees (with around 70 per cent drawn from local government and 30 per cent from other organizations). Though local government organization changed substantially in the 1990s and 2000s, the extent of the re-scaling of the planning system can be gauged from the fact that (using data for 2009) there were only nine regions in England, but fifty-six Unitary Authorities and twenty-seven shire counties.
62. Minute 6, RCEP thirteenth meeting 2001, 1–2 November.
63. The Commission was keen to influence forthcoming legislation, and therefore to publish the environmental planning report before the end of the consultation period for the planning Green Paper (that is, by 18 March 2002). But at the

December meeting in 2001, when it had been hoped that the report could be signed off, there was 'still a great deal of work to do on the text' and some members were unwilling to sign the report before going through it in detail again (Minute 10, fourteenth meeting 2001, 6–7 December). In the event, the Commission submitted a response to the consultation in time to meet the deadline, citing its forthcoming report, which was actually published a few days later.

64. *Today* Programme, 15 March 2002: presenter Alan Little. The Chair's interview (conducted by Roger Harrabin) had been pre-recorded; the Minister was invited to comment live. A Parliamentary select committee also complained, later that year, that there was 'a "business" agenda running through much of the [planning] Green Paper' and that environmental issues had been relatively neglected (House of Commons Transport, Local Government and the Regions Committee 2002: para 103). The Committee had taken evidence from the Royal Commission.

65. The Commission recognized that it could have handled the press more skilfully on this occasion (minute 10, RCEP fifth meeting 2002, 4–5 April).

66. One factor in the lukewarm response—and indeed in the rather strained relationship with government during preparation of the report—might have been that planning (in the sense of Town and Country planning) no longer fell within the remit of the environment department, where there had been a planning directorate for around three decades, but had been removed to a separate department responsible for transport, local government, and the regions (DTLR) (subsequently, without the transport portfolio, the Office of the Deputy Prime Minister [ODPM], then the Department for Communities and Local Government [CLG]—see Appendix 2). Possibly, the Commission was perceived as a 'Defra body', acting outside its remit—a forgetfulness of its cross-governmental status to which we return in Chapter 7.

67. The proposal was to combine Regional Spatial Strategies with the Regional Economic Strategies prepared by Regional Development Agencies, and to incorporate certain other plans—for example, for housing and for sport and culture.

68. Undated memorandum from Professor Sir Tom Blundell to Members and Secretariat. Similar claims were made in the Commission's review of activities for 2001–3 (RCEP undated).

69. Minutes, Annex B, RCEP sixth meeting 2002, 2–3 May; minute 95, RCEP eighth meeting 2002, 3–5 July.

70. In this respect, the guidance (Planning Policy Statement 1) echoed thinking in the revised national sustainability strategy (UK Government 2005). The Commission observed, in its review of activities for 2001–3, that in the aftermath of its report, '[t]he strong emphasis on business efficiency in the Green Paper has ... been moderated to recognise more explicitly the role of planning in protecting the environment' (RCEP undated: 12).

71. It is interesting to trace how the proposals for regional strategies evolved in response to consultation (CLG and BERR 2008, 2009; CLG and BIS 2009). One effect was that issues such as climate change and biodiversity, and the concept of environmental limits (neglected in the Sub-National Review), gained ground. Another was the sharing of responsibility for producing the strategies between RDAs and 'Local Authority Leaders' Boards'—the latter comprised of elected members drawn from

participating authorities—to provide 'a mechanism for democratic input' (CLG and BIS 2009: para. A3.3). The *Local Democracy, Economic Development and Construction Act* imposed a duty on both bodies to consult widely (with explicit reference to environmental organizations) and provided for early input from an Examination in Public panel. While the Royal Commission had advocated a single lead body for its Integrated Spatial Strategies, it had declined to identify a candidate. It would not have been happy, however, with this role being conferred on the RDAs alone.

72. Revocation was announced in a written Ministerial Statement by the Secretary of State for Communities and Local Government (Eric Pickles): HC Deb 6 July 2010, vol. 513, no. 57, cc4–5WS (the quote is from c5WS); the Minister had the necessary powers to do this under s79(6) of the *Local Democracy, Economic Development and Construction Act* 2009.

73. There was, nevertheless, one important area in which the planning report had tangible impact, in that it helped—obliquely but in a way that is traceable—to bring about the instigation of a specialist Environmental Tribunal in England and Wales. Such an institution had been mooted, in various guises, for several decades and, in the planning report, the Commission advocated a tribunal as a more coherent way of handling appeals under environmental legislation (RCEP 2002a). Through a complex series of developments, including commissioning by the government of further work on this subject by Professor Richard Macrory (member of the Commission 1992–2003), and more general moves to reduce administrative burdens on business, an Environmental Tribunal eventually came into being in 2010. A full account of these developments is given by Macrory (2013).

74. Interviews (March and January 1997), civil servants from environment department (1980s), the first of whom had moved to HMIP (later the Environment Agency).

75. The twenty-third report did find an audience amongst academics with interests in ecological and spatial planning; see, for example, Allmendinger (2011); Harrison (2006); Jackson and Illsley (2007); Miller and Wood (2007). Interestingly, Jackson and Illsley cite the Commission's argument that Sustainability Appraisal (assessing economic, social, and environmental impacts together) risked marginalizing the environment (RCEP 2002a: para. 7.47) and go on to note the Scottish Executive's decision not to require inclusion of social and economic factors in environmental reports prepared under the *Environmental Assessment (Scotland) Act* 2005. However, in stating its own view that such inclusion would risk 'obscuring' environmental considerations, the Executive made no explicit reference to the Commission's twenty-third report (Scottish Executive Environment Group 2005: para. 3.41).

76. As noted earlier, there was some interaction with officials in two departments while the *Planning and Compulsory Purchase* Bill was being drafted. Other activities by members included organization of a seminar (July 2002) by Professor Richard Macrory, an environmental lawyer (minute 91, RCEP eighth meeting 2002, 3–5 July), and publication of an article by the author in the professional journal, *Town and Country Planning* (Owens 2002). The possibility of placing a Parliamentary question was considered (minute 96, RCEP eighth meeting 2002, 3–5 July), but it was not clear who would ask it and the idea was not pursued. Members gave evidence to the House of Commons Transport, Local Government and the Regions

Endnotes

Committee (2002) and the ODPM (Office of the Deputy Prime Minister): Housing, Planning, Local Government and the Regions Committee (2003). In April 2003, the Chair and two members met the Minister for Housing, Planning and Regeneration (then Tony McNulty), who agreed that 'there was considerable scope for the Commission's report to inform the further development of planning policy, including the revision of planning guidance following the [Planning and Compulsory Purchase] Bill' (RCEP undated: 12). On this occasion, 'the Commission offered to work with the government in identifying ways in which environmental planning can be strengthened under the government's proposed Regional Spatial Strategies' (ibid.: 12), but no such collaboration actually took place.

77. Minute 3, RCEP tenth meeting 2007, 6–7 December. The author had prepared a short paper, 'Planning Matters' (4 December 2007), tabled at this meeting. Members expressed concern about the proposals in the Review and asked the Secretariat to find out more about the consultation process. During 2008, however, when the consultation took place, the Commission was fully engaged with both the new materials study (RCEP 2008) and a shorter report on light pollution (RCEP 2009). Even internally, then, the planning issue fell by the wayside, perhaps also because whatever institutional memory remained of the original study was not entirely positive.

78. By this stage, as with the study seeking to elucidate the concept of BPEO (RCEP 1988), members had to scrutinize the original report to see exactly what their predecessors had said.

79. Responses to the consultation on the Sub-National Review were placed in the public domain. Sixteen responses, primarily from environmental NGOs, were kindly made available to the author by the Department of Communities and Local Government.

80. There are dissenting views. Smith (2000), for example, argues that the changes to the pollution control regime institutionalized in the 1990 *Environmental Protection Act* should not be seen as paradigmatic. As well as providing a statutory basis for integrated pollution control, the *Act* envisaged that HMIP would adopt an 'arms' length approach' in its regulatory duties. Smith argues that the latter was difficult to sustain 'because IPC required the same exchange of regulatory resources as under the previous air regime' (ibid.: 108); in the 1990s, 'the rules of the game concerning consensus regulation' were re-established (ibid.:109). But even if a relatively closed policy network continued to exist, as Smith maintains, it is difficult to deny that both the conceptual framing and the operation of the system of pollution control underwent profound change.

Chapter 6

1. These well-known reflections on ideas and 'human affairs' appear in various discussions of policy and politics; see, for example, Radaelli (1995: 159), who cites Mill in this context in his paper on the role of knowledge in the policy process.
2. HL Deb 23 March 1983, vol. 440, c1210. Lord Flowers, a former Chair of the Royal Commission, was commenting on the government's failure to implement the key recommendations of its fifth report (RCEP 1976a).

Endnotes

3. Interview (November 1996), Deputy Secretary, environment department (1990s).
4. Interview (November 1996), Deputy Secretary, environment department (1990s).
5. Interview (May 1997), senior scientific officer, environment department (1980s–90s).
6. Many of the Commission's recommendations on radioactive waste were quickly accepted by the government of the day: these were 'no brainers', according to a civil servant who had been involved in drafting the response to the sixth report (cited in Rough 2011: 204). It is worth noting, however, that despite its lasting impact on policy discourse, the report neither halted the construction of nuclear power stations nor terminated the controversy about radioactive waste management. Interestingly, almost thirty years later, Lord Flowers himself seemed to declare that the Flowers criterion had been met: 'a method to ensure safe disposal for the indefinite future—namely, underground storage—has been demonstrated beyond reasonable doubt in other countries, especially Finland. The same can be done here, now that we know the method exists' (HL Deb 12 January 2005, vol. 668, c331).
7. Interview, June 1996. The ninth report was an 'argument stopper', to use the words of a civil servant in the environment department (1980s), reflecting on ways in which the Commission had exercised authority (interview, January 1997). It scored a 'direct hit' in the sense that the government's stance on a highly contentious issue was affected in an immediate and visible way. It should be noted, however, that the transition to lead-free petrol, which required agreement within Europe, was not in itself a particularly rapid process.
8. A comprehensive early history of the landfill tax, which explicitly mentions the Royal Commission's recommendation, can be found in Seely (2009).
9. The policy was subsequently strengthened, as outlined in Chapter 2.
10. Government responses to Royal Commission reports (see Appendix 1) are a source of considerable detail on the fate of individual recommendations. Independent evaluations of later reports, though small in number, are also helpful (Barron 2001; Farmer et al. 2005; Fergusson, Skinner, and Keles 2002; Fergusson et al. 2005). It should be remembered, too, that not all influence was manifest through policy and legislation, or at least not directly. As Radaelli (1995: 162) notes, 'decision-makers use knowledge to make choices, to implement decisions, and to develop standard operating procedures as part of a process of institutionalization'. In the latter sense, the Commission's advice had effect on research or monitoring on numerous occasions over its lifetime. After the third report (RCEP 1972a), for example, a number of organizations, including the Natural Environment Research Council, devoted more attention to estuaries, and the sixth report (RCEP 1976b) affected ways in which epidemiologists looked at radiation hazards. In the Commission's last decade, the report on bystander exposure to pesticides (RCEP 2005) produced a commitment to revisit the exposure model on which the risk assessment had been based; and the report on nanotechnologies (RCEP 2008) stimulated research at international level (via OECD) on potentially harmful effects, as well as UK government action to address a shortage of toxicologists (UK Government 2010: para. 47, Action 3.10 and para. 48)

Endnotes

11. At the end of 1970, for example, a junior minister (Lord Sandford) anticipated the Commission's first report in a House of Lords debate on pollution and the environment: 'We look forward keenly to reading that Report, and when we get it will consider [the Commission's] advice with urgency' (HL Deb 9 December 1970, vol. 313, c973).
12. Minute 57, RCEP third meeting 1988, 3–4 March.
13. Interview, December 1996.
14. HL Deb 18 May 1990, vol. 519, c484; HC Deb 30 April 1990, vol. 171, c820.
15. Interview (November 1996), chief scientific officer, environment department (1970s–80s).
16. Interview (March 1996), member involved in the lead study.
17. Specifically, the principle of 'contraction and convergence', under which per capita allowances for greenhouse gas emissions in more and less developed countries would gradually converge (decreasing in the former while increasing in the latter), and would thereafter contract together (RCEP 2000: para. 4.49). The Global Commons Institute, which promoted this principle, had given evidence to the Commission—a case of the latter acting as a 'conduit'. Interestingly, in evidence ten years later to the Environmental Audit Committee (on the subject of carbon budgets, as set by the Committee on Climate Change), the Institute claimed that 'the origins of the advice from the Committee on Climate Change could be traced back to the Royal Commission on Environmental Pollution's advocacy of contraction and convergence in their report, "Energy—the Changing Climate"' (EAC 2010b: para. 23). It should be noted, however, that in accepting the Commission's primary recommendation on emissions reduction, the government had *not* accepted the 'contraction and convergence' principle as a rationale, arguing that this approach was 'only one of a number of potential models for global agreement on addressing greenhouse gas emissions' (UK Government 2003b: 3).
18. Minute 8, RCEP second meeting 1999, 3–5 February.
19. A speech by the Prime Minister on the day of the White Paper's publication attracted commentary for apparently signalling a distancing from the US position on climate change (see Owens 2010).
20. The initial acceptance came in an announcement by the Secretary of State for the Environment (then Kenneth Baker) (ENDS 1985b).
21. Other important influences included the House of Lords Select Committee on Science and Technology (1981), the House of Commons Trade and Industry Committee (1984), and a series of reports from the Hazardous Waste Inspectorate (soon to be absorbed into HMIP); all of these were drawn upon in the Commission's eleventh report.
22. Like a number of other measures waiting for implementation, the 'duty of care' found its way into the omnibus Environmental Protection Bill during the period of heightened green awareness at the end of the 1980s (see Chapter 2). Interestingly, in moving the Second Reading of the Bill in the House of Lords, Lord Hesketh said that the proposals in Part II (on waste management) 'spring in the first instance from a report of the Select Committee on Science and Technology under the chairmanship of the noble Lord, Lord Gregson [the 1981 report on hazardous

Endnotes

waste—see note 21]. They also closely reflect the findings of the RCEP's eleventh report' (HL Deb 18 May 1990, vol. 519, c481). In the run up to the new legislation, the House of Commons Environment Committee (1989) had also reported on toxic waste (drawing extensively on the Royal Commission's report). These are interesting examples of interactions between the Royal Commission and select committees, considered again in Chapter 7.

23. 'Commission decides to go for the burn', News, *The Waste Manager* June 1993, p. 9.
24. For further detail, see RCEP (1993a), para. 9.39 and Appendix B, and DoE (1994), paras 3.1 and 3.2. The environment department had asked the Commission's consultants to conduct further work.
25. Interview (July 1996), civil servant in Waste Management Division, environment department (1990s), which coordinated the response to the incineration report.
26. The environment department had hoped that the delayed responses to the fifth and seventh reports might be published simultaneously. But towards the end of 1982, when MAFF were 'still being awkward about the pesticides part of RCEP7', it was felt that the response to the fifth report should not be further delayed (memorandum to ministers from Assistant Secretary, Central Directorate on Environmental Pollution, 24 September 1982 [CDEP 10/29/Part 5]). In a letter to the Minister of State for Agriculture, Fisheries and Food (4 October 1982), the Secretary of State for the Environment noted that the Chair of the Commission had 'felt obliged to criticise [the environment department] in acid terms for the lack of a response to RCEPs 5 and 7', and that an extended delay would cause 'further embarrassment'. (CDEP 10/29/Part 5.)
27. Interview (August 1997). A significant factor was MAFF's own need to reposition itself on matters concerning the countryside and the environment, given the growing recognition that the post-war agricultural regime—and the CAP in particular—would have to change.
28. The Minister of State at MAFF (Lord Belstead) explained to the House of Lords that the non-statutory control system had come under pressure because of the importation of uncleared pesticides and because the European Commission had been critical about the trade implications. The government, he said, had taken the opportunity to review the 'whole scope' of pesticide controls: 'Following the debate on this issue prompted by the Seventh Report of the Royal Commission on Environmental Pollution, we are committed to ensuring that United Kingdom pesticide use is the minimum necessary to ensure the efficient production and distribution of food and to safeguard human health' (HL Deb 22 November 1984, vol. 457, cc700–1).
29. Interview (May 1997), division head, environment department (1980s–90s), emphasis added.
30. Interview (March 1997), division head, environment department (1980s–90s), emphasis added. The 'request' mentioned at the beginning of this extract refers to measures that the Commission had called for in its first and fourth reports, as the interviewee goes on to explain.
31. Interview (December 1996), member (1970s–80s).

Endnotes

32. The package brought in over the period 1989–91 included new regulations on farm waste, a new farm and conservation grant scheme, a revised and more readily available code of good agricultural practice to protect water, and a more stringent approach to prosecution by the National Rivers Authority (with higher fines allowed for in the *Environmental Protection Act* 1990). For further detail, see Lowe et al. (1997).
33. Interview (July 1996), civil servant, Waste Management Division, environment department (1990s).
34. Interview (May 1997). The individual had been Director of the Central Unit on Environmental Pollution in the UK environment department in the 1970s and Head of DG Environment at the European Commission in the 1980s.
35. RWMAC (1978–2004) reported to the Secretary of State for the Environment.
36. The formal response to the sixth report procrastinated on the idea of an independent waste disposal body: 'The Government see the force of this proposal, but do not think that they need to come to a decision at this stage' (UK Government 1977: para. 20). The diaries of Tony Benn, then Secretary of State for Energy, reveal conflict over this issue not only at Cabinet level but between Benn himself (who was sympathetic to the Commission's proposal) and civil servants in his own Department (Benn 1990). According to one well-placed official in the environment department (late 1970s), previously with the Atomic Energy Authority, the (then Labour) government would probably have implemented the Commission's recommendation in the end (interview, April 1997). He suggested, however, that the political climate had become less favourable after the 1979 election, when the incoming Conservative government 'listened to the industry and industry said, "we can do it ourselves"'. The Nuclear Industry Radioactive Waste Executive (NIREX), was eventually established in 1982, keeping the waste management function 'in house'. It is interesting that ten years after the Commission's report, the House of Commons Environment Committee (1986: para. 253) could still complain that 'fundamental aspects of radioactive waste policy had remained immune from much needed scrutiny'.
37. The Commission was concerned to ensure that existing frameworks, such as the European REACH Regulation for chemicals (see Chapter 2), were 'fit for purpose' in terms of nanomaterials, and that gaps in regulatory arrangements would be identified and closed (though it acknowledged that such measures in themselves would be insufficient, as discussed in Chapter 4). One direct result—promised in the government's response to the report (UK Government 2009)—was the publication of a nanotechnologies strategy (UK Government 2010). However, while there was action on some of the Commission's recommendations (see note 10), the regulatory commitments in the strategy were weak. A Nanotechnology Strategy Forum was also established, 'to facilitate discussion and engagement between Government and key stakeholders on strategic issues for the responsible advancement of the UK's nanotechnologies industries'; at the time of writing (mid-2014) the Forum has held four meetings, the minutes of which suggest a primary focus on technology development (<https://www.gov.uk/government/groups/nanotechnology-strategy-forum>, accessed 17 July 2014). At European level, progress with

Endnotes

amendments to REACH were similarly slow in the years following the Royal Commission's report (see, for example, BiPRO and Öko-Institut e.V 2013; European Commission 2012). It should be noted, too, that this was another area in which the Royal Commission was contributing to wider expressions of concern; its report had been preceded by a landmark contribution from the Royal Society and Royal Academy of Engineering (2004) and was followed by a report on nanotechnologies and food from the House of Lords Select Committee on Science and Technology (2010), which endorsed several of the Commission's recommendations.

38. The standards report is best remembered for its argument that public values should be incorporated from the earliest stages into the processes of setting environmental standards—a position that was broadly welcomed by government (UK Government 2000), and recognized by some observers as being of 'enormous' significance (Weale 2001: 362). The Commission's ideas came to be reflected in advice on risk assessment (DETR, Environment Agency, and IEH 2000), in the reports of Parliamentary committees (for example, EAC 1999; House of Lords Select Committee on Science and Technology 2000), and in the argumentation of NGOs (for example, <http://www.greenpeace.org.uk/about/a-brief-history-of-science-and-society> [accessed 11 June 2014]). The standards study was also drawn upon in a broader sense by the Environment Agency, in developing its own framework for the setting and use of standards (Farmer et al. 2005). And it is interesting that consultees in the second Financial Management and Policy Review of the Commission, conducted in 2007–8, were still citing the standards report as one of the Commission's most influential, though in general they had nominated more recent contributions (see Chapter 7). Wider engagement remained challenging in implementation, however, and the Commission itself came to believe that discerning and incorporating public perspectives was far from straightforward, conceptually or practically (RCEP 2008).

39. The Commission was of the view that pressures for disclosure would increase. It is interesting that it cited the second report of the US Council on Environmental Quality (US CEQ 1971), which had contained a section on 'the citizen's right to know' (see RCEP 1972b: para. 8)—a further example of transnational cross-fertilization.

40. Interview (May 1997), campaigner who had worked for Social Audit, which had been active in the campaign against the Alkali Inspectorate in the 1970s, and later for the Freedom of Information Campaign (set up by Des Wilson in 1984).

41. The individual had given evidence during the air pollution study. Of course, the Commission's tone was more measured, and in significant respects it commended the Alkali Inspectorate's work. But it considered the Inspectorate's policy on releasing information to be 'misguided' (RCEP 1976a: para. 125), and was adamant that 'the public should have the right to know the state of the air they breathe and the amounts of pollution emitted by both registered and non-registered industry' (ibid.: para. 235). It recommended that both the Alkali Inspectorate and local authorities should be able to release data on emissions to the public and that copies of 'consents' as well as monitoring information should be kept on publicly accessible registers (ibid.: paras 215, 218). These recommendations went much further

than the provisions for public registers in the *Control of Pollution Act* 1974, which in any case had not yet been implemented.
42. The endurance of this principle is illustrated by its being paraphrased in HMIP's guidance for its inspectors (cited in Allott 1994) and cited in full, even twelve years after its appearance, in reports of the House of Lords Select Committee on the European Communities (1989: paras 16, 50, 1996: para. 46). See also Ross and Rowan-Robinson (1994).
43. Frankel was with the Campaign for Freedom of Information and became its Director in 1987. He had previously worked for Social Audit.
44. Interview, June 1996.
45. Interview (March 1996), Chair (1980s).
46. Interview (June 1996), Secretary (1980s).
47. Interview (May 1997), emphasis added. The campaigner specifically mentioned pesticides in this context: 'the Royal Commission said information should not be unnecessarily withheld, especially from those engaged in research on pesticides. Four years later, [there were still] no proposals for reviewing the confidentiality of agreements—"confidentiality is essential to the success of these arrangements".' But greater transparency was eventually forthcoming, as we have seen.
48. Interview (March 1997), Deputy Chief Scientific Officer (1980s), emphasis added.
49. To the examples of diffuse influence considered here we might add one of a more generic character. The majority of Commission members were academics, and many of them drew upon the various studies in their graduate and undergraduate classes. One member (1980s) emphasized this point in relation to the Centre for Environmental Technology at Imperial College, London, where a number of Commission members had been based: it wasn't going too far, he felt, to say that the students there 'were in part educated by the Royal Commission' (interview, June 1996). And such influence didn't stop in the universities: its greater significance was that concepts emanating from the Commission 'educated' numerous cohorts of graduates destined for government, industry, the media, academia, and the environmental professions, thus providing a powerful mechanism for the transmission of new ideas. Nor was the pedagogic effect confined to the UK; Barron (2001) records, for example, that the Chair of the European Environment Agency's Scientific Committee made the soils report (RCEP 1996) a set text in his university courses in Brussels.
50. Interview (November 1996), Deputy Secretary, environment department (1990s).
51. Interviews (November 1996, August 1997, and August 1996), Deputy Secretary (1990s); member (1970s–80s); and member (1990s).
52. Here I am indebted to Elizabeth Rough, whose doctoral research in The National Archives (TNA) revealed fascinating material on UK nuclear power policy formation, including the role of the Royal Commission. The quote is from a Cabinet Office Study, *Radioactive Waste*, K. P. Pagett, 13 August 1974 (TNA, Kew, AB 38/799), cited in Rough (2011: 202). Indeed, the very formation of the Commission had led some in the Atomic Energy Authority (AEA) to initiate a review of internal plans, so that it could get its 'own house in order' (letter from H. J. Dunster, Radiological Protection Division, UKAEA Health and Safety Branch, to T. Marsham, Deputy

Endnotes

Managing Director, Reactor Group, 26 March 1970 [TNA ibid.], cited in Rough [2011: 203]).
53. Interview, August 1996.
54. Interview (March 1996), member who served for a lengthy period spanning the 1980s and 1990s.
55. Interview (March 1997), division head, environment department, who had been involved with the fourteenth report. A similar point about practicality was made by a Special Advisor to the Commission, who had been engaged with the study of GMOs (RCEP 1989). However, the latter also made the interesting observation that GENHAZ 'wasn't rejected, it was just not taken up', and suggested that the thinking behind the concept had nevertheless had some influence on subsequent guidelines (interview, February 1997).
56. The author was a member at this time. There was never great enthusiasm within the Commission for the contract idea, and nor was there any particular expertise in public administration among its members (a limitation of the 'committee of experts' model in this case). But members were convinced that simply repeating past prescriptions would hold out little hope of improving urban environments, and some felt that a form of contract might at least be a way of stimulating action in the face of complexity and policy inertia. The latter argument prevailed. The government accepted that there was 'a clear benefit to having an agreed, shared understanding between central and local government of the most important environmental objectives'. However, it maintained that this would be achieved by the *Joint Environmental Prospectus*—an agreement by Defra and the Local Government Association to work together (Defra and LGA 2007)—and by Local Area Agreements, which had been set out in the 2006 Local Government White Paper and were broader in intent than anything that the Commission had in mind (Defra and Department of the Environment Northern Ireland 2008). The *Prospectus* postdated publication of the urban environment report by several months, but made no reference to the Commission's work.
57. Among other difficulties, there were questions about the weight that should be afforded to 'anecdotal' evidence. In concluding that a link between bystander exposure and chronic ill-health was plausible, even if causality had not been established, the Commission drew on its visits to those who believed that they had been affected as well as on its understanding of the biological mechanisms that might be involved (RCEP 2005: para. 6.20). In response, however, the government insisted that there was 'no scientific basis for additional precaution' (Defra 2006b: para. 18). This raises interesting issues. The Commission, as one senior civil servant observed, 'was doing exploratory epistemic work at the furthest boundaries of...scientific advice giving' (personal communication, June 2014), while the department was disinclined to compromise on what it regarded as 'sound science'. Note that, in this case, the Commission had not managed to produce a 'serviceable truth' (Jasanoff 1990: 250), as it had done in the cases of lead in petrol (RCEP 1983) and carbon dioxide emissions reductions (RCEP 2000).

Endnotes

58. In 2005–6, the department—by this stage responsible for environment, food, and rural affairs—was facing criticism over its implementation of a new European system of single farm payments; delays and poor communication had resulted in difficulties for a significant minority of farmers (NAO 2006). Politically, therefore, it was not a good time to endorse a measure to which the agricultural community had expressed firm opposition (see also Fisk 2007).

Chapter 7

1. Interview (May 1997), Director of Social Audit (1970s–80s), and afterwards of the Freedom of Information Campaign.
2. Interview (November 1996), Deputy Secretary, environment department (1990s).
3. This has been a recurrent theme in commentaries on, and reviews of, the Commission (for example, Defra 2007b; DETR 2000a; Lowe 1975a, b; Williams and Weale 1996); and see Owens (2011a, 2012; Owens and Rayner 1999). It was also a point made by many interviewees.
4. Chapter 5, for example, shows how governments of the day responded very differently to the fifth and twenty-third reports (RCEP 1976a, 2002a), which argued for integrated pollution control and Integrated Spatial Strategies respectively. In neither case were the main recommendations accepted, but there is a sense that governments in the 1970s and 1980s felt that they needed to take more trouble to say no. A reading of government responses in the Commission's last decade or so suggests that they became more perfunctory—even 'dismissive', as one reviewer said of the response on crop spraying (Fisk 2007: 289). This chimes with the views of a senior civil servant in the environment department (2000s), who observed a greater 'general impatience' with the need for civil servants to follow formal process in such ways (personal communication, March 2014).
5. For example, the outgoing President of the Royal Society, commenting in 2011 on the abolition of numerous quangos, included the Commission when expressing concern that the 'independence of scientific advice' would be compromised: <http://www.bionews.org.uk/page_83514.asp> (accessed 12 February 2014). Latterly, however, the environment department may not have been so convinced—a point to be considered later.
6. As Chapter 4 has shown, the Commission became more reflective and less confident about the possibility of such separation in some of its later reports.
7. Interviews (February and April 1996), members (1990s–2000s and 1970s–80s respectively).
8. The term was used in a written submission for this study (21 June 1997) (in lieu of an interview) by a member who served in the 1970s.
9. Interview (June 1996), member (1980s). Interestingly, Brian Flowers, when Chair of the Commission, told his audience at a National Society of Clean Air (NSCA) lecture that 'the Royal Commission is essentially a lay body. Whereas each of its members may be expert in something or other... it is not intended to be expert in any particular topic' (Flowers 1974; copy tabled at RCEP fifth meeting 1974, 4 July).

Endnotes

10. Interview (June 1997), member (1990s). The 'hallowed halls' were those of the Royal Society, where the interview took place.
11. Interview (August 1996), member (1990s).
12. Interviews (May and March 1996), members who served in the 1970s and 1980s–90s respectively.
13. Interview (February 1997), special advisor (1980s). The advisor felt that he, too, had benefited from the experience: 'I learnt how important it was to understand another point of view . . . I had to put a lot of effort into explaining things to people from very disparate backgrounds, which helped me enormously in better understanding them myself . . . that really sharpened my ideas up a lot.'
14. Interview (July 1996), member (1980s).
15. Interviews (March and August 1997, March and April 1996): Chair (1980–90s); member (1970s–80s); and two members (1970s), respectively. In the author's own experience, such learning was one of the most stimulating aspects of being a member of the Commission.
16. The phrase 'the independence thing' was used, in commenting on the Commission, by a division head in the environment department (1980s–90s) (interview, March 1997).
17. The distinction draws on Isaiah Berlin's 'two concepts of liberty' (Berlin 1969). It is noticeable, however, that the Commission's positive freedom did not extend to choosing its sponsoring department.
18. And also, of course, for fruitful interaction, requiring the delicate balance discussed in Chapter 3.
19. Interview (March 1997), Head of Pollution Policy, HMIP (later in the Environment Agency) (1990s), with previous experience in the environment department.
20. In the author's experience, the Commission tended to look all the more keenly at topics that the environment department had counselled against—at least, the department's reasons for not pursuing such topics would be closely scrutinized. It is possible, of course, that governments exercised an influence so subtle that the Commission was unaware of it, but the choice of subjects does not readily support this interpretation.
21. The phrase was used by Lord Ashby, the Commission's first Chair, during a debate on its tenth report in the House of Lords (HL Deb 29 October 1984, vol. 456, c354).
22. Interviews (June 1996 and March 1997), Secretary of the Commission (1980s) and division head in the environment department (1980s–90s).
23. The concept of functional independence featured in correspondence in November 1980 between the Chair (Professor Sir Hans Kornberg) and the Prime Minister (the Rt Hon. Margaret Thatcher), at a time when Vote responsibility for the Commission was being transferred from the Civil Service Department to the environment department (see Chapter 3).
24. Interview (May 1997), Deputy Secretary (1980s).
25. Interview (June 1996), member (1980s–90s).
26. Interview, July 1997.
27. Interview (February 1997), special advisor (1980s).
28. Interview, August 1998. The individual held several ministerial positions in the environment department in the 1980s.

Endnotes

29. The term was used by a member of the Secretariat (1990s) in interview (August 1996).
30. It might be said that the Secretariat came with baggage too: as noted in Chapter 3, there was something of a revolving door between the Commission and the environment department at this level.
31. Interview (December 1996), member, emphasis added.
32. For example, a former Chair, speaking of 'certain people who came through from MAFF [the Ministry of Agriculture, Fisheries and Food]' recalled having to say to one member, 'you are not here representing MAFF'. But he added that, as time went on, the same member had become 'very much one of the establishment [i.e. the Commission]' (interview, March 1997).
33. Interview (February 1996), member (1990s–2000s).
34. Personal communication (March 2014). This corroborates the view of a member from the 1980s—the Commission, he said, was 'far from being regarded as at the banner-waving end of the march' (interview, December 1996).
35. The same, as we have seen in Chapter 3, could be said of the Commission's treatment of evidence.
36. This comment on the Commission's report (RCEP 2004) was made by Lady Saltoun in a Lords debate on the marine environment (HL Deb13 December 2004, vol. 667, c1159). Measures advocated by the Commission (including a precautionary approach, Marine Protected Areas, and a system of marine spatial planning) featured soon afterwards in proposals for a Marine Bill (Defra 2006a; see also EAC 2006: Ev 61–2; Lawton 2007), and later in the *Marine and Coastal Access Act* 2009. But there had been a spate of reviews and reports on the marine environment, and this might be regarded as another case where the government's ideas were developing in line with those of a number of other bodies. The Commission's contribution, though it attracted ministerial, public, and Parliamentary attention, was never strongly associated with the subsequent policy developments—in contrast, for example, to the 'duty of care' in waste management (RCEP 1985), also advocated by a broad coalition, yet retaining the Commission's imprint as it came to be enacted in legislation.
37. EEAC continues at the time of writing (early 2014), without the Royal Commission. Some of its member bodies are similar in constitution and role to the Commission, others very different. The closest is probably the German Advisory Council on the Environment (SRU), with which the Commission held bilateral meetings from time to time—indeed, it was partly as a result of such interaction that the idea of a Europe-wide network was conceived (Macrory and Niestroy 2004). Joint perspectives are identified within EEAC's working groups, which periodically publish position statements and sometimes arrange meetings with European officials. EEAC's publications, and other information, can be found at: <http://www.eeac.eu/> (accessed 17 February 2014).
38. As, for example, in the early meeting between the inaugural Chair of the Commission and the Director of the environment department's Central Unit on Environmental Pollution, at which the status of the Secretariat was discussed (Chapter 3).
39. Lord Flowers recounted how the Benns came to dinner at that time: 'we sat in the kitchen and of course we talked, I don't know what about but it would have been

Endnotes

about nuclear things, I'm sure. He [Benn] was exceedingly troubled about it all, and the nuclear industry were infuriated by him because he would not make up his mind' (interview, August 1997). Benn similarly remembers Flowers coming to his office one evening in 1977: '[he] stayed for an hour and a half; I really enjoyed it and I somewhat poured my heart out to him. "Look, I really want your help because the whole nuclear thing is getting out of control. The lobby has got me by the neck; I don't know what to do about it"' (Benn 1990: 40).

40. Southwood was a Fellow of Merton College; Waldegrave of All Souls. Southwood recalled that sometimes they would meet informally in Oxford and discuss environmental matters (interview, March 1996).
41. At the time of the author's own membership, for example, meetings with ministers (usually involving the Chair with the Secretary and a small group of members) were arranged from time to time, and occasionally a minister (or sometimes a member of the Opposition) would attend part of a Commission meeting or join the members for dinner. Exceptionally, as in the case of the energy and aviation reports (RCEP 2000, 2002b)—both connecting closely with climate change—meetings were arranged to allow the Chair to brief the prime minister.
42. Interview (April 1996), member (1970s).
43. The phrase was used by a member of the Commission (1980s–90s) who served whilst also sitting in the House of Lords, where (as he said) he was 'quite happy to be "a good old boy"' (interview, March 1996). The last Peer to serve in both institutions was Lord Selborne, whose term on the Commission ended in 1998. Beyond the turn of the century, interventions in the Lords from former members of the Commission on matters with which it was concerned became less frequent, even when its reports were being cited or discussed.
44. For example, HL Deb 23 March 1983, vol. 440, cc1204–22; HL Deb 29 October 1984, vol. 456, cc354–404. In the latter debate, which was about the Commission's tenth report (RCEP 1984), Baroness Birk (a former junior minister in the environment department) said: 'probably these commission reports would never be discussed if it were not for this House' (c364). While her speculation may have been unduly pessimistic, it serves to emphasize the significance of the Commission's presence in the House of Lords.
45. Lord Hesketh, on moving the second reading of the Bill in 1990, told peers: 'I am pleased that our debates on this part of the Bill stand to benefit from the counsel of present and former members of the commission who are also Members of your Lordships' House' (HL Deb 18 May 1990, vol. 519, c484). In a later debate, Lord Nathan moved amendments and both he and Lord Lewis pressed for such measures as a statutory release committee, public access to information about releases, and a register of those who would be releasing GMOs (HL Deb 2 July 1990, vol. 520 c1964, cc1977–8, c1993). Their interventions were not wholly successful, but nevertheless made a substantial difference to the legislation.
46. For example, Ashby (RCEP 1970–3), Cranbrook (RCEP 1981–92), Lewis (RCEP 1986–92), Nathan (RCEP 1979–89), Selborne (RCEP 1993–8), and White (RCEP 1974–81) were at various times members or chairs of the European Communities Committee and its sub-committees. Both Flowers (RCEP 1973–6) and Selborne

chaired the Science and Technology Committee, of which Flowers was also a member for many years. The inter-connectedness was emphasized in a debate on water pollution in 1985, when Baroness White said of Lord Ashby: 'As the first chairman of the Royal Commission... he led the way in the early 1970s in considering how best we in the United Kingdom should tackle the problems of water quality and pollution control. He subsequently took the lead in the relevant subcommittees of the Select Committee on the European Communities... in scrutinizing the scientific justification for the proposals... emanating from the Commission in Brussels' (HL Deb 21 November 1985, vol. 468, c697).

47. HL Deb 21 November 1985, vol. 468, cc691–715. In the debate both Lord Nathan and Baroness White (RCEP 1974–81) cited the tenth report (c692), as did other peers, and lamented the absence of Lord Ashby who was unable to be present on that occasion (see also ENDS 1985a).
48. Former civil servant, environment department (personal communication, June 2014).
49. HL Deb 2 December 1993, vol. 550, cc674–8.
50. For example, Professor Richard Macrory (RCEP 1992–2003) and Professor Paul Ekins (RCEP 2002–8) acted as Special Advisors to the Environment Committee and the Environmental Audit Committee respectively. Ekins, whilst serving on the Commission, was also a Specialist Advisor to the Joint Parliamentary Committee on the Climate Change Bill.
51. In a letter to the chair of the Committee (Sir Hugh Rossi), after publication of its report on toxic waste (House of Commons Environment Committee 1989), a member of the Commission expressed gratification that the Committee had endorsed views expressed in the Commission's eleventh report (RCEP 1985). Noting that these views were already 'most ably and eloquently represented in the Upper House', the member urged Sir Hugh and his colleagues to do whatever they could in the House of Commons to rectify what the Commission saw as shortcomings relating to implementation: 'I trust that when the Green [Environmental Protection] Bill comes to be debated you and your colleagues on the Select Committee may be able to correct some of these deficiencies. Indeed it would be splendid if it proved possible to influence the production of the Bill itself' (letter, 15 March 1989, kindly made available by its author). During passage of the Bill, Sir Hugh Rossi was able to observe that the provisions on waste management were 'very much along the lines' of his Committee's recommendations, and that the concept of the 'duty of care', in particular, was of 'outstanding importance' (HC Deb 15 January 1990, vol. 165, c60).
52. Interview (August 1997), Commission Secretary (1990s–2000s).
53. Minute 9.2, RCEP first meeting 1970, 25 February.
54. 'Both sides' here may be taken to mean the worlds of science on the one hand and of politics and policy-making on the other. The Commission was not consciously established as a body to include both scientists and policy-makers, but in practice it involved 'both sides' in a variety of different ways. It included members of the Upper House of the British Parliament, and its chairs and members (including the scientists) were sometimes independently active within policy networks (or indeed

Endnotes

in other boundary organizations—Sir John Houghton, for example, chaired the Scientific Assessment Working Group of the Intergovernmental Panel on Climate Change when he was Chair of the Royal Commission). The Commission's membership also, on occasion, included retired senior civil servants (Martin Holdgate, for example, former Deputy Secretary and Chief Scientist in the environment department, served from 1994 to 2002), and its Secretariat was linked into the bureaucracy of which it formally remained a part (in the sense that most members of the Secretariat were civil servants).

55. Interview, March 1997. This official had experience of the Commission's contributions from senior positions in the environment department (1980s) and later in HMIP and the Environment Agency (1990s).
56. Member (2000s), personal communication (March 2011).
57. Interview (March 1996), member (1980s–90s).
58. One former Chair said '[the Commission] picks a field, it looks at it and shakes it around and makes some recommendations and then goes away and does something else', though he did acknowledge that some members would 'agitate a bit' and that chairs would try to 'move [government] along' (interview, August 1997). Similarly, a former Secretary expressed the view that 'it wasn't [the Commission's] responsibility to lobby against the government—its responsibility was to give advice...and once it had done that it had done its job' (interview, August 1996). Bijker, Bal, and Hendriks (2009) record similar sentiments among some members of the *Gezondheidsraad* (the Health Council of the Netherlands). Examples of issues that the Commission dropped after initial interest include lead in drinking water (RCEP 1983) and Integrated Spatial Strategies (RCEP 2002a)—see Chapters 4 and 5.
59. Interview (April 1996), member (1970s–80s).
60. Correspondence was not infrequent, as exemplified during and after the Commission's study of waste in the mid-1980s. An extensive correspondence between the Chair (Sir Richard Southwood) and the Secretary of State for the Environment (The Rt Hon. Patrick Jenkin MP) is reproduced in the report itself (RCEP 1985: Appendix 4); it concerned the government's proposed abolition of the Greater London and Metropolitan County Councils, and the Commission's anxieties about the implications for waste management. Later, following the government's response to the waste report (DoE 1986b), a letter from the Chair identified a number of issues on which the Commission felt that the government had not dealt satisfactorily with its recommendations. Interestingly, the Chair was at pains to point out that these areas of concern were 'to be distinguished from circumstances in which recommendations have been rejected for clearly stated reasons, where it would be unprofitable (and contrary to the Commission's normal practice) to enter into a debate with the Government' (letter from Professor Sir Jack Lewis to The Rt Hon. Nicholas Ridley MP, 18 November 1986). Further examples of correspondence are mentioned elsewhere in this book.
61. Interview (June 1996), Secretary (1980s).
62. HC Deb 22 July 2010, vol. 514, c32WS. See Chapter 3.
63. Bodies abolished in Europe included the Swedish Environmental Advisory Council (MVB, 1968–2011) and the Dutch Advisory Council for Research on Spatial

Planning, Nature and the Environment (RMNO, 1981–2009). As noted in Chapter 1, the US OTA was an independent agency of Congress, 1973–95 (see Bimber 1996; Blair 2011).

64. One further factor relating to the 'particular moment' was suggested by a former senior official from the environment department in the course of a discussion: that whereas in times of healthy majorities, 'criticism or goading is often quietly welcomed', coalition governments might be over-sensitive about independent reports with the potential to 'split the seam' (personal communication, April 2012). It is difficult to tell whether such concern had any influence on the decision to abolish, taken at time when the Coalition government was still very new.

65. The standards report (RCEP 1998b: 2–3) included a section on the '[c]hanging nature of environmental problems', and the urban report (RCEP 2007: 4–5) characterized urban environmental management as a 'wicked problem' (Rittel and Webber 1973), noting that such problems are intractable 'precisely because they lack a clear set of alternative solutions. They cannot be solved indefinitely, but rather must be managed for better or worse' (RCEP 2007: para. 1.15).

66. Written submission for the study (22 May 1996) from a member of the Commission (1980s–90s).

67. A former Secretary confirmed that problems with one of the later reports had arisen 'not least because so much of the detail we had to cover [was] different in each of the devolved administrations' (personal communication, May 2011). Similarly, the second Financial Management and Policy Review (FMPR) acknowledged that, while the UK-wide remit was important, devolution posed 'an ongoing relationship challenge' for the Commission (Defra 2007b: 12).

68. It is interesting that Sir Martin Holdgate, in his final years of service (early 2000s), found himself worrying about the Commission as a body: 'Was it taking on impossibly wide and awkward topics? Was it going too far from the central theme of pollution? Was it becoming rather like Antarctic exploration now that all the territory was mapped?' (Holdgate 2003: 343). A member who served in the final decade felt that, towards the end, 'it seemed as if the Commission was struggling to find issues, and too often choosing the wrong ones' (personal communication, March 2011); and, as if to emphasize this point, a senior civil servant in the department suggested that when it came to the Commission's last two reports, 'no-one was waiting for the answers' (personal communication, July 2014).

69. This might be said of reports on environmental planning (RCEP 2002a), the urban environment (RCEP 2007), climate change adaptation (2010), and demographic change (2011), though the last was published after the decision to abolish. It should be noted, however, that earlier reports, including those on nuclear power, lead, and GMOs, were seen by some critics as 'unscientific' in certain respects (though they certainly had scientific content), and some (such as the transport reports of the 1990s) dealt predominantly with non-scientific material. There was, however, a more noticeable tendency to choose broad-based topics in the later years, and the membership itself had a wider range of disciplinary backgrounds.

70. This possible reason for dissatisfaction was suggested, on different occasions, by two senior civil servants from the environment department (personal communication,

Endnotes

February and July 2014). Since the Commission had trodden with impunity on Whitehall toes before (one thinks of nuclear energy, transport, and agriculture, for example), any truth in this suggestion would serve only to emphasize the Commission's increasing exposure during the last few years of its life. (For further discussion of changing Whitehall structures, see Chapters 1 and 5, and Appendix 2.) In a more general sense, government had increasingly adhered to the view that policy was a matter for its own determination; while outside bodies could advise and inform, they should not be seen to be formulating policy—this, too, may have had a subtle effect on the reception of the Commission's advice.

71. A memorandum from the Chair to members in 2004 is perhaps indicative of the change. He explained that he had requested more frequent meetings at Director General level in the department, to restore an earlier position. In addition, he observed that, while the Commission had been 'encouraged to keep in close contact with Ministers', making such connections had been far from easy (memorandum by e-mail, 16 March 2004).

72. As suggested by a member who served during the 1990s (interview, August 1996). In the event, apart from a few voices regretting the Commission's passing, there was very little protest at all—a further sign, perhaps, of the diminution of its influence by that time. Its final Chair (Sir John Lawton) wondered, in retrospect, whether he should have done more to keep it alive; as he put it, a few years later, he still had 'niggling, "if only" moments'. When the new government came into office, he had felt reassured (naïvely, as he later realized), that the Commission was reasonably safe. Afterwards, he wondered whether, not being part of the 'London establishment', he had failed to rally sufficient support for the Commission (personal communication, July 2014). This interesting reflection reinforces the points made earlier about networks, though it seems unlikely, given the combination of factors involved, that any action by the Chair would have been sufficient to reprieve the Commission.

73. That is, a legitimacy deriving from 'superior authority' and accepted almost without question (Clark and Majone 1985: 16).

74. I am grateful to Albert Weale for this suggestion (personal communication, 5 December 2008). Weale uses the term 'practical public reasoning' in the sense of reasoning that 'involves the good or interest of the public', and which conforms to 'minimal standards of meaningfulness, inference and evidence' (Weale 2010: 275). See also Fischer (2009); Jasanoff (2013).

75. Boundary work of the kind that endowed the Commission with scientific authority can be distinguished from another form in which the Commission tried to hold a more general line between scientific evidence and values—and urged others to make such a distinction when conducting analysis to inform policy. Significantly, however, the Commission rarely assumed that science would 'settle' any issue that it was investigating, and in its last decade it became more circumspect about the possibility of separating the scientific and the political. This ambiguity about boundaries is especially evident in its treatment of environmental risk—see Chapter 4.

76. The term 'hybrid' characterizes the Commission in two senses: it 'mix[ed] elements from scientific and political forms of life' (Miller 2001: 480); and it was a hybrid within the typology of advisory bodies—with something of the traditional royal commission and something of the scientific advisory committee.
77. This relates to the point made in Chapter 1—in the context of Sheila Jasanoff's (1990) research on Federal advisory committees—that boundary work can be a flexible and valuable process. Its positive contributions have been confirmed in other in-depth studies. Bijker, Bal, and Hendriks (2009: 149), for example, emphasize the importance of 'coordination', in which the separating and connecting of social worlds are different sides of the same coin: 'It is in the act of coordinating over boundaries that distinctions are drawn anew.' In their useful table of coordinating and bridging mechanisms in the case of the *Gezondheidsraad* (ibid.: 150–1), there is much in common with the practices of the Royal Commission, as identified in the current study.
78. 'Quirkish things' could be 'immensely important' in policy evolution, according to a former Head of the Central Unit on the Environment in the department (1980s) (interview, January 1997).
79. This point is not to be confused with relativism, or with any tolerance of the wilful abuse and distortion of science on the part of politicians, or vice versa (for an interesting discussion, see *Nature* 2012).
80. The list could be extended to include statutory agencies, consultancies, think tanks, and other bodies, but the point remains essentially the same. It might be argued, as well, that different sources of advice are not mutually exclusive, and that the Commission was well placed to act as a synthesist and catalyst—indeed, these were amongst the most important roles that it performed.
81. Interview (February 1996), former Chair.
82. At the time of writing (August 2014), it is uncertain whether Scotland will choose to remain part of the UK: a referendum on this question is planned for September 2014.
83. Interview (March 1996), member (1980s–90s).
84. Personal communication, December 2008.
85. The analogy was drawn by a senior civil servant with long experience in the environment department (interview, June 1996).

Appendix 5

1. Lists of affiliations had become very long by this time. They are edited here to include only those current at the time and the first three of any previous roles listed.

References

ACP (Advisory Committee on Pesticides) (2005) *Crop Spraying and the Health of Residents and Bystanders: A Commentary on the Report published by the Royal Commission on Environmental Pollution in September 2005*, London: ACP, December. Available at <http://www.pesticides.gov.uk/acp.asp?id=2349> (accessed 17 October 2013).

Adams, J. (1995) *Risk*, London: University College Press.

AEBC (Agriculture and Environment Biotechnology Commission) (2001) *Crops on Trial*, London: AEBC.

Alkali Inspectorate (1888) *Annual Report for 1887*, London: Alkali Inspectorate.

Allmendinger, P. (2011) *New Labour and Planning: From New Right to New Left*, London: Routledge.

Allott, K. (1994) *IPC: The First Three Years*, London: Environmental Data Services.

Anderson, K., Bows, A., and Mander, S. (2008) 'From long term targets to cumulative emission pathways: reframing UK climate policy', *Energy Policy* 36, 10: 3714–22.

Argyris, C. and Schön, D. (1996) *Organizational Learning II: Theory, Method and Practice*, Reading, MA: Addison-Wesley.

Ashby, E. (1978) *Reconciling Man with the Environment*, London: Oxford University Press.

Ashby, E. and Anderson, M. (1981) *The Politics of Clean Air*, Oxford and New York: Oxford University Press.

Ashworth, J. (1997) 'Science, policy and risk: introduction', in Royal Society, *Science, Policy and Risk: a Discussion Meeting held at The Royal Society on Tuesday 18th March 1997, London*, London: The Royal Society, 3–5.

Baker, R. (1988) 'Assessing complex technical issues: public inquiries or commissions?', *Political Quarterly* 59, 2: 178–89.

Bal, R., Bijker, W., and Hendriks, R. (2004) 'Democratisation of scientific advice', *British Medical Journal* 329: 1339–441.

Ball, S. and Bell, S. (1994) *Environmental Law*, 2nd edn, London: Blackstone Press Limited.

Barker, A. (1993) 'Patterns of decision advice processes: a review of types and a commentary on some recent British practices', in B. Guy Peters and A. Barker (eds.) *Advising West European Governments: Inquiries, Expertise and Public Policy*, Edinburgh: Edinburgh University Press, 20–36.

Barker, A. and Peters, B. Guy (eds.) (1993) *The Politics of Expert Advice*, Edinburgh: Edinburgh University Press.

Barker, K. (2004) *Review of Housing Supply. Delivering Stability: Securing our Future Housing Needs. Final Report—Recommendations*, London: HMSO.

References

Barker, K. (2006) *Barker Review of Land Use Planning. Final Report—Recommendations*, London: HMSO.

Barron, E. M. (2001) *A Review of the Royal Commission on Environmental Pollution's Nineteenth Report: Sustainable Use of Soil*, London: RCEP.

Barry, J. and Patterson, M. (2003) 'The British state and the environment: New Labour's ecological modernisation strategy', *International Journal of Environment and Sustainable Development* 2, 3: 237–49.

Barry, J. and Patterson, M. (2004) 'Globalisation, ecological modernisation and New Labour', *Political Studies* 52: 767–84.

Baumgartner, F. R. (2006) 'Punctuated equilibrium theory and environmental policy', in R. Repetto (ed.) *Punctuated Equilibrium and the Dynamics of US Environmental Policy*, New Haven CT: Yale University Press, 24–46.

Baumgartner F. R. and Jones B. D. (1991) 'Agenda dynamics and policy subsystems', *The Journal of Politics* 53, 4: 1044–74.

Baumgartner F. R. and Jones B. D. (1993) *Agendas and Instability in American Politics*, Chicago: University of Chicago Press.

Baumgartner F. R. and Jones B. D. (2002) *Policy Dynamics*, Chicago: University of Chicago Press.

Beck, S. (2012) 'Between tribalism and trust: the IPCC under the "Public Microscope"', *Nature and Culture* 7, 2: 151–73.

Beck, U. (1992) *Risk Society: Towards a New Modernity*, London: Sage.

Beck, U., Giddens, A., and Lash, S. (1991) *Reflexive Modernization: Politics, Tradition and Aesthetics in the Modern Social Order*, Stanford: Stanford University Press.

Beddington, J. (undated) Food, energy, water and the climate: a perfect storm of global events? London: Government Office for Science. Available at: <http://webarchive.nationalarchives.gov.uk/20121212135622/http://www.bis.gov.uk/assets/goscience/docs/p/perfect-storm-paper.pdf> (accessed January 2015).

Beddington, J. (2009) Speech delivered to Sustainable Development UK, 19 March: <http://www.govnet.co.uk/news/govnet/professor-sir-john-beddingtons-speech-at-sduk-09> (accessed 29 January 2014).

Beloff, N. (1963) *The General Says No: Britain's Exclusion from Europe*, Harmondsworth, Middlesex: Penguin Books Ltd.

Benn, T. (1990) *Conflicts of Interest: Diaries, 1977–80*, edited by Ruth Winstone, London: Hutchinson.

Bennett, C. J. and Howlett, M. (1992) 'The lessons of learning: Reconciling theories of policy learning and policy change', *Policy Sciences* 25: 275–94.

Berkhout, F. (1991) *Radioactive Waste: Politics and Technology*, London and New York: Routledge.

Berlin, I. (1969) 'Two concepts of liberty', in *Four Essays on Liberty*, Oxford: Oxford University Press, 118–72.

BERR (Department of Business, Enterprise and Regulatory Reform) (2008) *Meeting the Energy Challenge: a White Paper on Nuclear Power*, Cm 7296, London: TSO.

Berry, C. (2000) *Marine Health Check 2000: A Report to Gauge the Health of the UK's Sealife*, London: WWF-UK, <http://www.wwf.org.uk/filelibrary/pdf/mhcr.pdf> (accessed 3 March 2014).

References

Bijker, W. E., Bal, R., and Hendriks, R. (2009) *The Paradox of Scientific Authority: The Role of Scientific Advice in Democracies*, Cambridge, Massachusetts and London: The MIT Press.

Bimber, B. (1996) *The Politics of Expertise in Congress: the Rise and Fall of the Office of Technology Assessment*, Albany, NY: State University of New York Press.

BiPRO (*Beratugsgesellschaft für integrierte Problemlösungen*) and Öko-Institut e.V (*Institut für Angewandte Ökologie*: Institute for Applied Ecology) (2013) *Examination and Assessment of Consequences for Industry, Consumers, Human Health and the Environment of Possible Options for Changing the REACH Requirements for Nanomaterials*, Final Report, IHCP/2011/1/05/27/OC, Ispra, Italy: European Commission, Joint Research Centre, Institute for Health and Consumer Protection.

Blair, P. D. (2011) 'Scientific advice for policy in the United States: lesson from the National Academies and the former Congressional Office of Technology Assessment', in J. Lentsch and P. Weingart (eds.) *The Politics of Scientific Advice: Institutional Design for Quality Assurance*, Cambridge, UK: Cambridge University Press, 297–333.

Blowers, A. (1986) *Something in the Air: Corporate Power and the Environment*, London: Harper and Row.

Boehmer-Christiansen, S. (1995) 'Reflections on scientific advice and EC transboundary pollution policy', *Science and Public Policy* 22, 3: 195–203.

Boholm, Å. (1998) 'Comparative studies of risk perception: a review of twenty years of research', *Journal of Risk Research* 1, 2: 135–63.

Booth, T. (1990) 'Researching policy research: issues of utilization in decision making', *Science Communication* 12, 1: 80–100.

Bugler, J. (1972) *Polluting Britain: A Report*, Harmondsworth, Middlesex: Penguin.

Bulkeley, H. (2005) 'Reconfiguring environmental governance: towards a politics of scales and networks', *Political Geography* 24, 8: 875–902.

Bulmer, M. (1980a) 'Introduction', in M. Bulmer (ed.) *Social Research and Royal Commissions*, London: George Allen and Unwin, 1–8.

Bulmer, M. (ed.) (1980b) *Social Research and Royal Commissions*, London: George Allen and Unwin.

Bulmer, M. (1980c) 'The Royal Commission on the Distribution of Income and Wealth', in M. Bulmer (ed.) *Social Research and Royal Commissions*, London: George Allen and Unwin, 158–79.

Bulmer, M. (1983a) 'An Anglo-American comparison: does social science contribute effectively to the work of governmental commissions?', *American Behavioural Scientist* 26, 5: 643–68.

Bulmer, M. (1983b) 'Commissions as instruments for policy research', *American Behavioral Scientist* 26, 5: 559–67.

Bulmer, M. (1993) 'The royal commission and departmental committee in the British policy-making process', in B. Guy Peters and A. Barker (eds.) *Advising West European Governments: Inquiries, Expertise and Public Policy*, Edinburgh: Edinburgh University Press, 37–49.

Burke, T. (2008) 'Things didn't only get better', in *ENDS at 30: How Green Has Britain Gone since 1978?* ENDS 30th Anniversary Supplement, May, 6–9.

References

Burnett, H. Sterling (2009) 'Understanding the precautionary principle and its threat to human welfare', *Social Philosophy and Policy* 26, 2: 378–410.

Cabinet Office (2002) *Improving Government's Capability to Handle Risk and Uncertainty*, Full Report: A Source Document, London: Cabinet Office Strategy Unit.

Cabinet Office (2006a) *Making and Managing Public Appointments: A Guide for Departments*, 4th edn, London: The Cabinet Office, February. Available at: <http://www.civilservice.gov.uk/wp-content/uploads/2011/09/public_appt_guide-pdf_tcm6-3392.pdf> (accessed 7 February 2014).

Cabinet Office (2006b) *Public Bodies: A Guide for Departments: Overview of Guidance Documents*, London: The Cabinet Office. <https://www.gov.uk/government/uploads/system/uploads/attachment_data/file/80077/PublicBodiesGuide2006_overview_0.pdf> (accessed 7 February 2014).

Cabinet Office (2009) *Public Bodies 2009*, London: The Cabinet Office.

Cabinet Office (2012) *Public Bodies 2012*, London: The Cabinet Office. Also at <http://www.civilservice.gov.uk/about/resources/ndpb> (accessed 22 May 2014).

Carter, N. and Lowe, P. (1994) 'Environmental politics and administrative reform', *Political Quarterly* 65, 3: 263–75.

Cartwright, T. J. (1975) *Royal Commissions and Departmental Committees in Britain: A Case-Study in Institutional Adaptiveness and Public Participation in Government*, London: Hodder and Stoughton.

Castells, M. (1996) *The Rise of the Network Society*, Oxford: Blackwell.

CEC (Commission of the European Communities) (2000) *Communication from the Commission on the Precautionary Principle*, COM(2000)1 final, 2 February, Brussels: The European Commission.

CEC (Commission of the European Communities) (2001) *A Sustainable Europe for a Better World: A European Union Strategy for Sustainable Development*, Communication from the Commission (Commission's Proposal to the Gothenburg European Council), COM(2001)264, 15 May, Brussels: The European Commission.

CEC (Commission of the European Communities) (2002) *Communication from the Commission on the Collection and Use of Expertise by the Commission: Principles and Guidelines*, COM(2002)713 final, 11 December, Brussels: The European Commission.

CEC (Commission of the European Communities) (2007a) *Green Paper on Market-based Instruments for Environment and Related Policy Purposes*, COM(2007)140 final, 28 March, Brussels: The European Commission.

CEC (Commission of the European Communities) (2007b) *Towards an Improved Policy on Industrial Emissions*, Communication from the Commission to the Council, the European Parliament, the Economic and Social Committee and the Committee of the Regions, COM(2007)843 final, 21 December, Brussels: The European Commission.

Chapman, R. A. (1973a) *Commissions in Policy-Making*, in R. A. Chapman (ed.) *The Role of Commissions in Policy-Making*, London: George Allen and Unwin Ltd., 174–88.

Chapman, R. A. (ed.) (1973b) *The Role of Commissions in Policy-Making*, London: George Allen and Unwin Ltd.

Christie, I., Southgate, M., and Warburton, D. (2002) *Living Spaces: A Vision for the Future of Planning*, Sandy, Bedfordshire: Royal Society for the Protection of Birds.

References

Christoff, P. (1996) 'Ecological modernisation, ecological modernities', *Environmental Politics* 5, 3: 476–500.

Clark, W. C. and Majone, G. (1985) 'The critical appraisal of scientific inquiries with policy implications', *Science, Technology, and Human Values* 10, 3: 6–19.

Clarke, R. H. and Valentin, J. (2005) 'A history of the International Commission on Radiological Protection', *Health Physics* 88, 5: 407–22.

Clarke, R. H. and Valentin, J. (2009) 'The history of the ICRP and the evolution of its policies', *Annals of the ICRP* 39, 1: 75–110.

CLG (Department for Communities and Local Government) (2012) *National Planning Policy Framework*, London: The Stationery Office Ltd., March.

CLG (Department for Communities and Local Government) and BERR (Department for Business, Enterprise and Regulatory Reform) (2008) *Prosperous Places: Taking Forward the Review of Sub-National Economic Development and Regeneration*, The Government Response to Public Consultation, London: CLG and BERR, 25 November.

CLG (Department for Communities and Local Government) and BERR (Department for Business, Enterprise and Regulatory Reform) (2009) *Local Democracy, Economic Development and Construction Bill: Policy Document on Regional Strategies*, London: CLG, January.

CLG (Department for Communities and Local Government) and BIS (Department for Business, Innovation and Skills) (2009) *Policy Statement on Regional Strategies and Guidance on the Establishment of Leaders' Boards*, Consultation, London: CLG and BIS, August.

Cohen, M. D., March, J. G., and Olsen, J. P. (1972) 'A garbage can model of organizational choice', *Administrative Quarterly* 17 (March): 1–25.

Collingridge, D. (1980) *The Social Control of Technology*, New York: Frances Pinter.

Collingridge, D. and Reeve, C. (1986) *Science Speaks to Power: The Role of Experts in Policy Making*, London: Frances Pinter.

Collins, H. and Evans, R. (2002) 'The Third Wave of science studies: studies of expertise and experience', *Social Studies of Science* 32: 235–96.

Collins, H. and Evans, R. (2007) *Rethinking Expertise*, Chicago and London: University of Chicago Press.

Collins, H., Weinel, M., and Evans, R. (2010) 'The politics and policy of the Third Wave: new technologies and society', *Critical Policy Studies* 4, 2: 185–201.

Commissioner for Public Appointments (2012) *Code of Practice for Ministerial Appointments to Public Bodies*, London: Commissioner for Public Appointments. <http://publicappointmentscommissioner.independent.gov.uk/the-code-of-practice/> (accessed 25 February 2014).

Committee on Climate Change (2008) *Building a Low-Carbon Economy: the UK's Contribution to Tackling Climate Change*, First Report, London: TSO.

Committee on Standards in Public Life (Nolan Committee) (1995) *Standards in Public Life: First Report*, Cm 2850-I, London: HMSO.

Cook, C. (1976) 'How a state probe split the atom chiefs', *The Guardian*, 23 September: 11.

CoRWM (Committee on Radioactive Waste Management) (2006) *Managing our Radioactive Waste Safely: CoRWM's Recommendations to Government*, CoRWM doc 700, London: CoRWM.

References

COT/COC (Committees on Toxicity and Carcinogenicity of Chemicals in Food, Consumer Products and the Environment) (2006) *Statement on Royal Commission on Environmental Pollution: Crop Spraying and the Health of Residents and Bystanders*, COT/06/05 and COC/06/S1, London: COT and COC, April.

Council of Europe Consultative Assembly (1950) *Official Report of Debates*, Vols 1–3, Strasbourg: Parliamentary Assembly of the Council of Europe.

Council of the European Communities (1983) *Action Programme of the European Communities on the Environment (1982 to 1986)* (Annex to Resolution of 7 February 1983 'on the continuation and implementation of a European Community policy and action programme on the environment 1982 to 1986'); OJ C46, 17.2.83.

Council of the European Communities (1987) *EEC Fourth Environmental Action Programme (1987–1992)* (Annex to Resolution of 19 October 1987 'on the continuation and implementation of a European Community policy and action programme on the environment 1987 to 1992'); OJ C328, 7.12.87.

Council of the European Communities (1993) *Towards Sustainability. A European Community Programme of Policy and Action in relation to the Environment and Sustainable Development* (Fifth Environmental Action Programme, 1993–2000), COM(92)23, vol 2; OJ C138/7, 17.5.93.

Council of the European Union (2009) 'Council adopts climate-energy legislative package', Press Release, 8434/09 (Presse 77), Brussels, Council of the EU.

Council of the European Union (2012) Consolidated versions of the Treaty on European Union and the Treaty on the Functioning of the European Union and the Charter of Fundamental Rights of the European Union, 6655/7/08 REV 7, 12 November, Brussels: CEU.

Cowell, R. and Owens, S. (2006) 'Governing space: planning reform and the politics of sustainability', *Environment and Planning C: Government and Policy* 24: 403–21.

CPRE (Council for the Protection of Rural England) (2002) *Planning to Improve*, London: CPRE.

Crenson, M. A. (1971) *The Un-Politics of Air Pollution: A Study of Non-Decisionmaking in the Cities*, Baltimore: Johns Hopkins University Press.

Czarniawska, B. (2000) *A City Reframed: Managing Warsaw in the 1990s*, Amsterdam: Harwood Academic Publishers.

Dake, K. (1992) 'Myths of nature: culture and the social construction of risk', *Journal of Social Issues* 48, 4: 21–37.

Dammann, S. and Gee, D. (2011) 'Science into policy: The European Environment Agency', in J. Lentsch and P. Weingart (eds.) *The Politics of Scientific Advice: Institutional Design for Quality Assurance*, Cambridge, UK: Cambridge University Press, 238–58.

de Wit, B. (2004) *Methodology of Boundary Work*, The Hague: RMNO (Dutch Advisory Council for Research on Spatial Planning, Nature and the Environment, PO Box 93051, 2509 AB The Hague, The Netherlands).

de Wit, B. (2011) 'RMNO and quality control of scientific advice to policy', in J. Lentsch and P. Weingart (eds.) *The Politics of Scientific Advice: Institutional Design for Quality Assurance*, Cambridge, UK: Cambridge University Press, 139–56.

References

Defra (Department for Environment, Food and Rural Affairs) (2004a) *Quality of Life Counts: 2004* Update, London: Defra.

Defra (Department for Environment, Food and Rural Affairs) (2004b) *The Royal Commission on Environmental Pollution Report on Chemicals in Products: Government Response*, Cm 6300, London: TSO.

Defra (Department for Environment, Food and Rural Affairs) (2004c) *The Royal Commission on Environmental Pollution's Special Report on Biomass as a Source of Renewable Energy—Government Response*, London: Defra.

Defra (Department for Environment, Food and Rural Affairs) (2006a) *A Marine Bill: A Consultation Document*, March, London: Defra.

Defra (Department for Environment, Food and Rural Affairs) (2006b) *The Royal Commission on Environmental Pollution Report on Crop Spraying and the Health of Residents and Bystanders: Government Response*, London: Defra.

Defra (Department for Environment, Food and Rural Affairs) (2006c) *The UK Government Response to the Royal Commission on Environmental Pollution's Twenty-fifth Report, Turning the Tide—Addressing the Impact of Fisheries on the Marine Environment*, Cm 6845, London: TSO.

Defra (Department for Environment, Food and Rural Affairs) (2007a) *A Sea Change: a Marine Bill White Paper*, Cm 7047, London: TSO.

Defra (Department for Environment, Food and Rural Affairs) (2007b) *Review of the Royal Commission on Environmental Pollution: Final Report*, prepared for Defra by PricewaterhouseCoopers, London: Defra, May. Available at: <http://archive.defra.gov.uk/corporate/about/with/rcep/documents/report.pdf> (accessed 13 January 2014).

Defra (Department for Environment, Food and Rural Affairs) (2007c) *Waste Strategy for England 2007*, Cm 7086, London: TSO.

Defra (Department for Environment, Food and Rural Affairs) (2008) *Government Response to the Review of the Royal Commission on Environmental Pollution*, London: Defra, June.

Defra (Department for Environment, Food and Rural Affairs) (2010a) *Measuring Progress: Sustainable Development Indicators 2010*, London: Defra.

Defra (Department for Environment, Food and Rural Affairs) (2010b) *The Royal Commission on Environmental Pollution (RCEP) Report on Artificial Light in the Environment—Government Response* (England), London: Defra.

Defra (Department for Environment, Food and Rural Affairs) (2011) *Government Review of Waste Policy in England*, PB 13540, London: Defra.

Defra and Cranfield University (2011) *Guidelines for Environmental Risk Assessment and Management: Green Leaves III*, London: Defra.

Defra (Department for Environment, Food and Rural Affairs) and Department of the Environment Northern Ireland (2008) *The Royal Commission on Environmental Pollution's Report on the Urban Environment, Government Response*, London: Defra.

Defra (Department for Environment, Food and Rural Affairs), Department of the Environment Northern Ireland, Scottish Executive, and Welsh Assembly Government (2003) *The Air Quality Strategy for England, Scotland, Wales and Northern Ireland: Addendum*, London: Defra.

References

Defra (Department for Environment, Food and Rural Affairs), Department of the Environment Northern Ireland, Scottish Executive, and Welsh Assembly Government (2007) *The Air Quality Strategy for England, Scotland, Wales and Northern Ireland*, Cm 7169, NIA 61/06–07, London: TSO.

Defra (Department for Environment, Food and Rural Affairs), Department of the Environment Northern Ireland, Scottish Government, and Welsh Assembly Government (2010) *Air Pollution: Action in a Changing Climate*, PB 13378, London: Defra.

Defra (Department for Environment, Food and Rural Affairs) and LGA (Local Government Association) (2007) *Joint Environmental Prospectus*, London: Defra, July.

Demmke, C., Bovens, M., Henökl, T., van Lierop, K., Moilanen, T., Pikker, G. and Salminen, A. (2007) *Regulating Conflicts of Interest for Holders of Public Office in the European Union: A Comparative Study of the Rules and Standards of Professional Ethics for the Holders of Public Office in the EU-27 and EU Institutions*, European Institute of Public Administration in co-operation with Utrecht School of Governance, University of Helsinki, and University of Vaasa, for European Commission, Bureau of European Policy Advisors. Available at: <http://ec.europa.eu/dgs/policy_advisers/publications/docs/hpo_professional_ethics_en.pdf> (accessed 25 February 2014).

Department of the Environment Northern Ireland (2006) *Towards Resource Management: the Northern Ireland Waste Management Strategy 2006–2020*, Belfast: Department of the Environment.

DETR (Department of the Environment, Transport and the Regions) (1998) *A New Deal for Transport: Better for Everyone*, Cm 3950, London: HMSO.

DETR (Department of the Environment, Transport and the Regions) (1999) *Quality of Life Counts: Indicators for a Strategy for Sustainable Development for the United Kingdom: a Baseline Assessment*, London: DETR.

DETR (Department of the Environment, Transport and the Regions) (2000a) *Financial Management and Policy Review of the Royal Commission on Environmental Pollution*, London: DETR, available from British Library Document Supply Centre, Boston Spa, West Yorkshire.

DETR (Department of the Environment, Transport and the Regions) (2000b) *Waste Strategy 2000 for England and Wales*, London: The Stationery Office Ltd.

DETR (Department of the Environment, Transport and the Regions) (2001) *Framework Document: The Royal Commission on Environmental Pollution and the Department of the Environment, Transport and the Regions*, London: DETR, February.

DETR (Department of the Environment, Transport and the Regions) with Environment Agency and IEH (Institute for Environment and Health) (2000) *Guidelines for Environmental Risk Assessment and Management*, London: The Stationery Office Ltd.

DETR (Department of the Environment, Transport and the Regions), Scottish Executive, The National Assembly for Wales, and Department of the Environment Northern Ireland (2000a) *Climate Change: the UK Programme*, Cm 4913 (SE 2000/209. NIA 19/00), London: The Stationery Office Ltd.

DETR (Department of the Environment, Transport and the Regions), Scottish Executive, The National Assembly for Wales, and Department of the Environment Northern Ireland (2000b) *The Air Quality Strategy for England, Scotland, Wales and Northern

References

Ireland: Working Together for Clean Air, Cm 4548 (SE2000/3, NIA 7), London: The Stationery Office Ltd.

DfT (Department for Transport) (2003) *The Future of Air Transport*, White Paper, Cm 6046, London: TSO.

DHSS (Department of Health and Social Security) (1980) *Lead and Health: the Report of a DHSS Working Party on Lead in the Environment* (the Lawther Report), London: HMSO.

DoE (Department of the Environment) (1974) *Lead in the Environment and its Significance to Man*, Pollution Paper no. 2, Central Unit on Environmental Pollution, London: HMSO.

DoE (Department of the Environment) (1975) *Controlling Pollution: A Review of Action Related to Recommendations by the Royal Commission on Environmental Pollution*, Pollution Paper no. 4, Central Unit on Environmental Pollution, London: HMSO.

DoE (Department of the Environment) (1982) *Air Pollution Control: The Government Response to the Fifth Report of the Royal Commission on Environmental Pollution*, Pollution Paper no. 18, Central Directorate on Environmental Pollution, London: HMSO.

DoE (Department of the Environment) (1983a) *Agriculture and Pollution: The Government Response to the Seventh Report of the Royal Commission on Environmental Pollution*, Pollution Paper no. 21, Central Unit of Environmental Protection, London: HMSO.

DoE (Department of the Environment) (1983b) *Lead in the Environment: The Government Response to the Ninth Report of the Royal Commission on Environmental Pollution*, Pollution Paper no. 19, Central Directorate of Environmental Protection, London: HMSO.

DoE (Department of the Environment) (1983c) *Oil Pollution of the Sea: The Government Response to the Eighth Report of the Royal Commission on Environmental Pollution*, Pollution Paper no. 20, Central Directorate of Environmental Protection, London: HMSO.

DoE (Department of the Environment) (1984) *Controlling Pollution: Principles and Prospects. The Government's Response to the Tenth Report of the Royal Commission on Environmental Pollution*, Pollution Paper no. 22, Central Directorate of Environmental Protection, London: HMSO.

DoE (Department of the Environment) (1986a) *Assessment of Best Practicable Environmental Options (BPEOs) for Management of Low- and Intermediate-level Solid Radioactive Waste*, Report by the Radioactive Waste (Professional) Division of the Department of the Environment, London: HMSO.

DoE (Department of the Environment) (1986b) *Managing Waste: The Duty of Care. The Government's Response to the Eleventh Report of the Royal Commission on Environmental Pollution*, Pollution Paper No. 24, Central Directorate of Environmental Protection, London: HMSO.

DoE (Department of the Environment) (1986c) *Public Access to Environmental Information: Report of an Interdepartmental Working Party on Public Access to Information held by Pollution Control Authorities*, Pollution Paper No. 23, Central Directorate of Environmental Protection, London: HMSO.

DoE (Department of the Environment) (1988) *Integrated Pollution Control: a Consultation Paper*, London: DoE.

References

DoE (Department of the Environment) (1992) *Government Response to the Twelfth Report of the Royal Commission on Environmental Pollution—Best Practicable Environmental Option*, London: DoE.

DoE (Department of the Environment) (1993a) *Government Response to the Thirteenth Report of the Royal Commission on Environmental Pollution—the Release of Genetically Engineered Organisms to the Environment*, London: DoE.

DoE (Department of the Environment) (1993b) *Preliminary Government Response to the Fourteenth Report of the Royal Commission on Environmental Pollution—GENHAZ: A System for the Critical Appraisal of Proposals to Release Genetically Modified Organisms to the Environment*, London: DoE.

DoE (Department of the Environment) (1994) *Government Response to the Seventeenth Report of the Royal Commission on Environmental Pollution: Incineration of Waste*, London: DoE.

DoE (Department of the Environment) (1995a) *A Guide to Risk Assessment and Risk Management for Environmental Protection*, London: HMSO.

DoE (Department of the Environment) (1995b) *Freshwater Quality: Government Response to the Sixteenth Report of the Royal Commission on Environmental Pollution*, London: DoE.

DoE (Department of the Environment) (1996) *Indicators of Sustainable Development for the United Kingdom*, London: HMSO.

DoE (Department of the Environment) (1997) *Sustainable Use of Soil: Government Response to the Nineteenth Report of the Royal Commission on Environmental Pollution*, London: DoE.

DoE (Department of the Environment) and Department of Employment (1986) *Inspecting Industry: Pollution and Safety*, Efficiency scrutiny commissioned by the Secretaries of State for Environment and Employment, London: HMSO.

DoE (Department of the Environment) and The Scottish Office (1997) *The United Kingdom National Air Quality Strategy*, Cm 3587, London: The Stationery Office Ltd.

Doherty, B. and de Geus, M. (eds.) (1996) *Democracy and Green Political Thought: Sustainability, Rights and Citizenship*, London and New York: Routledge.

Donnison, D. (1980) 'Committees and committeemen', in M. Bulmer (ed.) *Social Research and Royal Commissions*, London: George Allen and Unwin, 9–17.

DoT (Department of Transport) (1989) *National Road Traffic Forecasts (Great Britain)*, London: HMSO.

DoT (Department of Transport) (1992) *Emissions from Heavy Duty Diesel Engined Vehicles—The Government's Response to the Fifteenth Report of the Royal Commission on Environmental Pollution*, Vehicle Standards and Engineering Division, London: HMSO.

Doubleday, R. and Wilsdon, J. (eds.) (2013) *Future Directions for Scientific Advice in Whitehall*, Cambridge: University of Cambridge, Centre for Science and Policy. Available at: <http://www.csap.cam.ac.uk/events/future-directions-scientific-advice-whitehall/> (accessed 10 June 2014).

Doubleday, R. and Wynne, B. (2011) 'Despotism and democracy in the UK: experiments in reframing relations between the state, science and citizens', in S. Jasanoff

References

(ed.) *Reframing Rights: Bioconstitutionalism in the Genetic Age*, Cambridge MA.: MIT Press, 239–62.

Douglas, M. (1992) *Risk and Blame: Essays in Cultural Theory*. London: Routledge.

Douglas, M. and Wildavsky, A. (1982) *Risk and Culture: An Essay on the Selection of Technical and Environmental Dangers*, Berkeley: University of California Press.

Downs, A. (1972) 'Up and down with ecology: the issue attention cycle', *Public Interest* 28: 38–50.

Downs, G. (2010) 'Why I'm taking my campaign to protect the public from pesticides to Europe', *The Guardian* (online) 25 January. Available at: <http://www.guardian.co.uk/environment/cif-green/2010/jan/25/georgina-downs-pesticides> (accessed 21 January 2014).

Dreyfus, H. L. and Rabinow, P. (1982) *Michel Foucault: Beyond Structuralism and Hermeneutics*, Brighton, UK: The Harvester Press Ltd.

Dryzek, J. S. (1997) *The Politics of the Earth: Environmental Discourses*, Oxford: Oxford University Press.

Dryzek, J. S., Downes, D., Hunold, C., Schloberg, D., and Hernes, H. (2003) *Green States and Social Movements: Environmentalism in the United States, United Kingdom, Germany and Norway*, Oxford: Oxford University Press.

DTI (Department of Trade and Industry) (2003) *Our Energy Future: Creating a Low Carbon Economy*, Cm 5761, London: TSO.

DTI (Department of Trade and Industry) (2006) *The Energy Challenge. Energy Review: a Report*, Cm 6887, London: TSO.

DTI (Department of Trade and Industry) (2007) *Meeting the Energy Challenge: A White Paper on Energy*, Cm 8124, London: TSO.

DTLR (Department for Transport, Local Government and the Regions) (2001) *Planning: Delivering Fundamental Change* (Green Paper), London: DTLR.

EAC (House of Commons Environmental Audit Committee) (1999) *Genetically Modified Organisms and the Environment: Coordination of Government Policy*, Fifth Report Session 1998-9, HC 384-I, London: The Stationery Office Ltd.

EAC (House of Commons Environmental Audit Committee) (2003) *Budget 2003 and Aviation*, Ninth Report Session 2002-3, HC 672, London: The Stationery Office Ltd.

EAC (House of Commons Environmental Audit Committee) (2004a) *Aviation: Sustainability and the Government Response*, Seventh Report Session 2003-4, HC 623, London: The Stationery Office Ltd.

EAC (House of Commons Environmental Audit Committee) (2004b) *Aviation: Sustainability and the Government's Second Response*, Eleventh Report Session 2003-4, HC 1063, London: The Stationery Office Ltd.

EAC (House of Commons Environmental Audit Committee) (2004c) *Pre-Budget Report 2003: Aviation Follow-up*, Third Report Session 2003-4, HC 233, London: The Stationery Office Ltd.

EAC (House of Commons Environmental Audit Committee) (2006) *Proposals for Draft Marine Bill*, Eighth Report Session 2005-6, HC 1323, London: The Stationery Office Ltd.

EAC (House of Commons Environmental Audit Committee) (2010a) *Air Quality*, Fifth Report Session 2009-10, HC 229-I, London: The Stationery Office Ltd.

References

EAC (House of Commons Environmental Audit Committee) (2010b) *Carbon Budgets*, Third Report Session 2009–10, HC 228-I, London: The Stationery Office Ltd.

EAC (House of Commons Environmental Audit Committee) (2011) *Air Quality: A Follow up Report*, Ninth Report Session 2010–12, HC 1024-I, London: The Stationery Office Ltd.

EEA (European Environment Agency) (2005) *Environmental Policy Integration in Europe: State of Play and an Evaluation Framework*, Copenhagen: EEA.

EEAC (European Environment and Sustainable Development Advisory Councils) (2004) *Towards a European Marine Strategy*, Statement, The Hague, The Netherlands: EEAC.

EEAC (European Environment and Sustainable Development Advisory Councils) (2006) *Impact Assessment at the EU Level: Achievements and Prospects*, Statement by EEAC Working Group on Governance, available at: <http://eeac.eu/documents/statements> (accessed 3 March 2014).

Ecologist Magazine (1972) *Blueprint for Survival*, 2, 1.

Ecologist Magazine (1994) 'Wrong question', 24, 1: 19.

Eddington, R. (2006) *The Eddington Transport Report. The Case for Action: Sir Rod Eddington's Advice to Government*, London: The Stationery Office Ltd (with agreement of HM Treasury).

Elkington, J. and Hailes, J. (1988) *The Green Consumer Guide*, London: Victor Golancz Ltd.

ENDS (Environmental Data Services) (1983) 'How the Royal Commission swung the case against lead in petrol', *ENDS Report* 99 (April): 9–11.

ENDS (Environmental Data Services) (1985a) 'Lords look for a compromise on EEC water pollution policy', ENDS Report 127 (August): 11–12.

ENDS (Environmental Data Services) (1985b) 'New Act promised to tighten controls on waste disposal', *ENDS Report* 131 (December): 3.

ENDS (Environmental Data Services) (1985c) 'Raising the status of waste management: a prescription from the Royal Commission', *ENDS Report* 131 (December): 9–13.

ENDS (Environmental Data Services) (1985d) 'Time for action on waste', *ENDS Report* 131 (December): 2.

ENDS (Environmental Data Services) (1991) 'Regulating hazardous waste incineration: yesterday's failures and tomorrow's aspirations', *ENDS Report* 202 (November): 16–20.

ENDS (Environmental Data Services) (1993) 'RCEP's incineration report turn the heat up on landfills', *ENDS Report* 220 (May): 13–15.

ENDS (Environmental Data Services) (1996) 'Ten years ago this month: Government move on environmental disclosure', *ENDS Report* 255 (April): 51.

ENDS (Environmental Data Services) (2002) 'Putting environmental sustainability into planning', *ENDS Report* 326 (March): 22–4.

ENDS (Environmental Data Services) (2008) 'The key bodies setting the green agenda', in *ENDS at 30: How Green Has Britain Gone since 1978?* ENDS 30th Anniversary Supplement (May): 24–30.

ENDS (Environmental Data Services) (2010) 'RCEP closure puts an end to "awkward advice"', *ENDS Report* 428 (September): 8–9.

References

Environment and Heritage Service Northern Ireland (EHNI) (2005) *Best Practicable Environmental Option for Waste Management in Northern Ireland: Guidance Document, Final Report on Municipal Solid Waste, Commercial & Industrial and Construction Sector Wastes*, Belfast: EHNI, June.

Epstein, S. (2011) 'Misguided boundary work in studies of expertise: time to return to the evidence', *Critical Policy Studies* 5, 3: 323–8.

European Commission (2002) *Environment 2010: Our Future, Our Choice*, Sixth EU Environment Action Programme, Luxembourg: Office for Official Publications of the European Communities.

European Commission (2007) *Reach in Brief*, DG Enterprise and Industry and DG Environment, Brussels: The European Commission.

European Commission (2009) *EU Action Against Climate Change: Leading Global Action to 2020 and Beyond*, 2009 edn, Luxembourg: Office for Official Publications of the European Communities.

European Commission (2012) *Second Regulatory Review on Nanomaterials*, Communication from the Commission to the European Parliament, the Council and the European Economic and Social Committee (Text with EEA relevance), COM(2012)572 final, {SWD(2012)288 final}, 3 October, Brussels: the European Commission.

Everest, D. A. (1990) 'The provision of expert advice to government on environmental matters: the role of advisory committees', *Science and Public Affairs* 4: 17–40.

Expert Group on Science and Governance (2007) *Taking European Knowledge Society Seriously*, Report of the Expert Group on Science and Governance to the Science, Economy and Society Directorate, Directorate-General for Research, European Commission, EUR 22700, Luxembourg: Office for Official Publications of the European Communities.

Ezrahi, Y. (1980) 'Utopian and pragmatic rationalism: The political context of scientific advice', *Minerva* 18, 1: 111–31.

Falconer, C. (Lord Falconer of Thoroton) (2002) Speech at launch of Royal Society for the Protection of Birds report, *Living Spaces*, 18 March.

Farmer, A., Bowyer, C., Kekki, M., Fergusson, M., and Skinner, I. (2005) *Evaluation of the 21st Report of the Royal Commission on Environmental Pollution, 'Setting Environmental Standards'*, Final Report, London: Institute for European Environmental Policy.

Feindt, P. H. and Cowell, R. (2010) 'The recession, environmental policy and ecological modernization: What's new about the Green New Deal?', *International Planning Studies* 15, 3: 191–211.

Fergusson, M., Skinner, I., and Keles, P. (2002) *Review of the Royal Commission on Environmental Pollution's 20th Report on Transport and the Environment—Developments Since 1994*, A Report to the Royal Commission on Environmental Pollution, London: Institute for European Environmental Policy.

Fergusson, M., Skinner, I., Keles, P., Herodes, M., Kekki, M., and Bowyer, C. (2005) *Evaluation of the 22nd Report of the Royal Commission on Environmental Pollution, 'Energy: the Changing Climate'*, Final Report, London: Institute for European Environmental Policy.

Fiorino, D. (2001) 'Environmental policy as learning: a new view of an old landscape', *Public Administration Review* 61, 3: 322–34.

References

Fischer, F. (1990) *Technocracy and the Politics of Expertise*, Newbury Park, CA: Sage.

Fischer, F. (1993) 'Policy discourse and the politics of Washington think tanks', in F. Fischer and F. Forester (eds.) *The Argumentative Turn in Policy Analysis and Planning*, Durham, NC and London: Duke University Press, 21–42.

Fischer, F. (2000) *Citizens, Experts and the Environment*, Durham, NC and London: Duke University Press.

Fischer, F. (2009) *Democracy and Expertise: Re-orienting Policy Inquiry*, Oxford and New York: Oxford University Press.

Fischer, F. (2011) 'The "policy turn" in the Third Wave: return to the fact–value dichotomy?', *Critical Policy Studies* 5, 3: 311–16.

Fischer, F. and Forester, F. (eds.) (1993) *The Argumentative Turn in Policy Analysis and Planning*, Durham, NC and London: Duke University Press.

Fischhoff, B. (2000) Review of *Setting Environmental Standards* (Royal Commission on Environmental Pollution, Twenty-first Report, Cm 4053, 1988), *Journal of Risk Research* 3, 2: 180–2.

Fischhoff, B., Slovic, P., and Lichtenstein, S. (1982) 'Lay foibles and expert fables in judgement about risks', *The American Statistician* 36, 3: 240–55.

Fisk, D. (1998) 'Environmental science and environmental law', *Journal of Environmental Law* 10, 1: 3–8.

Fisk, D. (2007) 'Review. *Crop Spraying and the Health of Residents and Bystanders, Special Report*, by Royal Commission on Environmental Pollution (RCEP); *UK Government Response to RCEP Special Report*, by Department for Environment, Food and Rural Affairs (Defra)', *Journal of Environmental Law* 19, 2: 288–9.

Flood, M. and Grove-White, R. (1976) *Nuclear Prospects: A Comment on the Individual, the State and Nuclear Power*, London: Friends of the Earth.

Flowers, B. (1974) 'The work of the Royal Commission on Environmental Pollution', National Society for Clean Air lecture, 3 July (News Release, Imperial College, London SW7 2AZ).

Flowers, B. (1995) 'The Commission's achievement', in Royal Commission on Environmental Pollution *Environmental Policy into the 21st Century*, A Seminar held on 28 March 1995 to mark the first 25 years of the Commission at Church House, Westminster, London: RCEP.

Flyvbjerg, B. (1998) *Rationality and Power: Democracy in Practice*, Chicago: University of Chicago Press.

Forsyth, T. (2011) 'Expertise needs transparency not blind trust: a deliberative approach to integrating science and social participation', *Critical Policy Studies* 5, 3: 317–22.

Foucault, M. (1972) *The Archaeology of Knowledge*, translated by A. M. Sheridan Smith, London: Tavistock Publications.

Foucault, M. (1979) *Discipline and Punish: The Birth of the Prison*, translated by A. Sheridan, Harmondsworth: Penguin.

Foucault, M. (1991a) 'Governmentality', in G. Burchell, C. Gordon, and P. Miller (1991) *The Foucault Effect: Studies in Governmentality*, London: Harvester Wheatsheaf, 87–104.

References

Foucault, M. (1991b) 'Questions of method' (interview), in G. Burchell, C. Gordon, and P. Miller (1991) *The Foucault Effect: Studies in Governmentality*, London: Harvester Wheatsheaf, 73–86.

Frankel, M. (1974) 'The Alkali Inspectorate: the control of industrial air pollution', *Social Audit* 1, 4 (Special Report) London: Social Audit Ltd.

Frankel, M. (1984) 'How secrecy protects the polluter', in D. Wilson (ed.) *The Secrets File: The Case for Freedom of Information in Britain Today*, London: Heinemann Educational, 22–58.

Friends of the Earth (1990) *Stealing our Future: Friends of the Earth's Critique of 'This Common Inheritance', the Government's White Paper on the Environment*, London: Friends of the Earth, September.

Funtowicz, S. and Ravetz, J. (1985) 'Three kinds of risk assessment: a methodological analysis', in C. Whipple and V. Covello (eds.), *Risk Analysis in the Private Sector*, New York: Plenum Press, 217–31.

Funtowicz, S. and Ravetz, J. (1990) *Uncertainty and Quality in Science for Policy*, Dordrecht, The Netherlands: Kluwer Academic Publishers.

Funtowicz, S. and Ravetz, J. (1993) 'Science for the post-normal age', *Futures* 25, 7: 739–55.

Gaffney, J. (1991) 'Political think tanks in the UK and Ministerial Cabinets in France', *West European Politics* 14, 1: 1–17.

Gaskell, G. (2004) 'Science policy and society: the British debate over GM agriculture', *Current Opinion in Biotechnology* 15: 241–5.

Gibbons, M., Limoges, C., Nowotny, H., Schwartzman, S., Scott, P., and Trow, M. (1994) *The New Production of Knowledge: the Dynamics of Science and Research in Contemporary Societies*, London: Sage.

Giddens, A. (1990) *The Consequences of Modernity*, Cambridge: Polity, in association with Blackwell.

Giddens, A. (1999) 'Risk and responsibility', *The Modern Law Review* 62, 1: 1–10.

Gieryn, T. (1983) 'Boundary work and the demarcation of science from non-science: strains and interests in professional ideologies of scientists', *American Sociological Review* 48, 6: 781–95.

Gieryn, T. (1995) 'Boundaries of science', in S. Jasanoff, G. E. Markle, J. C. Petersen, and T. Pinch (eds.) *Handbook of Science and Technology Studies*, Thousand Oaks, London and New Delhi: Sage, 393–443.

Glasbergen, P. (1996) 'Learning to manage the environment', in W. Lafferty and J. Meadowcroft (eds.) *Democracy and the Environment: Problems and Prospects*, Cheltenham, UK: Edward Elgar, 175–93.

Glickman, T. S. and Gough, M. (eds.) (1990) *Readings in Risk*, Washington DC: Resources for the Future.

GM Nation Steering Board (Department of Trade and Industry) (2003) *GM Nation? The Findings of a Public Debate*, London: HMSO.

Goodwin, P., Hallett, S., Kenny, F., and Stokes, G. (1991) *Transport: The New Realism*, Report to the Rees Jeffreys Road Fund. Oxford: University of Oxford, Transport Studies Unit.

References

Gottweis, H. (1998) *Governing Molecules*, Cambridge, MA.: MIT Press.

Gottweis, H. (2003) 'Theoretical strategies of poststructuralist policy analysis: towards an analytics of government', in M. Hajer and H. Wagenaar (eds.) *Deliberative Policy Analysis: Understanding Governance in the Network Society*, Cambridge: Cambridge University Press, 247–65.

Government Office for Science (UK) (2007) *Code of Practice for Scientific Advisory Committees*, URN 07/1570, London: Department for Innovation, Universities and Skills.

Government Office for Science (UK) (2011) *Code of Practice for Scientific Advisory Committees*, URN 11/1382, London: Department for Business, Innovation and Skills.

Government Office for Science (UK) (2013) *Review of Science Advisory Councils 2013*, URN 13/850, London: Government Office for Science. Available at: <https://www.gov.uk/government/uploads/system/uploads/attachment_data/file/278421/13-850-science-advisory-council-review-2013-1.pdf> (accessed 30 July 2014).

Gray, T. (ed.) (1995) *UK Environmental Policy in the 1990s*, Basingstoke: Palgrave Macmillan.

Greenpeace (1993) *Royal Commission gives Licence to Burn: Greenpeace Responds*, Press Release 20 May, London: Greenpeace.

Grove-White, R. (2001) 'New wine, old bottles? Personal reflections on the new biotechnology commissions', *Political Quarterly* 72, 4: 466–72.

Grove-White, R., Kapitza, S., and Shiva, V. (1992) 'Public awareness, science and the environment', in J. C. I. Doodge, J. W. M. I. Rivière, J. Morton-Lefèvre, T. O'Riordan, and F. Praderie (eds.) *An Agenda of Science for Environment and Development into the 21st Century*, Cambridge: Cambridge University Press, 239–54.

Grove-White, R., Macnaghten, P., Mayer, S., and Wynne, B. (1997) *Uncertain World: Genetically Modified Organisms, Food and Public Attitudes in Britain*, Lancaster: University of Lancaster, Centre for the Study of Environmental Change.

Guston, D. H. (1999) 'Stabilizing the boundary between US politics and science: the role of the Office of Technology Transfer as a boundary organization', *Social Studies of Science* 29,1: 87–112.

Guston, D. H. (2001) 'Boundary organizations in environmental policy and science: an introduction', *Science, Technology, and Human Values* 26, 4: 399–408.

Haas, P. M. (1990) *Saving the Mediterranean: The Politics of International Environmental Cooperation*, New York: Columbia University Press.

Haas, P. M. (1992) 'Introduction: epistemic communities and international policy co-ordination', *International Organisation* 46, 1: 1–35.

Haigh, N. (1984) *EEC Environmental Policy and Britain*, London: Environmental Data Services.

Haigh, N. (1986) 'Public perceptions and international influences', in G. Conway (ed.) *The Assessment of Environmental Problems*, London: Centre for Environmental Technology, Imperial College, 73–83.

Haigh, N. (1987) *EEC Environmental Policy and Britain*, 2nd edn, Harlow, UK: Longman.

Haigh, N. (1992) *Manual of Environmental Policy: The EC and Britain*, London: Longman with Institute for European Environmental Policy (loose-leaf, updated to Release 15, 1999).

References

Haigh, N. (1995) 'Environmental protection in the DoE (1970–1995) or one and a half cheers for bureaucracy', in Department of the Environment, *A Perspective for Change*, proceedings of a conference to review the Department after twenty-five years, London, 30 October, London: DoE, 7–16.

Haigh, N. (1998) 'Roundtable 4: challenges and opportunities for IEA—science–policy interactions from a policy perspective', *Environmental Modelling and Assessment* 3, 3: 135–42.

Haigh, N. and Irwin, F. (eds.) (1990) *Integrated Pollution Control in Europe and North America*, Washington DC: Conservation Foundation and Bonn: Institute for European Environmental Policy.

Hajer, M. (1993) 'Discourse coalitions and the institutionalization of practice: the case of acid rain in Britain', in F. Fischer and J. Forester (eds.) *The Argumentative Turn in Policy Analysis and Planning*, Durham NC and London: Duke University Press, 43–76.

Hajer, M. (1995) *The Politics of Environmental Discourse: Ecological Modernization and the Policy Process*, Oxford: Clarendon Press.

Hajer, M. (1996) 'Ecological modernization as cultural politics', in S. Lash, B. Szerszynski, and B. Wynne (eds.) *Risk, Environment and Modernity: Towards a New Ecology*, London: Sage, 246–68.

Hajer, M. (2003a) 'A frame in the fields: policymaking and the reinvention of politics', in M. Hajer and H. Wagenaar (eds.) *Deliberative Policy Analysis: Understanding Governance in the Network Society*, Cambridge: Cambridge University Press, 88–110.

Hajer, M. (2003b) 'Policy without polity? Policy analysis and the institutional void', *Policy Sciences* 36: 175–95.

Hajer, M. and Laws, D. (2006) 'Ordering through discourse', in M. Moran, M. Rein, and R. E. Goodin (eds.) *The Oxford Handbook of Public Policy*, Oxford: Oxford University Press, 251–68.

Hajer, M. and Versteeg, W. (2005a) 'A decade of discourse analysis of environmental politics: achievements, challenges, perspectives', *Journal of Environmental Policy and Planning* 7, 3: 175–84.

Hajer, M. and Versteeg, W. (2005b) 'Performing governance through networks', *European Political Science* 4, 3: 340–7.

Hajer, M. and Wagenaar, H. (eds.) (2003a) *Deliberative Policy Analysis: Understanding Governance in the Network Society*, Cambridge: Cambridge University Press.

Hajer, M. and Wagenaar, H. (2003b) 'Introduction', in M. Hajer and H. Wagenaar (eds.) *Deliberative Policy Analysis: Understanding Governance in the Network Society*, Cambridge: Cambridge University Press, 1–32.

Hall, P. A. (1993) 'Policy paradigms, social learning, and the state: the case of economic policymaking in Britain', *Comparative Politics* 25, 3: 275–96.

Ham, C. and Hill, M. (1993) *The Policy Process in the Modern Capitalist State*, 2nd edn, Hemel Hempstead, UK: Harvester Wheatsheaf.

Harrison, C. (2006) 'Future ecological planning for London', Gresham College Lecture 23 October, London; available at: <http://www.gresham.ac.uk/lectures-and-events/london's-ecology-future-ecological-planning> (accessed 28 January 2014).

Harrison R. M. (1993) 'A perspective on lead pollution and health 1972–1992', *Journal of the Royal Society of Health* 113, 3: 142–8.

References

Hart, D. M. (2014) 'An agent, not a mole: assessing the White House Office of Science and Technology Policy', *Science and Public Policy* 41, 4: 411-18 (online 25 August 2013, doi:10.1093/scipol/sct061).

Hawes, D. (1993) *Power on the Backbenches? The Growth of Select Committee Influence*, Bristol, UK: University of Bristol, School for Advanced Urban Studies.

Hawkes, N. (1976) 'A warning light on the nuclear road', *The Observer* 26 September: 9.

Hawkesworth, M. E. (1988) *Theoretical Issues in Policy Analysis*, Albany, NY: State University of New York Press.

Heath, E. (1970) 'A policy for the environment', *The Spectator* no. 7398 (11 April): 472–3.

Heclo, H. (1974) *Modern Social Politics in Britain and Sweden*, New Haven: Yale University Press.

Heclo, H. (1978) 'Issue networks and the executive establishment', in A. King (ed.) *The New American Political System*, Washington DC: American Enterprise Institute, 87–124.

Helm, D. (2004) *Energy, the State and the Market: British Energy Policy since 1979*, revised edn, Oxford: Oxford University Press.

Hendriks, R., Bal, R., and Bijker, E. (2004) 'Beyond the species barrier: the Health Council of The Netherlands, legitimacy and the making of objectivity', *Social Epistemology* 18, 2–3: 271–99.

Hennessy, P. (1989) *Whitehall*, London: Secker & Warburg.

Herbert, A. P. (1961) 'Anything but action? A study of the uses and abuses of committees of inquiry', in R. Harris (ed.) *Radical Reaction: Essays in Competition and Affluence*, 2nd edn, London: Hutchinson for the Institute of Economic Affairs, 249–302.

Hey, C., Jacob, K., and Volkery, A. (2006) Better Regulation by New Governance Hybrids? Governance Models and the Reform of European Chemicals Policy, FFU Report 02-2006, Berlin: *Forschungsstelle für Umweltpolitik, Freie Universität Berlin*, Ihnestr 22, 14195, Berlin.

Hilgartner, S. (2000) *Science on Stage: Expert Advice as Public Drama*, Stanford, CA: Stanford University Press.

Hinrichsen, D. (1990) 'Integrated permitting and inspection in Sweden', in N. Haigh and F. Irwin (eds.) *Integrated Pollution Control in Europe and North America*, Washington DC: The Conservation Foundation and London: Institute for European Environmental Policy, 147–68.

Hisschemöller, M., Hoppe, R., Dunn, W. N., and Ravetz, J. R. (2001) *Knowledge, Power and Participation in Environmental Policy Analysis* (Policy Studies Review Annual), Piscataway, NJ: Transaction Publishers.

HM Treasury, BERR (Department for Business, Enterprise and Regulatory Reform), and CLG (Department for Communities and Local Government) (2007) *Review of Sub-National Economic Development and Regeneration*, London: HM Treasury, July.

Holdgate, M. (2003) *Penguins and Mandarins*, Spennymoor, County Durham: The Memoir Club.

Hoppe, R. (2005) 'Re-thinking the science-policy nexus: from knowledge utilization and science technology studies to types of boundary arrangements', *Poiesis Prax* 3: 199–215.

References

Hoppe, R., Wesselink, A., and Cairns, R. (2013) 'Lost in the problem: the role of boundary organisations in the governance of climate change', *WIREs Climate Change* 4, 4: 283–300.

Houghton, J. (2013) *In the Eye of the Storm: The Autobiography of Sir John Houghton* (with G. Tavner), Oxford: Lion.

Houghton, J., Jenkins, G. J., and Ephraums, J. J. (eds.) (1990) *Climate Change: The IPCC Scientific Assessment*, Cambridge: Cambridge University Press.

House of Commons Environment Committee (1986) *Radioactive Waste*, First Report Session 1985–6, HC 191 (3 vols), London: HMSO.

House of Commons Environment Committee (1987) *Pollution of Rivers and Estuaries*, Third Report Session 1986–7, HC 183 (2 vols), London: HMSO.

House of Commons Environment Committee (1989) *Toxic Waste*, Second Report Session 1988–9, HC 22, London: HMSO.

House of Commons Innovation, Universities, Science and Skills Committee (2009) *Putting Science and Engineering at the Heart of Government Policy*, Eighth Report Session 2008–9, vol. 1, HC 168-I, London: The Stationery Office Ltd., July.

House of Commons ODPM (Office of the Deputy Prime Minister): Housing, Planning, Local Government and the Regions Committee (2003) *Planning for Sustainable Housing and Communities: Sustainable Communities in the South East*, Eighth Report Session 2002–3, vol. 1, HC 77-I, London: The Stationery Office Ltd., April.

House of Commons Public Administration Select Committee (2001) *Mapping the Quango State*, HC 367, Fifth Report Session 2000–1, London: The Stationery Office Ltd.

House of Commons Science and Technology Committee (2010) *The Disclosure of Climate Data from the Climatic Research Unit at the University of East Anglia*, Eighth Report Session 2009–10, HC 387-I, London: The Stationery Office Ltd.

House of Commons Trade and Industry Committee (1984) *The Wealth of Waste*, Fourth Report Session 1983–4, HC 640-478, London: HMSO.

House of Commons Transport Committee (2009) *The Future of Aviation*, First Report Session 2009–10, HC 125-I, London: The Stationery Office Ltd.

House of Commons Transport, Local Government and Regions Committee (2002) *Planning Green Paper*, Thirteenth Report Session 2001–2, HC 476-I, London: The Stationery Office Ltd.

House of Lords Select Committee on the European Communities (1985) *The Discharge of Dangerous Substances*, Fifteenth Report Session 1984–5, HL 227, London: HMSO.

House of Lords Select Committee on the European Communities (1989) *Freedom of Access to Information on the Environment*, First Report Session 1989–90, HL 2, London: HMSO.

House of Lords Select Committee on the European Communities (1996) *Freedom of Access to Information on the Environment*, First Report Session 1996–7, HL 9, London: HMSO.

House of Lords Select Committee on Science and Technology (1981) *Hazardous Waste Disposal*, First Report Session 1980–1, HL 273, London: HMSO.

References

House of Lords Select Committee on Science and Technology (1993) *Regulation of the United Kingdom Biotechnology Industry and Global Competitiveness*, Seventh Report Session 1992–3, HL 80, London: HMSO.

House of Lords Select Committee on Science and Technology (2000) *Science and Society*, Third Report 1999–2000, HL 38, London: The Stationery Office Ltd.

House of Lords Select Committee on Science and Technology (2010) *Nanotechnologies and Food*, First Report, 2009–10, HL 22, London: The Stationery Office Ltd.

House of Lords Select Committee on Science and Technology (2012) *The Role and Function of Departmental Chief Scientific Advisors*, Fourth Report 2010–12, HL 264, London: The Stationery Office Ltd.

Huber, J. (1982) *Die Verlorene Unschuld der Ökologie: Neue Technologien und Superindustrielle Entwicklung (The Lost Innocence of Ecology. New Technologies and Superindustrialized Development)*, Frankfurt am Main: Fischer Verlag.

Huber, J. (1985) *Die Regenbogengesellschaft. Ökologie und Sozialpolitik (The Rainbow Society: Ecology and Social Policy)*, Frankfurt am Main: Fischer Verlag.

Huber, J. (1991) '*Ecologische modernisering: Weg van schaarste, soberheid en bureaucratie?* (Ecological modernization: abandoning scarcity, sobriety, and bureaucracy?)', in A. P. J. Mol, G. Spaargaren, and A. Klapwijk (eds.) *Technologie en Milieubeheer: Tussen Sanering en Ecologische Modernisering (Technology and Environmental Policy: Between Remediation and Ecological Modernization)*, Den Haag, The Netherlands: *Staats Drukkerüen Ultseverü* (SDU).

Hulme, M. (2009) *Why We Disagree About Climate Change*, Cambridge: Cambridge University Press.

Hulme. M. (2010) The IPCC on Trial: Experimentation Continues, available at <http://environmentalresearchweb.org/cws/article/opinion/43250> (accessed 14 March 2014).

Hunt, J. (1994) 'The social construction of precaution', in T. O'Riordan and J. Cameron (eds.) *Interpreting the Precautionary Principle*, London: Cameron May, 117–25.

Hyam, R. (2007) *Britain's Declining Empire: The Road to Decolonisation, 1918–1968*, Cambridge: Cambridge University Press.

IAC (Inter-Academy Council) (2010) *Climate Change Assessments: Review of the Processes and Procedures of the IPCC*, Amsterdam: IAC.

in 't Veld, R. (ed.) (2009) *Willingly and Knowingly: the Roles of Knowledge about Nature and the Environment in Policy Processes*, 2nd edn, Paper V15e, The Hague, The Netherlands: RMNO.

in 't Veld, R. and de Wit, A. (2009) 'Clarifications', in R. in 't Veld (ed.) *Willingly and Knowingly: the Roles of Knowledge about Nature and the Environment in Policy Processes*, 2nd edn, Paper V15e, The Hague, The Netherlands: RMNO, 147–57.

IPCC (Intergovernmental Panel on Climate Change) (1992) *Climate Change: The 1990 and 1992 IPCC Assessments*. IPCC First Assessment Report Overview and Policy Maker Summaries and 1992 IPCC Supplement. Geneva, Switzerland: IPCC Secretariat.

IPCC (Intergovernmental Panel on Climate Change) (1995) *IPCC Second Assessment: Climate Change 1995*, Geneva, Switzerland: IPCC Secretariat.

IPCC (Intergovernmental Panel on Climate Change) (1996) *The Science of Climate Change 1995. Summary for Policymakers*. Cambridge University Press, Cambridge, UK.

References

IPCC (Intergovernmental Panel on Climate Change) (2001) *Intergovernmental Panel on Climate Change Third Assessment: Climate Change 2001* (four vols), Geneva, Switzerland: IPCC Secretariat. Also available at: <http://www.ipcc.ch/ipccreports/assessments-reports.htm> (accessed 14 March 2014).

IPCC (Intergovernmental Panel on Climate Change) (2007) *Intergovernmental Panel on Climate Change Fourth Assessment: Climate Change 2007* (four vols), Geneva, Switzerland: IPCC Secretariat. Also available at: <http://www.ipcc.ch/ipccreports/assessments-reports.htm> (accessed 14 March 2014).

Irwin, A. (2001) *Sociology and the Environment: A Critical Introduction to Society, Nature and Knowledge*, Cambridge: Polity.

Irwin, A. (2008) 'STS perspectives on scientific governance', in: E. J. Hackett, O. Amsterdamska, M. Lynch, and J. Wajcman (eds.), *The Handbook of Science and Technology Studies*, 3rd edn, Cambridge MA: MIT Press, 583–607.

Jachtenfuchs, M. (1996) *International Policy-making as a Learning Process? The European Union and the Greenhouse Effect*, Aldershot, Hants, UK: Avebury.

Jackson, T. and Illsley, B. (2007) 'An analysis of the theoretical rationale for using strategic environmental assessment to deliver environmental justice in the light of the Scottish Environmental Assessment Act', *Environmental Impact Assessment Review* 27, 7: 607–23.

Jacobs, M. (1991) *The Green Economy*, London: Pluto Press.

Jäger, J. and O'Riordan, T. (1996) 'The history of climate change and politics', in T. O'Riordan and J. Jäger (eds.) *Politics of Climate Change: A European Perspective*, London: Routledge: 1–31.

James, S. (2000) 'Influencing government policymaking', in D. Stone (ed.) *Banking on Knowledge: The Genesis of the Global Development Network*, London: Routledge, 162–79.

Jänicke, M. (1985) Preventive Environmental Policy as Ecological Modernisation and Structural Policy, Discussion Paper IIUG dp 85-2, Berlin: *Internationales Institut Für Umwelt und Gessellschaft, Wissenschaftszentrum Berlin Für Socialforschung* (WZB).

Jänicke, M. (1988) '*Ökologische Modernisierung: Optionen und restriktionen präventiver umweltpolitik*' (Ecological modernisation: options for and restrictions on preventive environmental policy), in U. Simonis (ed.) *Präventive Umweltpolitik* (*Preventive Environmental Policy*), Frankfurt am Main: Campus, 13–26.

Jänicke, M. and Weidner, H. (eds.) (1997) *National Environmental Policies: A Comparative Study of Capacity Building*. Berlin: Springer.

Jasanoff, S. (1987) 'Cultural aspects of risk assessment in Britain and the United States', in B. B. Johnson and V. T. Covello (eds.) *The Social and Cultural Construction of Risk*, Dordrecht, Boston: Reidel, 359–97.

Jasanoff, S. (1990) *The Fifth Branch: Science Advisers as Policy Makers*, Cambridge MA.: Harvard University Press.

Jasanoff, S. (1995) 'Product, process or programme: three cultures and the regulation of biotechnology', in M. Bauer (ed.) *Resistance to New Technology: Nuclear Power, Information Technology and Biotechnology*, Cambridge: Cambridge University Press, 311–31.

References

Jasanoff, S. (1996) 'Beyond epistemology: relativism and engagement in the politics of science', *Social Studies of Science* 26, 2: 393–418.

Jasanoff, S. (1997) 'Civilisation and madness: the great BSE scare of 1996', *Public Understanding of Science* 6, 3: 221–32.

Jasanoff, S. (2003a) 'Breaking waves in science studies: comment on H.M. Collins and Robert Evans, "The Third Wave of science studies"', *Social Studies of Science* 33, 3: 389–400.

Jasanoff, S. (2003b) 'Technologies of humility: citizen participation in governing science', *Minerva* 41: 223–44.

Jasanoff, S. (2005) *Designs on Nature: Science and Democracy in Europe and the United States*, Princeton, NJ.: Princeton University Press.

Jasanoff, S. (2006a) 'Ordering knowledge, ordering society', in S. Jasanoff (ed.), *States of Knowledge: The Co-production of Science and Social Order*. London and New York, Routledge, 13–45.

Jasanoff, S. (ed.) (2006b) *States of Knowledge: The Co-production of Science and Social Order*, London and New York: Routledge.

Jasanoff, S. (2006c) 'The Idiom of Co-Production', in S. Jasanoff (ed.) *States of Knowledge: The Co-Production of Science and Social Order*, London and New York: Routledge, 1–12.

Jasanoff, S. (2011) 'Quality control and peer review in advisory science', in J. Lentsch and P. Weingart (eds.) *The Politics of Scientific Advice: Institutional Design for Quality Assurance*, Cambridge, UK: Cambridge University Press, 19–35.

Jasanoff, S. (ed.) (2013) *Science and Public Reason*, London and New York: Routledge.

Jenkins-Smith, H. and Sabatier, P. A. (1994) 'Evaluating the advocacy coalition framework', *Journal of Public Policy* 14, 2: 175–203.

Johnson, S. (1973) *The Politics of Environment: The British Experience*, London: Tom Stacey Ltd.

Joint Review Committee on the Control of Pollution (Special Waste) Regulations (1985) *Report of a Review of the Control of Pollution (Special Waste) Regulations*, London: Department of the Environment.

Jordan, A. (1998) 'The impact on UK environmental administration', in P. Lowe and S. Ward (eds.) (1998) *British Environmental Policy and Europe*. London: Routledge, 173-94.

Jordan, A. (2002) *The Europeanization of British Environmental Policy: A Departmental Perspective*, Basingstoke, Hants: Palgrave MacMillan.

Jordan, A. and O'Riordan, T. (1995) 'The precautionary principle in UK environmental law and policy', in T. Gray (ed.) *UK Environmental Policy in the 1990s*, London: Macmillan, 57–84.

Jordan, A., Wurzel, R. K. W., Zito, R., and Brückner, L. (2003) 'Policy innovation or "muddling through"? New environmental policy instruments in the United Kingdom', *Environmental Politics* 12, 1: 179–98.

Jordan, A. G. and Richardson, J. J. (1983) 'Policy communities: the British and European policy style', *Policy Studies Journal* 11, 4: 603–15.

Kasperson, R. E., Renn, O., Slovic, P., Brown, H. S., Emel, J., Goble, R., Kasperson, J. X., and Ratick, S. (1988) 'The social amplification of risk: a conceptual framework', *Risk Analysis* 8, 2: 177–87.

References

Keeley, J. and Scoones, I. (2003) *Understanding Environmental Policy Processes: Cases from Africa*, London: Earthscan.

Kennet, W. (1970) *Controlling Our Environment*, Fabian Research Series 283, London: The Fabian Society.

Kennet, W. (1972) *Preservation*, London: Temple Smith.

Kennet, W. (2007) 'The Royal Commission on Environmental Pollution', letter to the editor, *Science in Parliament* 64, 1: 48.

Kingdon, J. (2003) *Agendas, Alternatives, and Public Policies*, 2nd edn, New York: Longman (Longman Classics in Political Science).

Lafferty, W. M. and Meadowcroft, J. (eds.) (2000) *Implementing Sustainable Development: Strategies and Initiatives in High Consumption Societies*, Oxford: Oxford University Press.

Lash, S., Szerszynski, B., and Wynne, B. (eds.) (1996) *Risk, Environment and Modernity: Towards a New Ecology*, London: Sage.

Lash, S. and Wynne, B. (1992) 'Introduction', in U. Beck, *Risk Society*, translated by M. Ritter, London: Sage, 1–8.

Lasswell, H. D. (1951) 'The policy orientation', in D. Lerner and H. D. Lasswell (eds.) *The Policy Sciences: Recent Developments in Scope and Method*, Stanford University Press: Stanford, CA.

Lasswell, H. D. (1971) *A Pre-View of Policy Sciences*, New York: American Elsevier.

Lasswell, H. D. (1976 edn) *Power and Personality*, New York: Norton.

Laws, D. and Rein, M. (2003) 'Reframing practice', in M. Hajer and H. Wagenaar (eds.) *Deliberative Policy Analysis: Understanding Governance in the Network Society*, Cambridge: Cambridge University Press, 172–206.

Lawton, J. (2007) 'Ecology, policy and politics', *Journal of Applied Ecology* 44, 3: 465–74.

Layard, A. (2002) 'Planning and environment at a crossroads', *Journal of Environmental Law* 14, 3: 401–17.

Lean, G. (2011) 'Axing this Commission is a right royal shame', *Telegraph Blogs*, 10 March. Available at: <http://blogs.telegraph.co.uk/news/geoffreylean/100079274/axing-this-commission-is-a-right-royal-shame/> (accessed 19 March 2014).

Lentsch, J. and Weingart, P. (eds.) (2011) *The Politics of Scientific Advice: Institutional Design for Quality Assurance*, Cambridge: Cambridge University Press.

Levidow, L., Carr, S., and Wield, D. (2000) 'Genetically modified crops in the European Union: regulatory conflicts as precautionary opportunities', *Journal of Risk Research* 3, 3: 189–208.

Levidow, L. and Tait, J. (1991) 'The greening of biotechnology: GMOs as environment-friendly products', *Science and Public Policy* 18, 5: 271–80.

Levidow, L. and Tait, J. (1992) 'Release of genetically modified organisms: precautionary legislation', *Project Appraisal*, 7, 2: 93–105.

Lewens, T. (ed.) (2007) *Risk: Philosophical Perspectives*, London: Routledge.

Liberatore, A. (1995) 'The social construction of environmental problems', in P. Glasbergen and A. Blowers (eds.) *Perspectives on Environmental Problems*, London: Arnold, 59–84.

Lindblom, C. E. (1959) 'The science of "muddling through"', *Public Administration Review* 19, 2: 79–88.

References

Litfin, K. (1994) *Ozone Discourses: Science and Politics in Global Environmental Cooperation*, New York: Columbia University Press.

Lovell, H., Bulkeley, H., and Owens, S. (2009) 'Converging agendas: energy and climate change policies in the UK', *Environment and Planning C: Government and Policy* 27, 1: 90–109.

Lowe, P (1975a) 'Science and government: the case of pollution', *Public Administration* 53, 3: 287–98.

Lowe, P (1975b) 'The Royal Commission on Environmental Pollution', *Political Quarterly* 46, 1: 87–93.

Lowe, P., Clark, J., Seymour, S., and Ward, N. (1997) *Moralizing the Environment*, London: UCL Press.

Lowe, P. and Flynn, A. (1989) 'Environmental politics and policy in the 1980s', in J. Mohan (ed.) *The Political Geography of Contemporary Britain*. Basingstoke: Macmillan, 255–79.

Lowe, P. and Goyder, J. (1983) *Environmental Groups in Politics*, London: Allen and Unwin.

Lowe, P. and Ward, S. (1998a) 'Britain in Europe: themes and issues in national environmental policy', in P. Lowe and S. Ward (eds.) (1998) *British Environmental Policy and Europe*. London: Routledge, 3–32.

Lowe, P. and Ward, S. (eds.) (1998b) *British Environmental Policy and Europe*, London: Routledge.

Maasen, S. and Weingart, P. (eds.) (2005) *Democratization of Expertise? Exploring Novel Forms of Scientific Advice in Political Decision-Making* (*Sociology of the Sciences Yearbook*, vol. 24), Dordrecht: Springer.

McCormick, J. (1991) *British Politics and the Environment*, London: Earthscan.

MacGillivray, B. H. (2014) 'Heuristics structure and pervade formal risk assessment', *Risk Analysis* 34, 4: 771–87.

McQuail, P. (1994) *Origins of the Department of the Environment*, London: Department of the Environment.

Macrory, R. (2013) 'The long and winding road—towards an environmental court in England and Wales', Editorial comment, *Journal of Environmental Law* 25, 3: 371–81.

Macrory, R. and Niestroy, I. (2004) 'Emerging transnational policy networks: the European Environmental Advisory Councils', in N. J. Vig and M. G. Faure (eds.) *Green Giants: Environmental Policies of the United States and the European Union*, Cambridge, MA and London: The MIT Press, 305–28.

Majone, G. (1989), *Evidence, Argument, and Persuasion in the Policy Process*, New Haven and London: Yale University Press.

Marsh, D. and Rhodes, R. A. W. (1992a) 'Policy communities and issue networks: beyond typology', in D. Marsh and R. A. W. Rhodes (eds.) *Policy Networks in British Government*, Oxford: Clarendon Press, 249–68.

Marsh, D. and Rhodes, R. A. W. (eds.) (1992b) *Policy Networks in British Government*, Oxford: Clarendon Press.

Mason, B. J. (1990) *The Surface Waters Acidification Programme*, Cambridge: Cambridge University Press (on behalf of the Royal Society).

Matthews, F. (2011) 'The capacity to co-ordinate: Whitehall, governance and the challenge of climate change', *Public Policy and Administration* 27, 2: 169–89.

May, P. (1992), 'Policy learning and failure', *Journal of Public Policy* 12, 4: 331–54.

Meadows, D. H., Meadows, D. L., Randers, J., and Behrens III, W. W. (1972) *The Limits to Growth*, New York: University Books.

Mill, J. S. (1859) *Dissertations and Discussions: Political, Philosophical and Historical, Vol. II*, London: John W. Parker and Son.

Miller, Clark (2001) 'Hybrid management: boundary organizations, science policy, and environmental governance in climate change', *Science, Technology, and Human Values* 26, 4: 478–500.

Miller, C. and Wood, C. (2007) 'The adaptation of UK planning and pollution control policy', *Town Planning Review* 78, 5: 597–618.

Mills, R. (1998) 'The IPPC Directive and the role of creative ambiguity', *UK CEED Bulletin* 52 (Winter 1997–8): 10–11.

Millstone, E. (1997) *Lead and Public Health: The Dangers for Children*, London: Earthscan.

Mitchell, H. (2002) 'UK planning: fit for the future?', *Conservation Planner* (Royal Society for the Protection of Birds), Edition 18, Autumn 2002: 3.

Mol, A. P. J. (1996) 'Ecological modernisation and institutional reflexivity: environmental reform in the late modern age', *Environmental Politics* 5, 2: 302–23.

Mol, A. P. J. and Sonnenfield, D. A. (2000) 'Ecological modernisation around the world: an introduction', *Environmental Politics* 9, 1: 1–14.

Morley, J. (1903) *The Life of William Ewart Gladstone*, vol. 2 (1859–80), London: MacMillan and Co. Ltd.

Morris, H. (2002) 'Study says what the government left out', Editorial, *Planning* (Royal Town Planning Institute), 29 March: 11.

NAO (National Audit Office, UK) (2006) *Regulatory Impact Assessments and Sustainable Development*, Briefing for the Environmental Audit Committee, 22 May, London: NAO.

Nathan, R. (2003) *The Spice of Life*, Spennymoor, County Durham: The Memoir Club.

Nature (2010) 'Closing the Climategate', Editorial, 468, 7322 (18 November): 345.

Nature (2012) 'Political Science', Editorial, 483, 7388 (8 March): 123–4.

Nelkin, D. (1975) 'The political impact of technical expertise', *Social Studies of Science* 5, 1: 35–54.

Nichols, R. W. (1972) 'Some practical problems of scientific advisers', *Minerva* 10, 4: 603–13.

Nilsson, M. and Eckerberg, K. (2007) *Environmental Policy Integration in Practice: Shaping Institutions for Learning*, London and Sterling, VA: Earthscan.

Nowotny, H., Scott, P., and Gibbons, M. (2001) *Re-thinking Science: Knowledge and the Public in an Age of Uncertainty*, Cambridge: Polity Press.

Nowotny, H., Scott, P., and Gibbons, M. (2003) 'Introduction. Mode 2 revisited: the new production of knowledge', *Minerva* 41, 3: 179–94.

Nye, J. S. (1987) 'Nuclear learning and US–Soviet security regimes', *International Organization* 41, 3: 371–402.

ODPM (Office of the Deputy Prime Minister) (2005) *Planning Policy Statement 1: Delivering Sustainable Development*, London: ODPM.

References

OECD (Organisation for Economic Cooperation and Development) (2000) *Framework for Integrating Socio-Economic Analysis in Chemical Risk Management Decision-Making*, ENV/JM/MONO(2000) 5, Paris: OECD.

Oltedal, S., Moen, B-E., Klempre, H., and Rundmo, T. (2004) *Explaining Risk Perception: An Evaluation of Cultural Theory*, Rotunde no 85, Trondheim, Norway: Norwegian University of Science and Technology.

OST (Office of Science and Technology [UK]) (1997) *The Use of Scientific Advice in Policymaking*, London: Department of Trade and Industry, March.

OST (Office of Science and Technology [UK]) (2000) *Guidelines 2000: Scientific Advice and Policymaking*, London: Department of Trade and Industry, July.

OST (Office of Science and Technology [UK]) (2001) *Code of Practice for Scientific Advisory Committees*, London: Department of Trade and Industry, December.

Owens, S. (1986) 'Environmental politics in Britain: new paradigm or placebo?', *Area* 18, 3: 195–201.

Owens, S. (1987) 'A rejoinder to David Pepper', *Area* 19, 1: 77–9.

Owens, S. (1989) 'Integrated pollution-control in the United Kingdom: prospects and problems', *Environment and Planning C: Government and Policy* 7, 1: 81–91.

Owens, S. (1990) 'The unified pollution inspectorate and best practicable environmental option in the United Kingdom', in N. Haigh and F. Irwin (eds.) *Integrated Pollution Control in Europe and North America*, Washington DC: The Conservation Foundation and London: Institute for European Environmental Policy, 169–208.

Owens, S. (1995) 'Twenty five years of the Royal Commission on Environmental Pollution: a time for reflection', *Environment and Planning A*, 27, 11: 1686–91.

Owens, S. (2002) 'Environmental planning: the need for fundamental change', *Town and Country Planning* 71, 7/8: 198–201.

Owens, S. (2003) 'The Royal Commission on Environmental Pollution', in G. Altner, H. Leitschuh-Fecht, G. Michelsen, U. E. Simonis, and E. U. von Weizsäcker (eds.), *Jarhbuch Ökologie 2004*, München: Verlag C. H. Beck, 96–103.

Owens, S. (2004) 'Siting, sustainable development and social priorities', *Journal of Risk Research* 7, 2: 101–14.

Owens, S. (2005) 'Making a difference? Some perspectives on environmental research and policy', *Transactions of the Institute of British Geographers* NS, 30, 3: 287–92.

Owens, S. (2006) 'Risk and precaution: changing perspectives from the Royal Commission on Environmental Pollution', *Science in Parliament* 63, 1: 16–17.

Owens, S. (2007) 'A balanced appraisal? Impact assessment of European Commission proposals', *European Law Network International (ELNI) Review* 1/07: 2–8.

Owens, S. (2010) 'Learning across levels of governance: expert advice and the adoption of carbon dioxide emissions reduction targets in the UK', *Global Environmental Change* 20, 3: 394–401.

Owens, S. (2011a) 'Knowledge, advice and influence: the role of the UK Royal Commission on Environmental Pollution, 1970–2009', in J. Lentsch and P. Weingart (eds.) *The Politics of Scientific Advice: Institutional Design for Quality Assurance*, Cambridge, UK: Cambridge University Press, 73–101.

Owens, S. (2011b) 'Three thoughts on the Third Wave', *Critical Policy Studies* 5, 329–33.

References

Owens, S. (2012) 'Experts and the Environment: The UK Royal Commission on Environmental Pollution, 1970–2011', *Journal of Environmental Law* 24, 1: 1–22.

Owens, S. (2013) 'Commentary: Johan Rockström, Will Steffen, Kevin Noone, et al., "A safe operating space for humanity (2009)"', in L. Robin, S. Sörlin, and P. Warde (eds.) *The Future of Nature: Documents of Global Change*, New Haven and London: Yale University Press, 502–5.

Owens, S. (2014) 'Learning about limits', in E. Kessler and A. Karlqvist (eds.) *A Changing World: Redrawing the Map—Climate, Human Migration and Food Security: The 11th Royal Colloquium May 2013*, Stockholm: Kessler and Karlqvist, 31–6.

Owens, S. and Cowell, R. (2011) *Land and Limits: Interpreting Sustainability in the Planning Process*, 2nd edn, London: Routledge.

Owens, S. and Rayner, T. (1998) 'The Role and Influence of the Royal Commission on Environmental Pollution: Final Report to the Royal Commission on Environmental Pollution', University of Cambridge, Department of Geography, February.

Owens, S. and Rayner, T. (1999) '"When knowledge matters": the role and influence of the Royal Commission on Environmental Pollution', *Journal of Environmental Policy and Planning* 1, 1: 7–24.

Owens, S., Rayner, T., and Bina, O. (2004) 'New agendas for appraisal: reflections on theory, practice and research', *Environment and Planning A* 36, 11: 1943–59.

Oxburgh, R. (2010) Report of the International Panel set up by the University of East Anglia to Examine the Research of the Climatic Research Unit, Norwich, UK: University of East Anglia, available at: <http://www.uea.ac.uk/mac/comm/media/press/crustatements/sap> (accessed 1 May 2014).

Paehlke, R. C. (1989) *Environmentalism and the Future of Progressive Politics*, New Haven and London: Yale University Press.

Pallemaerts, M. (1996) 'The proposed IPPC Directive: re-regulation or de-regulation?', *European Environmental Law Review* 5, 6: 174–9.

Parker, M. (2013) 'Making the most of scientists and engineers in government', in R. Doubleday and J. Wilsdon (eds.) *Future Directions for Scientific Advice in Whitehall*, Cambridge: University of Cambridge Centre for Science and Policy, 49–55. Available at: <http://www.csap.cam.ac.uk/events/future-directions-scientific-advice-whitehall/> (accessed 10 June 2014).

Parson, E. A. and Clark, W. C. (1991) *Learning to Manage Global Environmental Change: a Review of Relevant Theory*, Center for Science and International Affairs (CSIA) Discussion Paper 91-13, Kennedy School of Government, Harvard University. Available at: <http://belfercenter.hks.harvard.edu/files/disc_paper_91_13.pdf> (accessed 27 March 2014).

Parson, E. A. and Clark, W. C. (1997) 'Sustainable development as social learning: theoretical perspectives and practical challenges for the design of a research program', in L. H. Gunderson, C. S. Holling, and S. S. Light (eds.) *Barriers and Bridges to the Renewal of Ecosystems and Institutions*, New York: Columbia University Press, 428–60.

Parsons, W. (2000) 'When dogs don't bark', in Symposium on Paul Sabatier's *Theories of the Policy Process*, *Journal of European Public Policy* 7, 1: 122–40.

Paterson, M. (1996) *Global Warming and Global Politics*, London: Routledge.

References

Pearce, F. (2010) *The Climate Files: The Battle for the Truth about Global Warming*, London: Guardian Books.

Pepper, D. (1987) 'Environmental politics: who are the real radicals?' *Area* 19, 1: 75–7.

Peters, B. Guy and Barker, A. (eds.) (1993a) *Advising West European Governments: Inquiries, Expertise and Public Policy*, Edinburgh: Edinburgh University Press.

Peters, B. Guy and Barker, A (1993b) 'Introduction: Governments, information, advice and policy-making', in B. Guy Peters and A. Barker (eds.) *Advising West European Governments: Inquiries, Expertise and Public Policy*, Edinburgh: Edinburgh University Press, 1–19.

Phillips, N. (2000) *Report of the BSE Enquiry, Volume 1*, London: The Stationery Office Ltd.

Pielke, Roger A. Jr (2007) *The Honest Broker: Making Sense of Science in Policy and Politics*, Cambridge: Cambridge University Press.

PIU (Performance and Innovation Unit, UK Cabinet Office) (2002) *The Energy Review* London: Performance and Innovation Unit.

Podger, G. (2011) 'Quality control and the link between science and regulation from a national and EU administrator's perspective', in J. Lentsch and P. Weingart (eds.) *The Politics of Scientific Advice: Institutional Design for Quality Assurance*, Cambridge, UK: Cambridge University Press, 229–37.

Porritt, J. (1993) 'It may no longer be a bad thing to go for the burn', *The Daily Telegraph* ('Weekend'), 2 October 1993, III.

Porter, M. (1998) 'Waste management', in P. Lowe and S. Ward (eds.) *British Environmental Policy and Europe*, London: Routledge, 195–213.

Prime Minister's Strategy Unit (2004) *Net Benefits: A Sustainable and Profitable Approach for UK Fishing*, London: Cabinet Office, Prime Minister's Strategy Unit.

Prospect (2011) *Is this the Lightest Green Government Ever?* (Leaflet), London: Prospect.

Raab, C. and McPherson, A. (1988) *Governing Education: A Sociology of Policy Since 1945*, Edinburgh: Edinburgh University Press.

Radaelli, C. M. (1995) 'The role of knowledge in the policy process', *Journal of European Public Policy* 2, 2: 159–83.

Radaelli, C. M. (2000) 'Public policy comes of age', in Symposium on Paul Sabatier's *Theories of the Policy Process*, *Journal of European Public Policy* 7, 1: 122–40.

Ravetz, J. (1990) *The Merger of Knowledge with Power: Essays in Critical Science*, London and New York: Mansell Publishing Ltd.

Rayner, S. (2003) 'Democracy in the age of assessment: reflections on the roles of expertise and democracy in public-sector decision making', *Science and Public Policy* 30, 3: 163–70.

Rayner, S. and Cantor, R. (1987) 'How fair is safe enough? The cultural approach to societal technology choice', *Risk Analysis* 7, 1: 3–9.

Rayner, T. (2003) 'Constructing Sustainable Transport in the UK', unpublished PhD thesis, Department of Geography, University of Cambridge, UK.

RCEP (Royal Commission on Environmental Pollution) (undated) *Review of Activities, 2001–2003*, London: RCEP.

RCEP (Royal Commission on Environmental Pollution) (1971) *First Report*, Cmnd 4585, London: HMSO, February.

References

RCEP (Royal Commission on Environmental Pollution) (1972a) *Pollution in Some British Estuaries and Coastal Waters*, Third Report, Cmnd 5054, London: HMSO, September.

RCEP (Royal Commission on Environmental Pollution) (1972b) *Three Issues in Industrial Pollution*, Second Report, Cmnd 4894, London: HMSO, March.

RCEP (Royal Commission on Environmental Pollution) (1974) *Pollution Control: Progress and Problems*, Fourth Report, Cmnd 5780, London: HMSO, December.

RCEP (Royal Commission on Environmental Pollution) (1976a) *Air Pollution Control: an Integrated Approach*, Fifth Report, Cmnd 6371, London: HMSO, January.

RCEP (Royal Commission on Environmental Pollution) (1976b) *Nuclear Power and the Environment*, Sixth Report, Cmnd 6618, London: HMSO, September.

RCEP (Royal Commission on Environmental Pollution) (1979) *Agriculture and Pollution*, Seventh Report, Cmnd 7644, London: HMSO, September.

RCEP (Royal Commission on Environmental Pollution) (1981) *Oil Pollution of the Sea*, Eighth Report, Cmnd 8358, London: HMSO, October.

RCEP (Royal Commission on Environmental Pollution) (1983) *Lead in the Environment*, Ninth Report, Cmnd 8852, London: HMSO, April.

RCEP (Royal Commission on Environmental Pollution) (1984) *Tackling Pollution: Experience and Prospects*, Tenth Report, Cmnd 9149, London: HMSO, February.

RCEP (Royal Commission on Environmental Pollution) (1985) *Managing Waste: The Duty of Care*, Eleventh Report, Cmnd 9675, London: HMSO, December.

RCEP (Royal Commission on Environmental Pollution) (1988) *Best Practicable Environmental Option*, Twelfth Report, Cm 310, London: HMSO, February.

RCEP (Royal Commission on Environmental Pollution) (1989) *The Release of Genetically Engineered Organisms to the Environment*, Thirteenth Report, Cm 720, London: HMSO, July.

RCEP (Royal Commission on Environmental Pollution) (1990) 'A New Series of Studies', Press Release, 12 July, London: RCEP.

RCEP (Royal Commission on Environmental Pollution) (1991a) *Emissions from Heavy Duty Diesel Vehicles*, Fifteenth Report, Cm 1631, London: HMSO, September.

RCEP (Royal Commission on Environmental Pollution) (1991b) *GENHAZ: A System for the Critical Appraisal of Proposals to Release Genetically Modified Organisms into the Environment*, Fourteenth Report, Cm 1557, London: HMSO, June.

RCEP (Royal Commission on Environmental Pollution) (1992) *Freshwater Quality*, Sixteenth Report, Cm 1966, London: HMSO, June.

RCEP (Royal Commission on Environmental Pollution) (1993a) *Incineration of Waste*, Seventeenth Report, Cm 2181, London: HMSO, May.

RCEP (Royal Commission on Environmental Pollution) (1993b) *Release of Genetically Modified Organisms to the Environment*, Press Release, 29 November, London: RCEP.

RCEP (Royal Commission on Environmental Pollution) (1994) *Transport and the Environment*, Eighteenth Report, Cm 2674, London: HMSO, October.

RCEP (Royal Commission on Environmental Pollution) (1995a) *A Guide to the Commission and its Work*, London: RCEP.

RCEP (Royal Commission on Environmental Pollution) (1995b) *Environmental Policy into the 21st Century*, A Seminar held on 28 March 1995 to mark the first twenty-five years of the Commission at Church House, Westminster, London: RCEP.

References

RCEP (Royal Commission on Environmental Pollution) (1995c) *Transport and the Environment: the Royal Commission on Environmental Pollution's Report*, Oxford: Oxford University Press.

RCEP (Royal Commission on Environmental Pollution) (1996) *Sustainable Use of Soil*, Nineteenth Report, Cm 3165, London: HMSO, February.

RCEP (Royal Commission on Environmental Pollution) (1997a) *A Guide to the Commission and its Work*. London: RCEP.

RCEP (Royal Commission on Environmental Pollution) (1997b) Statement on sulphur content of motor fuels, in *Royal Commission Calls for Strict Limits on Sulphur in Motor Fuels*, Press Release 10 June 1997, London: RCEP.

RCEP (Royal Commission on Environmental Pollution) (1997c) *Transport and the Environment: Developments since 1994*, Twentieth Report, Cm 3762, London: The Stationery Office Ltd., September.

RCEP (Royal Commission on Environmental Pollution) (1998a) *Guidelines for the Conduct of Commission Studies*, London: RCEP, February.

RCEP (Royal Commission on Environmental Pollution) (1998b) *Setting Environmental Standards*, Twenty-first Report, Cm 4053, London: The Stationery Office Ltd., September.

RCEP (Royal Commission on Environmental Pollution) (1999) *Royal Commission to Study Environmental Planning*, Press Release 22 July, London: RCEP.

RCEP (Royal Commission on Environmental Pollution) (2000) *Energy: The Changing Climate*, Twenty-second Report, Cm 4749, London: The Stationery Office Ltd., June.

RCEP (Royal Commission on Environmental Pollution) (2002a) *Environmental Planning*, Twenty-third Report, Cm 5459, London: The Stationery Office Ltd.

RCEP (Royal Commission on Environmental Pollution) (2002b) *The Environmental Effects of Civil Aircraft in Flight*, Special Report, London: RCEP.

RCEP (Royal Commission on Environmental Pollution) (2003) *Chemicals in Products: Safeguarding the Environment and Human Health*, Twenty-fourth Report, Cm 5827, London: TSO.

RCEP (Royal Commission on Environmental Pollution) (2004) *Turning the Tide: Addressing the Impact of Fisheries on the Marine Environment*, Twenty-fifth Report, Cm 6392, London: TSO.

RCEP (Royal Commission on Environmental Pollution) (2005) *Crop Spraying and the Health of Residents and Bystanders*, Special Report, London: RCEP, September.

RCEP (Royal Commission on Environmental Pollution) (2006) *Response to Commentary of the Advisory Committee on Pesticides on the RCEP's Report on Crop Spraying and the Health of Residents and Bystanders*, London: RCEP, July.

RCEP (Royal Commission on Environmental Pollution) (2007) *The Urban Environment*, Twenty-sixth Report, Cm 7009, London: TSO.

RCEP (Royal Commission on Environmental Pollution) (2008) *Novel Materials in the Environment: the Case of Nanotechnology*, Twenty-seventh Report, Cm 7468, London: TSO.

RCEP (Royal Commission on Environmental Pollution) (2009) *Artificial Light in the Environment*, Special Report, London: RCEP, November.

RCEP (Royal Commission on Environmental Pollution) (2010) *Adapting Institutions to Climate Change*, Twenty-eighth Report, Cm 7843, London: TSO.

References

RCEP (Royal Commission on Environmental Pollution) (2011) *Demographic Change and the Environment*, Twenty-ninth Report, Cm 8001, London: TSO.

Rees, J. (1990) *Natural Resources: Allocation, Economics and Policy*, 2nd edn, London: Routledge.

Rein, M. and Schön, D. (1991) 'Frame-reflective policy discourse', in P. Wagner, C. Weiss, B. Wittrock, and H. Wollman (eds.) *Social Sciences and Modern States*, Cambridge: Cambridge University Press, 262–89.

Rein, M. and White, S. (1977) 'Policy research: belief and doubt', *Policy Analysis* 3, 2: 239–71.

Renn, O. (1995) 'Styles of using scientific expertise: a comparative framework', *Science and Public Policy* 22, 3: 147–56.

Rhodes, G. (1975) *Committees of Inquiry*, London: Allen and Unwin for Royal Institute of Public Administration.

Rhodes, R. A. W. and Marsh, D. (1992) 'Policy networks in British politics: a critique of existing approaches', in D. Marsh and R. A. W. Rhodes (eds.) *Policy Networks in British Government*, Oxford: Clarendon Press, 1–26.

Richardson, J. J. and Jordan, A. G. (1979) *Governing Under Pressure*, Oxford: Martin Robertson.

Rip, A. (2003) 'Constructing expertise: in a Third Wave of science studies?' *Social Studies of Science* 33, 3: 419–34.

Rittel, H. and Webber, M. (1973) 'Dilemmas in a general theory of planning', *Policy Sciences* 4, 2: 155–69.

Rockström, J., Steffen, W., Noone, K., Persson, Å., Chapin, F. S. III, Lambin, E., Lenton, T. M., Scheffer, M., Folke, C., Schellnhuber, H. J., Nykvist, B., de Wit, C. A., Hughes, T., van der Leeuw, S., Rodhe, H., Sörlin, S., Snyder, P. K., Costanza, R., Svedin, U., Falkenmark, M., Karlberg, L., Corell, R. W., Fabry, V. J., Hansen, J., Walker, B., Liverman, D., Richardson, K., Crutzen, P., and Foley, J. (2009) 'A safe operating space for humanity', *Nature* 461, 7263 (24 September): 472–5.

Roqueplo, P. (1995) 'Scientific expertise among political powers, administrations and public opinion', *Science and Public Policy* 22, 3: 175–82.

Rose, R. (1991) 'What is lesson-drawing?' *Journal of Public Policy* 11, 1: 3–30.

Ross, A. and Rowan-Robinson, J. (1994) 'Public registers of environmental information: an assessment of their role', *Journal of Environmental Planning and Management* 37, 3: 349–59.

Rough, E. (2011) 'Nuclear Narratives in UK Energy Policy, 1955–2008: Exploring the Dynamics of Policy Framing', unpublished PhD thesis, University of Cambridge, Department of Geography.

Rowcliffe, N. (2008) 'In from the cold', in *ENDS at 30: How Green Has Britain Gone since 1978*? ENDS 30th Anniversary Supplement, May, 3.

Royal Society (1983) *Risk Assessment: Report of a Royal Society Study Group*, London: The Royal Society.

Royal Society (1992) *Risk: Analysis, Perception and Management. Report of a Royal Society Study Group*, London: The Royal Society.

Royal Society (1997) *Science, Policy and Risk: A Discussion Meeting held at The Royal Society on Tuesday 18th March 1997, London*, London: The Royal Society.

References

Royal Society (2012) *People and the Planet*, London: The Royal Society.

Royal Society and Royal Academy of Engineering (2004) *Nanoscience and Nanotechnologies: Opportunities and Uncertainties*, RS Policy Document 19/04, London: The Royal Society.

Russel, D. (2005) 'Environmental Policy Appraisal in UK Central Government: A Political Analysis', unpublished PhD thesis, University of East Anglia, School of Environmental Sciences.

Russel, D. and Jordan, A. J. (2007) 'Gearing up governance for sustainable development: patterns of policy appraisal in central government', *Journal of Environmental Planning and Management* 50, 1: 1–21.

Russell, M. (2010) *The Independent Climate Change E-mails Review*, July. Available at: <http://www.cce-review.org/> (accessed 3 April 2014).

RWMAC (Radioactive Waste Management Advisory Committee) (1995) *Historical Review of Radioactive Waste Management Policy and Practices*, London: RWMAC.

Sabatier, P A. (1987) 'Knowledge, policy-oriented learning and policy change: an advocacy coalition framework', *Knowledge: Creation, Diffusion, Utilization*, 8, 4: 649–92.

Sabatier, P A. (1988) 'An advocacy coalition framework of policy change and the role of policy-oriented learning therein', *Policy Sciences* 21, 2/3: 129–68.

Sabatier, P A. (1998) 'The advocacy coalition framework: revisions and relevance for Europe', *Journal of European Public Policy* 5, 1: 98–130.

Sabatier, P A. (1999) 'The need for better theories', in P. A. Sabatier (ed.) *Theories of the Policy Process*, Boulder, Colorado: Westview Press, 3–17.

Sabatier, P A. (2000) 'Clear enough to be wrong', in Symposium on Paul Sabatier's *Theories of the Policy Process*, *Journal of European Public Policy* 7, 1: 122–40.

Sabatier, P. A. and Jenkins-Smith, H. (eds.) (1993) *Policy Change and Learning: An Advocacy Coalition Approach*, Boulder, CO: Westview Press.

Sandbach, F. (1980) *Environment, Ideology and Policy*, Oxford: Blackwell.

Sarewitz, D. (2010) 'Curing Climate Backlash', *Nature* 464, 7285 (4 March): 28.

Saward, M. (1992) 'The civil nuclear network in Britain', in D. Marsh and R. A. W. Rhodes (eds.) *Policy Networks in British Government*, Oxford: Clarendon Press, 75–99.

Schmant, J. (1984) 'Regulation and science', *Science, Technology, and Human Values* 9, 1: 23–38.

Schneider, S., Rosencranz, A., and Niles, J. (eds.) (2002) *Climate Change Policy: a Survey*, NY: Island Press.

Schön, D. A. and Rein, M. (1994) *Frame Reflection: Toward the Resolution of Intractable Policy Controversies*, New York: Basic Books.

Science and Public Policy (1995) Special Issue on *Scientific Expertise in Europe*, Vol. 22, no 3.

Scott, A. (2000) *The Dissemination of the Results of Environmental Research: A Scoping Report for the European Environment Agency*, Environmental Issues Series no. 15, Copenhagen: European Environment Agency.

Scottish Executive (2003) *Scottish Executive Response to the Royal Commission on Environmental Pollution's Twenty-third Report, Environmental Planning*, SE/2003/172, Edinburgh: Scottish Executive Development Department.

References

Scottish Executive (2006) *The Royal Commission on Environmental Pollution Report 'Turning the Tide'—Addressing the Impact of Fisheries on the Marine Environment, The Scottish Executive Response*, Edinburgh: Scottish Executive.

Scottish Executive Environment Group (2005) *The Environmental Assessment (Scotland) Bill: Summary of Consultation Comments and Scottish Executive Response*, Paper 2005/3, Edinburgh: Scottish Executive.

Scottish Government (2010) *Scotland's Zero Waste Plan*, Edinburgh: The Scottish Government.

SDC (Sustainable Development Commission) (2002) *Air Transport and Sustainable Development: A Submission from the SDC*, London: The Sustainable Development Commission.

SDC (Sustainable Development Commission) (2003) *UK Climate Change Programme: A Policy Audit*, SDC Report, London: Sustainable Development Commission.

SDC (Sustainable Development Commission) (2004) *Missed Opportunity: Summary Critique of the Air Transport White Paper*, London: Sustainable Development Commission.

SDC (Sustainable Development Commission) (2006) *Climate Change: The UK Programme 2006: SDC response*, Submission to the Environmental Audit Committee, London: Sustainable Development Commission, July; available at <http://www.sd-commission.org.uk/data/files/publications/SDC_submission_to_EAC_on_CCP_2006.pdf> (accessed 3 April 2014).

SDC (Sustainable Development Commission) (2007) *Sub-National Review of Economic Development and Regeneration: Preliminary Analysis from the Sustainable Development Commission*, London: SDC, November.

Seely, A. (2009) Landfill Tax: Introduction and Early History, Standard Note SN/BT/237, London: UK Parliament, House of Commons Library; available at <http://www.parliament.uk/briefing-papers/SN00237> (accessed 11 February 2014).

Shackley, S. and Wynne, B. (1996) 'Representing uncertainty in global climate change science and policy: boundary ordering devices and authority', *Science, Technology, and Human Values* 21, 3: 275–302.

Shrader-Frechette, K. (1995) 'Evaluating the expertise of experts', *Risk: Environment, Health, and Safety* 6, 2: 115–26.

Silberston, A. (1993) 'Economics and the Royal Commission on Environmental Pollution', *National Westminster Bank Quarterly Review*, February: 29–39.

Simon, H. A. (1947) *Administrative Behavior: A Study of Decision-Making Processes in Administrative Organization*, New York: Macmillan.

Skea, J. and Smith, A. (1998) 'Integrating pollution control', in P. Lowe and S. Ward (eds.) *British Environmental Policy and Europe*, London and New York: Routledge, 265–81.

Slovic, P. (1987) 'Perception of risk', *Science* 236, 4799: 280–5.

Slovic, P. (1993) 'Perceived risk, trust, and democracy', *Risk Analysis* 13, 6: 675–82.

Slovic, P. (2000) *The Perception of Risk*, London: Earthscan.

Slovic, P., Fischhoff, B., and Lichtenstein, S. (1980) 'Facts and fears: understanding perceived risk', in R. C. Schwing and A. Albers (eds.), *Societal Risk Assessment: How Safe is Safe Enough?* New York: Plenum Press, 181–216.

References

Smith, A. (1997) *Integrated Pollution Control: Change and Continuity in the UK Industrial Pollution Policy Network*, Aldershot, Hants: Avebury.

Smith, A. (2000) 'Policy networks and advocacy coalitions: explaining policy change and stability in UK industrial pollution policy?', *Environment and Planning C: Government and Policy* 18, 1: 95–114.

Smith, A. (2004) 'Policy transfer in the development of UK climate policy', *Policy and Politics* 32, 1: 79–93.

Solesbury, W. (1976) 'The environmental agenda: an illustration of how situations may become political issues and issues may demand responses from government: or how they may not', *Public Administration* 54, 4: 379–97.

Southwood, T. R. E. (1985) 'The roles of proof and concern in the Royal Commission on Environmental Pollution', *Marine Pollution Bulletin* 16, 9: 346–50.

Southwood, T. R. E. (1992) 'The environment: problems and prospects', in B. Cartledge (ed.) *Monitoring the Environment*, Oxford: Oxford University Press, 5–41.

Spiegelhalter, D. J. and Riesch, H. (2011) 'Don't know, can't know: embracing deeper uncertainties when analysing risks', *Philosophical Transactions of the Royal Society A* 369, 1956: 4730–50.

SRU (German Advisory Council on the Environment) (2012) *Environmental Report 2012: Responsibility in a Finite World*. Summary for policy makers, Berlin: SRU.

Star, S. L. and Griesemer, J. (1989) 'Institutional ecology, "translations" and coherence: amateurs and professionals in Berkeley's Museum of Vertebrate Zoology, 1907–39', *Social Studies of Science* 19, 3: 387–420.

Starr, C. (1969) 'Social benefit versus technological risk: what is our society willing to pay for safety?', *Science* 165, 3899: 1232–8.

Stern, N. (2006) *The Stern Review on the Economics of Climate Change*, London: HM Treasury.

Stirling, A. (2003) 'Risk, uncertainty and precaution: some instrumental implications from the social sciences', in F. Berkhout, M. Leach, and I. Scoones (eds.) *Negotiating Environmental Change: New Perspectives from Social Science*, Cheltenham, UK and Northampton, MA.: Edward Elgar, 33–76.

Stirling, A. (2007) 'Risk, precaution and science: towards a more constructive policy debate', *EMBO [European Molecular Biology Organization] Reports*, 8, 4: 309–15.

Stone D. (1996) *Capturing the Political Imagination: Think Tanks and the Policy Process*, London: Frank Cass.

Stone D. (2004) 'Introduction: think tanks, policy advice and governments', in D. Stone and A. Denham (eds.) *Think Tank Traditions: Policy Research and the Politics of Ideas*, Manchester and New York: Manchester University Press, 1–16.

Stone, D. and Denham, A. (2004) *Think Tank Traditions: Policy Research and the Politics of Ideas*, Manchester and New York: Manchester University Press.

Tait, J. and Levidow, L. (1992) 'Proactive and reactive approaches to risk regulation: the case of biotechnology', *Futures* 24, 3: 219–31.

Thompson, M., Ellis, R. J., and Wildavsky, A. (1990) *Cultural Theory*, Boulder, CO: Westview Press.

References

Thompson, M. and Rayner, S. (1998) 'Risk and governance part I: the discourse of climate change', *Government and Opposition* 33, 2: 139–66.

Tinker, J. (1972) 'Royal Commission barks and bites', *New Scientist* 53, 787, 16 March: 579.

Tinker, J. (1975) 'River pollution: the Midlands dirty dozen', *New Scientist* 65, 939, 6 March: 551–4.

Torgerson, D. (1985) 'Contextual orientation in policy analysis: the contribution of Harold D. Lasswell', *Policy Sciences* 18, 3: 241–61.

Torgerson, D. (1986) 'Between knowledge and politics: the three faces of policy analysis', *Policy Sciences* 19, 1: 33–59.

Torgerson, D. (2003) 'Democracy through policy discourse', in M. Hajer and H. Wagenaar (eds.) *Deliberative Policy Analysis: Understanding Governance in the Network Society*, Cambridge: Cambridge University Press, 113–38.

Townsend, A. (2009) 'Integration of economic and spatial planning across scales', *International Journal of Public Sector Management* 22, 7: 643–59.

Tribe, L. H. (1972) 'Policy science: analysis or ideology?' *Philosophy and Public Affairs* 2, 1: 66–110.

UK Government (1970) *The Protection of the Environment: The Fight Against Pollution*, Cmnd 4373, London: HMSO.

UK Government (1977) *Nuclear Power and the Environment. The Government's Response to the Sixth Report of the Royal Commission on Environmental Pollution (Cmnd. 6618)*, Cmnd 6820, London: HMSO.

UK Government (1990) *This Common Inheritance*, Cm 1200, London: HMSO.

UK Government (1994a) *Biodiversity: The UK Action Plan*, Cm 2428, London: HMSO.

UK Government (1994b) *Climate Change: The UK Programme*, Cm 2427, London: HMSO.

UK Government (1994c) *Sustainable Development: The UK Strategy*, Cm 2426, London: HMSO.

UK Government (1995) *Making Waste Work*, White Paper, Cm 3040, London: HMSO, December.

UK Government (1996) *Transport: The Way Forward—the Government's Response to the Transport Debate*, Green Paper, Cm 3234, London: The Stationery Office Ltd.

UK Government (1998) *Government Response to the Royal Commission on Environmental Pollution's Twentieth Report: Transport and the Environment—Developments since 1994*, Cm 4066, London: The Stationery Office Ltd.

UK Government (1999) *A Better Quality of Life: Strategy for Sustainable Development for the United Kingdom 1999*, Cm 4345, London: The Stationery Office Ltd.

UK Government (2000) *Government Response to the Royal Commission on Environmental Pollution's Twenty-first Report, Setting Environmental Standards*, Cm 4794, London: The Stationery Office Ltd.

UK Government (2003a) *The Government's Response to the Royal Commission on Environmental Pollution's Twenty-third Report, Environmental Planning, England*, Cm 5887, London: TSO.

References

UK Government (2003b) *The UK Government Response to the Royal Commission on Environmental Pollution's Twenty-second Report: Energy—The Changing Climate*, Cm 5766, London: TSO.

UK Government (2003c) *The UK Government's Response to the Royal Commission on Environmental Pollution's Twenty-third Report, Environmental Planning*, Cm 5888, London: TSO.

UK Government (2005) *Securing the Future: Delivering UK Sustainable Development Strategy*, Cm 6467, London: TSO.

UK Government (2006) *Climate Change: The UK Programme 2006*, Cm 6764, SE/2006/43, London: TSO.

UK Government (2009) *UK Government Response to the Royal Commission on Environmental Pollution (RCEP) Report 'Novel Materials in the Environment: the Case of Nanotechnology'*, Cm 7620, London: TSO.

UK Government (2010) *UK Nanotechnologies Strategy: Small Technologies, Great Opportunities*, URN 10/825, London: Department for Business, Innovation and Skills, <http://webarchive.nationalarchives.gov.uk/20121212135622/http://www.bis.gov.uk/assets/goscience/docs/u/10-825-uk-nanotechnologies-strategy.pdf> (accessed February 2015).

UN (United Nations) (1992) *United Nations Conference on Environment and Development, Rio de Janeiro, 3–14 June 1992, vol. I: Resolutions Adopted by the Conference, Resolution 1, Annex I: Rio Declaration on Environment and Development* (UN Publication E.93.I.8 and corrigenda), New York: UN; available at: <http://www.un.org/documents/ga/conf151/aconf15126-1> (accessed 21 May 2014).

US CEQ (Council on Environmental Quality) (1971) *Environmental Quality: The Second Annual Report of the Council on Environmental Quality*, Washington DC: US Government Printing Office.

US National Academies (2005) *Getting to Know the Committee Process*, Washington DC: the National Academies; available at: <http://www.national-academies.org> (accessed 8 January 2014).

US NRC (National Research Council) (1983) *Risk Assessment in the Federal Government: Managing the Process*, Washington DC: National Academy Press.

US NRC (National Research Council) (1989) *Field Testing Genetically Modified Organisms—Framework for Decisions*, Washington DC: National Academy Press.

van Hemmen, J. J. (2006) 'Pesticides and the residential bystander', *Annals of Occupational Hygiene* 50, 7: 651–5.

Vig, N. J. and Faure, M. G. (eds.) (2004) *Green Giants: Environmental Policies of the United States and the European Union*, Cambridge, MA and London: The MIT Press.

von Moltke, K. (1988) The *Vorsorgeprinzip* in West German Environmental Policy, London: Institute for European Environmental Policy; reproduced in RCEP (1988) *Best Practicable Environmental Option*, Twelfth Report, Cm 310, London: HMSO (Appendix 3).

von Moltke, K. (1995) *The Maastricht Treaty and the Winnipeg Principles on Trade and Sustainable Development*, Winnipeg, Manitoba: International Institute for Sustainable Development.

References

Ward, N. (1998) 'Water quality', in P. Lowe and S. Ward (eds.) *British Environmental Policy and Europe*, London and New York: Routledge, 244–64.

Warren, L. M. (2009) 'Healthy crops or healthy people? Balancing the needs for pest control against the effect of pesticides on bystanders', *Journal of Environmental Law* 21, 3: 483–99.

WCED (World Commission on Environment and Development) (1987) *Our Common Future*, Oxford: Oxford University Press.

Weale, A. (1992) *The New Politics of Pollution*, Manchester: Manchester University Press.

Weale, A. (1996) 'Grinding slow and grinding sure? The making of the Environment Agency', *Environmental Management and Health* 7, 2: 40–3.

Weale, A. (1997) 'Great Britain', in M. Jänicke and H. Weidner (eds.) *National Environmental Policies: A Comparative Study of Capacity Building*, Berlin: Springer, 89–108.

Weale, A. (2001) 'Can we democratize decisions on risk and the environment?', *Government and Opposition* 36, 3: 355–78.

Weale, A. (2002) (ed.) *Risk, Democratic Citizenship and Public Policy*, Oxford: Oxford University Press for the British Academy.

Weale, A. (2010) 'Political theory and practical public reasoning', *Political Studies* 58, 2: 266–81.

Weale, A., O'Riordan., T., and Kramme, L. (1991) *Controlling Pollution in the Round*, London: Anglo-German Foundation.

Weale, A., Pridham, G., Cini, M., Konstadakopulos, D., Porter, M., and Flynn, B. (2003) *Environmental Governance in Europe*, Oxford: Oxford University Press.

Weinberg, A. (1972) 'Science and trans-science', *Minerva* 10, 2: 209–22.

Weingart, P. (1999) 'Scientific expertise and political accountability: paradoxes of science in politics', *Science and Public Policy* 26, 3: 151–61.

Weiss, C. (1975) 'Evaluation research in the political context', in E. L. Struening and M. Guttentag (eds.) *Handbook of Evaluation Research*, vol. 1, Beverly Hills and London: Sage: 13–25.

Weiss, C. (1977) 'Research for policy's sake: the enlightenment function of social research', *Policy Analysis* 3, 4: 531–45.

Weiss, C. (1991) 'Policy Research: data, ideas, or arguments?', in P. Wagner, C. H. Weiss, B. Wittrock, and H. Wollman (eds.) *Social Sciences and Modern States. National Experiences and Theoretical Crossroads*, Cambridge: Cambridge University Press, 307–32.

Weiss, C. and Bucuvalas, M. J. (1980) *Social Science Research and Decision-making*, New York: Columbia University Press.

Welsh Assembly Government (2002) *Wise About Waste: The National Waste Strategy for Wales* Part I, Cardiff: Welsh Assembly Government.

Welsh Assembly Government (2003) *Government Response to the Royal Commission on Environmental Pollution's Twenty-third Report, Environmental Planning, Wales*, Cardiff: Welsh Assembly Government.

Wheare, K. C. (1955) *Government by Committee: An Essay on the British Constitution*, Oxford: Clarendon Press.

Wibberley, G. (1981) 'Strong agricultures but weak rural economies: the undue emphasis on agriculture in European rural development', *European Rural Economy* 8, 2–3: 151–70.

References

Wildavsky, A. and Dake, K. (1990) 'Theories of risk perception: who fears what and why?', *Daedalus* 119, 4: 41–60.

Wilks, S. (1999) *In the Public Interest: Competition Policy and the Monopolies and Mergers Commission*, Manchester: Manchester University Press.

Williams, A. and Weale, A. (1996) 'The UK's Royal Commission on Environmental Pollution after 25 years', *Environmental Management and Health* 7, 2: 35–9.

Williams, M. K. (1983) 'Lead pollution', letter to the Editor, *The Times*, April 27.

Williams, R. (1993) 'The House of Lords Select Committee on Science and Technology within British science policy and the nature of science policy advice', in B. Guy Peters and A. Barker (eds.) *Advising West European Governments: Inquiries, Expertise and Public Policy*, Edinburgh University Press: Edinburgh, 137–50.

Wilsdon, J. and Willis, R. (2004) *See-through Science: Why Public Engagement Needs to Move Upstream*, London: Demos.

Wilsdon, J., Wynne, B., and Stilgoe, J. (2005) *The Public Value of Science: Or How to Ensure that Science Really Matters*, London: Demos.

Wilson, D. (1983) *The Lead Scandal*, London: Heinemann Educational.

Wilson, H. (1971) *The Labour Government 1964–1970: A Personal Record*, London: Weidenfield and Nicholson and Michael Joseph.

Wolf, N. and Stanley, S. (2003) *Environmental Law*, 4th edn, London: Cavendish Publishing.

Wynne, B. (1975) 'The rhetoric of consensus politics: a critical review of technology assessment', *Research Policy* 4, 2: 108–58.

Wynne, B. (1992) 'Uncertainty and environmental learning: reconceiving science and policy in the preventive paradigm', *Global Environmental Change* 2, 2: 111–27.

Wynne, B. (1996) 'May the sheep safely graze?', in S. Lash, B. Szerszynski, and B. Wynne (eds.) *Risk, Environment and Modernity: Towards a New Ecology*, London: Sage, 44–83.

Wynne, B. (2002) 'Risk and environment as legitimatory discourses of technology: reflexivity inside out?', *Current Sociology* 50, 3: 459–77.

Wynne, B. (2003) 'Seasick on the Third Wave? Subverting the hegemony of propositionalism: response to Collins and Evans (2002)', *Social Studies of Science* 33, 3: 402–17.

Yanow, D. (2000) *Conducting Interpretive Policy Analysis*, Newbury Park, CA.: Sage.

Yearley, S. (1991) *The Green Case: A Sociology of Environmental Issues, Arguments and Politics*, London and New York: Routledge.

Zuckerman, S. (1988) *Monkeys, Men and Missiles: An Autobiography 1946–1988*, London: Collins.

Index

A
Aarhus Convention 1998 39, 139
accountability
 Alkali Inspectorate 104, 105
 audit 51–2, 68, 69, 162
 and autonomy 151, 170–2
 BPEO 111, 112
 demand for 19, 28, 162
 integrated pollution control 116, 117
Acheson, Donald 83
acid rain 16, 30, 86
ACP (Advisory Committee on Pesticides) 96–7
Action Programmes on the Environment (EC) 30
adaptive capacity (governance) 99
advisory bodies
 abolition in Europe 160
 analytic role 16
 autonomy 16
 inclusion of environmental groups 39
 manipulation of 8, 9
 multiple roles 17
 within networks 17–18
 positioning in theoretical perspectives 16–18
 purpose of 8
 reduction in 29
 selectivity of use of advice 8, 9
 technocratic function 16
Advisory Committee on Pesticides (ACP) 96–7
advocacy coalitions 10, 17, 129, 130
Agenda 21 35
agriculture 60, 133–5 *see also* pesticides
air pollution 104–11
 aim of control 78
 'best practicable means' 105
 climate change 41
 complexity of regulation 104
 European governance 37
 government guidance, RCEP study 50
 guidelines 106
 Macmillan and 8
 national strategies 36, 37
 public access to information 138
 RCEP report 123, 138
 study 59, 114
 urban 41
 see also Alkali Inspectorate; Campaign for Lead Free Air; Clean Air Council; Industrial Air Pollution Inspectorate; lead pollution
airports 41
Alkali Inspectorate 27, 28, 104, 105, 106, 107, 138
'anchoring' 62, 141, 155
anthropology 73
Ashby, Sir Eric (later Lord) 27, 48, 49, 66, 67, 156
Ashworth, John 71
assimilative capacities 26, 30, 76, 100
Atomic Energy Authority 80
'authentic experts' 64
aviation 37

B
Bank of England 40
Bankside Power Station 105
Bathing Waters Directive, EEC 28, 29
Baumgartner, F. R. 43, 109, 110
Baumgartner, F. R. and Jones, B. D. 11, 29
Beckerman, Wilfred 78n14
Beeching, Dr Richard 48
behavioural change 10, 41
Belstead, Lord 134n28
Benn, Tony 64, 128, 137n36, 155
'best practicable environmental option' (BPEO) 92, 104, 105, 108–9, 111–12, 114, 133
'best practicable means' (BPM) 27n16, 78, 79, 105, 106
Bijker, W. E., Bal, R. and Hendriks, R. 14, 15, 64
bio-accumulative pollutants 77
biodiversity action plan 35
Blair, Tony 36, 51, 131
Blueprint for Survival (*Ecologist*) 25
Blundell, Professor Sir Tom 60, 64, 118, 120
'boundary organization' 15, 145, 157, 167

Index

boundary work 29, 67, 72, 95, 96, 165–8
    and authority 148
    relaxation of 101
BPEO ('best practicable environmental option') 92, 104, 105, 108–9, 111–12, 114, 133
BPM ('best practicable means') 27n16, 78, 79, 105, 106, 111n31
Brent Spar North Sea Oil Platform 88n42
Britain, and Europe 24, 28, 31
British Petroleum 83
Brundtland Report 1987 33
Brussels 66
buffer zones 142
Bugler, Jeremy 26, 27, 28n21, 104
burden of proof 31, 72
Burke, Tom 147, 160

C
Cameron, David 41n78
Campaign for Lead Free Air (CLEAR) 82, 83, 130
Cancun conference 2009 41
carbon budgets 40, 41
carbon dioxide emissions 36, 128, 129, 130, 131, 154
carbon trading system 39
'Cardiff process' 38
Cartwright, T. J. 12, 151
causality 72, 96, 125, 129
CBI (Confederation of British Industry) 110
Central Council for Science and Technology 46
Central Electricity Generating Board 32
Central Unit on Environmental Pollution (CUEP) 27, 28, 30, 47, 49–50
Chapman, R. A. 12
chemical substances 26, 37, 76, 95, 142
Chernobyl disaster 30
Chief Scientific Advisors 46, 47, 171
children, impact of lead in petrol 81, 83
civic engagement 117
'classical–modernist government' 18, 160
Clayton, Professor Barbara 83
Clean Air Council 29
CLEAR (Campaign for Lead-free Air) 82, 83, 130
climate change 35–41, 42, 60, 130, 131
Climate Change Act 2008 36, 40
'climate change levy' 39
'Climategate' 41
Clinton-Davis, Lord 34
coalition government 41, 69, 121
cognitive models 9, 10–12, 13, 15, 17, 165–6, 168
Collingridge, D. 88, 98
Collingridge, D. and Reeve, C. 2
Command Papers 57, 65

Commission on Energy and the Environment 29
Committee on Science and Technology (House of Lords) 156
competitiveness 29, 40, 41, 121, 143
Comprehensive Spending Review 70
Confederation of British Industry (CBI) 110
confidentiality 64, 138, 157
Conservation Society 26n9
Conservative government 4, 31, 39, 47, 107
constructivism 75, 87, 100–1
'control dilemma' 88, 98
Control of Pollution Act 1974 27, 127, 138, 158
Copenhagen conference 2008 41
co-production 13–17, 75, 101, 130, 145, 157, 165–8
cost-benefit analysis 78, 94n56, 143
cost-effectiveness 82, 172
Cranbrook, Lord 156
Creutzfeldt-Jakob Disease (vCJD) 19n51
Crosland, Anthony 46, 47, 48, 104, 105
cross-media approach 108, 109, 114, 133
CUEP (Central Unit on Environmental Pollution) 27, 28, 30, 47, 49–50

D
'dash for gas' 32, 36
decentralization 27
Defra bodies 69, 70, 160, 163
Deposit of Poisonous Waste Act 1972 27
devolution 24, 57, 161, 171
discourse coalitions 17, 18, 37, 82, 142, 154
'discourse institutionalization' 38
discursive model
    advisory bodies as autonomous 16
    impact of RCEP reports 131
    networks 17
    role of knowledge 9, 11, 12, 16, 42, 123
'doing good by stealth' 140–1, 155
Donnison, D. 61, 62, 63
Douglas, Mary 93
Douglas, Mary and Wildavsky, Aaron 73
Downs, Anthony 36
Durban Conference 2011 41
'duty of care' principle 131, 132, 154

E
Ecologist 91
ecomodernism (ecological modernization)
    British government 114
    development of concept 25
    Environmental Protection Act 112
    institutionalization 38
    integration (environmental regulation and policy) 103
    interventionism 135
    loosening of policy networks 39

302

Index

precautionary principle 71, 81
and RCEP 85, 101, 109, 140, 154
recession 41
Stern Review 36
Vorsorgeprinzip 92
Eddington, R. 143
EEA (European Environment Agency) 15
EEAC (European Environment and Sustainable Development Advisory Councils) 154–5
electricity generation 32, 37
Elizabeth II, Queen 52, 70
emission standards 27, 28, 30, 32, 106
endangered species 27
'end-of-pipe' approach 26, 38, 42
ENDS (journal) 146
energy security 36, 37, 131
Energy: the Changing Climate (RCEP) 128
'enlightenment' 13n38, 123, 134, 144
environmental policy
  abolition of advisory bodies 67
  controversies 16
  early concerns 25, 26–7
  and economics 36, 41, 77, 78, 94
  fragmentation of responsibilities 46
  functions of government departments 177app
  global issues 24
  growing importance of 68
  inspection 110
  integration 38
  as low government priority 29, 30, 40
  in politics 23–42, 43
  policy integration 30, 38, 161
  standards 93
  systemic approach to 30
Environment Act 1995 36, 38, 113
environmental groups 26, 30, 33, 39, 62n99, 153
environmental planning 115–21, 124
Environmental Protection Act 1990 34, 88, 112, 128, 137, 143
Environmental Protection Bill 33, 89, 129, 156
epistemic network 17, 147, 154, 155, 157, 164
ethical decisions 77–8, 89, 90, 93
EU (European Union)
  advisory bodies and RCEP 154
  British membership 24
  centralized pollution control 28–9
  devolution 36
  importance of 111
  integrated pollution control 113
  key driver environmental governance 36
  legislation 40
  policy making influenced by member states 24
  sustainable development 35

European Action Programmes 38
European Commission 30, 32, 39, 171
European Communities Committee (House of Lords) 156
European Council 41n81
European Court of Human Rights 97
European Economic Community (EEC) 28
European Environment Agency (EEA) 15
European Environment and Sustainable Development Advisory Councils (EEAC) 154–5
European Union (EU)
  advisory bodies and RCEP 154
  British membership 24
  centralized pollution control 28–9
  devolution 36
  importance of 111
  integrated pollution control 113
  key driver environmental governance 36
  legislation 40
  policy making influenced by member states 24
  sustainable development 35
European Year of the Environment 111
Euro-scepticism 31
evidence, taking of 61–2, 63, 64, 140
'evidence-based policy' 168
expertise
  deference 19, 39
  interrogation of 14
  loss of influence/deference 150
  network connectivity 157
  rival claims 88, 89
  role of 12

F
Factory Inspectorate 110
Fairclough, A. J. 49
Falconer, Lord 118, 120
fat solubility 76
financial crisis, global 40
Financial Management and Policy Reviews (FMPRs) 51–2, 64, 68, 69, 162
fines 27
Flowers, Sir Brian (later Lord)
  on BPM 105
  effect of deliberation 149
  friendship with Tony Benn 155
  increase in membership 48
  influence of RCEP 63, 125, 136, 140
  network structure 156
  and nuclear industry 153
  on radioactive waste 80, 128n6
  on RCEP as 'lay' body 149n9
'Flowers criterion' 77, 80, 128, 128n6
flue gas desulphurization technology 32
Flyvbjerg, Bent 16

303

Index

FMPRs (Financial Management and Policy Reviews) 51–2, 64, 68, 69, 162
Fogg, Gordon 104
Food and Environment Protection Act 1985 31n34, 134
Foucault, Michel 7n24
framework document 51, 52
Frankel, Maurice 139, 144
Freedom of Environmental Information Directive, 1990 39n71
'freedom of information' 137
Freedom of Information Act 2000 139
Freedom of Information Campaign 139
freshwater quality 60, 67, 86, 151
Friends of the Earth 26n9
'fuel duty escalator' 39

G

gas-fired generation 32, 36
general elections
   1983 31
   1997 37
   2010 159, 160
genetically modified organisms (GMOs) 88–90, 91
   as new area 129, 137
   empirical evidence 88
   learning within RCEP 149
   precautionary principle 128
   RCEP reports 72, 99, 142, 156, 166
GENHAZ risk assessment 142
Gezondheidsraad (the Health Council of the Netherlands) 14, 64, 171
Gieryn, Thomas 13, 148
Gladstone, William Ewart 8
GMOs (genetically modified organisms) 88–90, 91
   as new area 129, 137
   empirical evidence 88
   learning within RCEP 149
   precautionary principle 128
   RCEP reports 72, 99, 142, 156, 166
governance
   changes in 161
   and ethics 77
   European 171
   model 39
   multiple levels of 24, 36
   risk and 71–101, 102
   scientific advisors 15
Grant, Professor (later Sir) Malcolm 58n70
'Great and the Good' 52, 53, 54, 55
Green Alliance 129
Green Consumer Guide (Elkington and Hailes) 33
'green ministers' 38
Green Paper 118
Green Party 33
greenhouse gas emissions 35–6, 40
Greenpeace 62n99
Gummer, John 19n51, 24n3, 39n72
Guston, D. H. 157, 160, 166

H

Haigh, Nigel 25n6, 32, 36n52, 130
Hajer, Maarten 16, 17, 23, 35, 38, 86, 161
Hajer, Maarten and Wagenaar, H. 18, 160
Hall, Peter 11, 18, 26, 36, 42, 123, 157
Hawkes, N. 79
hazard identification 142
Health and Safety Executive (HSE) 107, 110
Health Council of the Netherlands (*Gezondheidsraad*) 14, 64, 171
Heath, Edward 46n5
Heclo, Hugh 9, 10, 10n32, 13, 15, 31, 165
Helm, D. 131
Hennessy, Peter 8n28, 52n32
Her Majesty's Inspectorate of Pollution (HMIP) 32, 111, 112–13, 122
'Her Majesty's Pollution Inspectorate' (HMPI) 105, 106, 107
herbicides 89
Hesketh, Lord 132n22, 156n45
Hilgartner, S. 141
Holdgate, Dr (later Sir) Martin 28, 35n48, 46n4, 47, 48, 162n68
Houghton, J. 152
House of Commons 156, 157
House of Lords 19, 29, 68, 90, 155, 156, 163
HSE (Health and Safety Executive) 107, 110

I

IAPI (Industrial Air Pollution Inspectorate) 34
IEEP (Institute for European Environmental Policy) 92, 114
incineration 87, 88, 91–2, 64, 73, 132, 136, 140
inclusiveness 18, 39, 94, 100, 137, 172
indicators of sustainability 38
Industrial Air Pollution Inspectorate (IAPI) 34
industrial interests, involvement of 27
influence 125–45, 167
   delayed impact 132, 133–6
   gradual 136–9, 140
   short-term responses 127–32
information, access to 28, 39, 137, 138–9
Inspectorate of Pollution 38
Institute for European Environmental Policy (IEEP) 92, 114
institutionalization 5, 38, 40
in 't Veld, R. and de Wit, A. 2
integrated pollution control (IPC) 104–5, 107–8, 110, 113–15, 122, 133
   RCEP report 119–20

304

Index

Integrated Pollution Prevention and Control (IPPC) Directive 38, 39, 135
Integrated Spatial Strategies 143
integration 103–24
 environmental planning 115–21
 institutions 104–14, 115
intensive livestock units, waste disposal 134–5, 136
Intergovernmental Panel on Climate Change (IPCC) 34, 35, 114, 115
interviewees 20n53, 20n54, 179app–180app
IPCC (Intergovernmental Panel on Climate Change) 34, 35, 114, 115
IPPC (Integrated Pollution Prevention and Control) Directive 38, 39, 135
Iraq War 131
irreversibility 76, 83, 89
'issue–attention cycle' 36

J
James, S. 13n38, 124
Jasanoff, Sheila 1, 13, 14, 14n39, 17, 75n8, 84, 89, 95, 100, 165, 166
Johnson, S. 47
Journal of Environmental Law 119

K
Kennet, Lord 46, 47
Kingdon, J. 11, 12, 17, 26, 42, 59, 81, 110, 125, 130, 142, 159
knowledge 165, 167–8
 and authority 147
 cognitive perspectives 9, 10–12, 13, 17
 'control dilemma' 88
 co-production and boundary work 13–16, 17, 75
 democratization of expertise 19
 EEAC 154, 155
 effectiveness 2
 epistemic network 17, 147, 154, 155, 157, 164
 forms of learning 10–13
 individual members 157
 limits of 149
 'natural' 101
 non-linear process 42
 and policy interface vi, 5–6, 9, 12–13, 16–17, 70, 145, 164, 168
 political rationality 8–9
 precautionary principle 72
 strategic model 142
 technical rationality 6, 7
 and uncertainty 98
Kornberg, Hans 158
Kyoto Protocol 35, 36, 41, 41n80

L
Labour government 107, 117, 151
Labour Party 39
land use ('Town and Country') planning 116
landfill 91, 109
landfill tax 39, 128, 131, 132, 136
L'Aquila summit 2009 40
Large Combustion Plant Directive 1988 32
Larminie, Geoffrey 83
Lasswell, Harold 7n21
Lawther Committee 81, 82, 83, 84
Lawton, Sir John 69, 163n72
Layard, A. 119
lead pollution 81–5
 blood lead concentration 83
 and children 81, 83
 and diversity of RCEP membership 150
 economics of 84, 85
 framing of enquiry 83–5
 government policy reversal 31, 128
 health benefits as not proven 85
 importance of 33
 influence of network 154
 public perceptions 87, 88
 RCEP influence 129, 130
 RCEP report 73
 reduction in levels 81, 83
 as subject for study 60
Lean, Geoffrey 45
legislation 181app–187app
legitimizing functions 15
Levidow, L. and Tait, J. 88
Lewis, Lord 156
Liberal Party 31
'limits' discourse 42
Limits to Growth, The (Meadows et al) 25
linear–rational model 6–9, 13, 17, 73, 144–5, 165
Litfin, Karen 1, 12n37, 15, 16
local authorities 106
Local Democracy, Economic Development and Construction Act 2009 121
localism 121, 161
Localism Act 2011 121
London 57
Lowe, P. 49
Lowe, P. and Flynn, A. 26n9, 67
Lowe, P. and Ward, S. 24

M
Maastricht Treaty 1992 35
Macmillan, Harold 8
Major, John 3n9
Marine and Coastal Access Act 2009 41
Marine Conservation Zones 41
marine environment 26, 60, 40, 41, 86
 RCEP report 59, 154, 155
 see also water quality

305

Index

Marine Protected Areas 41
market-based instruments 39
Marshall, Walter 80
McCormick, J. 29, 32
McQuail, Paul 47n9
methodology 18–19, 20
Mill, John Stuart 125
Miller, C. 166n76
Mills, R. 113
ministries, reorganization of 26
Mitchell, H. 118
Mol, A. P. J. 25n7
monitoring 76, 96, 106
Morris, H. 118

N

nanomaterials 97, 98–9, 137
nanotechnologies report (RCEP) 72
Nathan, Lord 156
National Academy of Sciences (US) 64
National Environmental Policy Act (US) 1969 26
National Planning Policy Framework 121
National Rivers Authority (NRA) 32, 38, 113, 135, 135n32
National Society for Clean Air 110
NDPBs ('non-departmental public bodies') 1n3
Nelkin, Dorothy 9
neo-liberalism 32
Netherlands 160, 171
networks 17–18, 126, 157, 163
neutrality 16n46, 75, 167
'New Labour' 36, 51
'new realist' discourse 37, 142, 143, 154
NIMBY ('not in my backyard') 87
NIREX (Nuclear Industry Radioactive Waste Executive) 137n36
Nolan, Lord 52
Nolan Committee 1995 3, 53
'Nolan principles' 3n9
non-action, legitimization of 128
'non-departmental public bodies' (NDPBs) 1n3
North Sea Conference 1987 32
Northern Ireland 24
Northern Rock (bank) 40
NRA (National Rivers Authority) 32, 38, 113, 135, 135n32
Nuclear Industry Radioactive Waste Executive (NIREX) 137n36
nuclear power 79–81
 competitiveness 41
 costs 32
 focus on 30
 learning within RCEP 149
 opening up of policy community 18
 RCEP influence 129
 RCEP report 72, 79, 127, 137, 153, 155
Nuclear Power and the Environment (RCEP) 79

O

Observer, The 26
Office of Technology Assessment (OTA) (US) 160
oil crisis 27, 29
oil pollution, marine 26, 60
openness 79
Owens, S. 149

P

Paget, Dame Shirley 48n14
Patten, Chris 33, 85n38
PCBs (polychlorinated biphenyls) 26
Peat Producers' Association 61n93
peer review 63
pesticides
 bystander exposure to 50, 59, 65, 68, 73, 95–7, 142, 163, 166
 RCEP report 95, 133, 134
 voluntary initiative 39
 and wildlife 26
Pesticides Safety Precautions Scheme (PSPS) 133, 134
Pielke, Roger 2, 12n37, 159
Planning (journal) 118
Planning Act 2008 41, 119
Planning and Compulsory Purchase Act 2004 121
planning system, devolution of 117, 119–20
pluralism 137
plutonium 79, 80
policy
 analysis 7, 16
 communities 18, 28
 development 42
 failure 34
 learning 34, 42
 opening up of process 137
 process 9, 11, 18, 21, 45, 101, 130
 theories 3
 use of term 6
'policy-oriented learning' 10
political fallout 39
political rationality model 8–9
'polluter pays' 28
Polluting Britain (Bugler) 26, 27
pollution
 controversies 30
 cross-media 32, 105
 early focus on 26
 integration of control 38
 as mainstream issue 32, 33
 new institutions 104–14, 15
 overview 86
 prevention 28
 traditional approach to control 27
 as worsening 34
polychlorinated biphenyls (PCBs) 26

306

Index

Porritt, Jonathon 92n51
'post-normal science' 13
post-positivism 7, 16
post-structuralism 7
precautionary principle 71, 72
  acid rain 86, 87
  addressed by RCEP 92
  and burden of proof 31, 85
  changing attitude 100–1
  and costs 94
  'cultural bias' 93, 94
  early RCEP reports 76
  GMOs 88, 89, 90
  pesticides 96
  proactive 95
  RCEP influence 136
prestige 61
prevention 79, 80, 83, 86, 99
PricewaterhouseCoopers 68, 68n129
privatization of utilities 32, 60, 67, 113, 135
proactivity 86, 90, 92, 94, 95
'process streams' 11
prosecutions 27
PSPS (Pesticides Safety Precautions Scheme) 133, 134
psychometric studies 74
public biotechnology commission 90
public engagement 31, 99
public perceptions 92, 100
'public risk perception' 73, 74, 92
'punctuated equilibrium' 11

Q
quality objectives 27
quangos 29, 67, 69

R
Raab, C. and McPherson, A. 21n55
Radaelli, C. M. 13, 128n10, 165
radioactive waste 77, 79, 80, 137, 140
Radioactive Waste Management Advisory Committee (RWMAC) 137
rational analytical model 152, 166
RDAs (Regional Development Agencies) 119
REACH (registration, evaluation, and authorization of chemical substances) 37, 39, 137n37
recession 27, 40–1, 42
'recommendation–acceptance–action' 128
Reconciling Man with the Environment (Ashby) 27
recycling 91
'red list' 32
'reflexive modernization' 75
Regional Development Agencies (RDAs) 119

registration, evaluation, and authorization of chemical substances (REACH) 37, 39, 137n37
regulatory approach 27, 28, 78, 88, 90, 98
regulatory science 13, 75, 86
Rein, M. and Schön, D. 11n35, 150
renewable energy 37, 41n83
resource-dependency 171
'responsible stewardship' 77
Rhodes, R. A. W. and Marsh, D. 17n49
Ridley, Nicholas 33
Rio Conference 1992 35
Rio Declaration 1992 71, 72
risk
  'acceptable' level 78
  assessment 73–5
  controversies 14, 16
  'cultural theory' 73, 74
  as social construct 74
risk governance 71–101, 102
  boundaries 101
  ecomodernism 81–5
  paradigms 73–5, 76
  RCEP 76–81
  wider engagement 75
  realist/constructivist perspectives 75
risk perception 79, 87, 88, 92, 138
road fuel duty 39n70
road transport 37, 142
role playing 14
Rose, R. 92
Rossi, Sir Hugh 157n51
Rough, E. 127
Round Table on Sustainable Development 39
Rowcliffe, Nick 23
Royal Commission on Environmental Pollution, The
  abolition 4, 69–70, 159, 160–3
  advice legitimizing shift of policy 85
  as ally of environmental groups 91
  appointments 52–5, 153
  audit 51
  authority 97, 144, 147, 148, 150, 164
  autonomy 49–51, 69, 150–3, 154, 155, 171
  budget 69
  camaraderie 56–7
  chairs 189app, 190app
  conceptualization of pollution 107
  on commercial pressure 30
  constitution 151
  continuity 21, 158–9
  cross-governmental remit 163
  deliberation and drafting 63–4, 65, 150
  demographic change 42
  diversity of membership 48, 55, 69, 84, 97, 148
  early reports 26, 76

Index

Royal Commission on Environmental Pollution, The (*cont.*)
 'educative function' 12
 'failures' 141, 142, 143
 'Financial Management and Policy Reviews' 51–2
 government 'interference' 50, 51, 54, 55
 hybridity 166, 169
 identity 56, 58, 147, 158, 169
 importance as standing body 48
 as interdisciplinary 148, 149–50
 involvement in policy change 43
 as knowledge broker 165, 166
 membership 190app–195app
 minority report 78
 networks 154–7, 158
 origins of 7, 46–8
 payments 56
 policy communities 18
 profile of 60, 82
 and public access to information 39, 138
 relationship with Whitehall 49–50, 118
 remit 4, 24, 170, 171
 reports 65–6
 as 'scientific' 6, 29, 148, 166
 selection of studies 58, 59–61, 151
 setting up as tactic 47
 'sponsorship' by government 50
 survival 66, 67–9
 working practices 57–64, 65
Royal Society for the Protection of Birds (RSPB) 118, 120
Royal Society 52, 55, 112
Royal Warrant 4, 59, 70, 151, 171
RWMAC (Radioactive Waste Management Advisory Committee) 137

S
Sabatier, Paul 10, 17, 21n56, 149
Saward, M. 18
Scandinavia 30
science
 authority 90
 and expertise 39, 93, 169
 interpretation of 96
 and objectivity 101
 and politics 95
 public attitude towards 19
 reduced deference towards 39
 and risk reduction 98
science and technology studies (STS) 9, 14, 75
Scotland 24, 57, 119, 161
SDC (Sustainable Development Commission) 39
SDP (Social Democratic Party) 31
Selborne, Lord 54, 156n43
select committees 29, 68, 156, 157, 169

Environment Committee (House of Commons) 132n22, 135, 137n36, 157, 157n50
Environmental Audit Committee (House of Commons) 38, 39n72, 130n17, 157n50
ODPM Housing, Planning, Local Government and the Regions Committee (House of Commons) 123n76
Select Committee on the European Communities (House of Lords) 138n42, 156
Select Committee on Science and Technology (House of Lords) 19, 90, 131n21, 132n22, 137n37, 156
Trade and Industry Committee (House of Commons) 131n21
Transport, Local Government and the Regions Committee (House of Commons) 118n64, 123n76
self-interest 89, 99
Sellafield reprocessing plant 30
Setting Environmental Standards, RCEP 71
Shrader-Frechette, K. 74, 93
Single European Act 1986 30
site visits 58
Sizewell B pressurized water reactor 30, 32
Slovic, P., Fischhoff, B., and Lichtenstein, S. 74
Smith, A. 85n38, 110, 123n80
smoke control, domestic 106
Social Democratic Party (SDP) 31
Solesbury, W. 26n13
source-based controls 77
sources 20–1
Southwood, Professor (later Sir) Richard
 on acid rain 86, 87
 Control of Pollution Act 1974 158
 as ecologist 83
 on influence of RCEP 128
 and ministers 50
 public access to information 139
 and RCEP profile 60, 82, 130
 Round Table 39n72
 and Waldegrave 155
special reports 57
Special Waste Regulations review 131
'standards' study 137
standing body 48, 158, 159
Stern Review 2006 36
Stockholm Conference on Human Environment 1972 26
Stone, Diane 127
strategic model 8–9, 17, 130, 142, 145, 165
STS (science and technology studies) 9, 14, 75
Sub-National Review 119, 121, 123
sulphur dioxide emissions 32
sustainable development 33, 35, 38

Index

Sustainable Development Commission (SDC) 39, 57n67, 70n134
Sweden 26n11, 160
systemic approach 30, 38, 39

T

Tackling Pollution: Experience and Prospects, RCEP 71
taxation 39, 128, 131–2, 136
technical rationality model 1, 6, 7, 8, 9, 12
technocracy 7, 75, 78, 79, 94
Thatcher, Margaret 1, 29, 32, 33, 54, 67, 112
think tanks 2, 127
Tinker, J. 66, 105
tokenism 38
Torrey Canyon 26, 46
'Town and Country' planning 116
toxic waste, illegal disposal 27
toxicological testing 76
trade unionism 54, 107
'traditional pragmatist' approach 27, 31, 82, 86
transnational organizations 161
transport 37n59, 60, 142–3
  integration of policy 115
  'new realism' 154
  public engagement 64
  RCEP report 159
  road 37, 142
'trans-scientific' issues 5

U

unified inspectorate 105, 133
United Nations Environment Programme (UNEP) 35
United States, financial crisis 40
urban environment 141, 142
US National Research Council 89
utilities privatization 32, 60, 67, 113, 135

V

von Moltke, K. 92

W

Waldegrave, William 32, 110, 111, 155
Wales 24, 57, 119, 136, 161
Walker, Peter 47
Warner, Sir Frederick 75n9
waste management 131–2, 136
  climate change 41
  devolved governance 37
  'duty of care' principle 128
  incineration 64, 73, 87, 88, 91–2, 132, 136, 140
  integration of 38
  landfill 91, 109
  minimization 91
  radioactive 111
Water Framework Directive 38
water industry, regulation 32
water privatization 32, 60, 67, 135
water quality 26, 28, 29, 76, 78, 127, 135
  see also marine environment
Weale, A. 8n25, 9, 10, 23, 34, 48, 52, 103, 166, 172
Weale, A., O'Riordan, T. and Kramme, L. 107
Weinberg, Alvin 5n15
Weiss, C. 13n38, 123, 134, 144
Wheare, Kenneth 8n28, 12, 15n44, 146, 148, 149
White, Baroness 80n19
wildlife 26, 27
Wilks, Stephen 45
Wilson, D. 82, 83
Wilson, Harold 4, 8, 46, 47, 48
World Meteorological Society (WMO) 34
World Summit on Sustainable Development, Johannesburg 2002 41n76
Wynne, B. 74, 75n8, 79, 80, 87, 88, 101

Z

Zuckerman, Sir Solly (later Lord) 46, 46n5, 47n6, 47n8, 48, 78n14